D0759318

NO LONGER PROPERTY OF
SEATTLE PUBLIC LIBRARY

Scientific Babel

Scientific Babel

HOW SCIENCE WAS DONE BEFORE AND AFTER GLOBAL ENGLISH

Michael D. Gordin

The University of Chicago Press Chicago and London

MICHAEL D. GORDIN is the Rosengarten Professor of Modern and Contemporary History at Princeton University and the author of *The Pseudoscience Wars*, also published by the University of Chicago Press.

The University of Chicago Press, Chicago 60637
The University of Chicago Press, Ltd., London
© 2015 by Michael D. Gordin.
All rights reserved. Published 2015.
Printed in the United States of America

24 23 22 21 20 19 18 17 16 15   1 2 3 4 5

ISBN-13: 978-0-226-00029-9 (cloth)
ISBN-13: 978-0-226-00032-9 (e-book)
DOI: 10.7208/chicago/9780226000329.001.0001

Library of Congress Cataloging-in-Publication Data
Gordin, Michael D., author.
Scientific Babel : how science was done before and after global English / Michael D. Gordin.
pages cm
Includes bibliographical references and index.
ISBN 978-0-226-00029-9 (cloth : alk. paper) —
ISBN 978-0-226-00032-9 (e-book)
1. Communication in science. 2. English language—Technical English. I. Title.
Q223.G67 2015
501'.4—dc23
2014032723

♾ This paper meets the requirements of ANSI/NISO Z39.48-1992 (Permanence of Paper).

To my language teachers

CONTENTS

INTRODUCTION

Talking Science

> Les savants des autres nations à qui nous avons donné l'exemple, ont cru avec raison qu'il écriraient encore mieux dans leur langue que dans la nôtre. L'Angleterre nous a donc imités; l'Allemagne, où le latin semblait s'être réfugié, commence insensiblement à en perdre l'usage: je ne doute pas qu'elle ne soit bientôt suivie par les Suédois, les Danois et les Russes. Ainsi, avant la fin du XVIIIe siècle, un philosophe qui voudra s'instruire à fond des découvertes de ses prédécesseurs, sera contraint de charger sa mémoire de sept à huit langues différentes; et après avoir consumé à les apprendre le temps le plus précieux de sa vie, il mourra avant de commencer à s'instruire.*
>
> JEAN LE ROND D'ALEMBERT[1]

You are able to read this sentence. That is obvious, but it is also quite an achievement. You read English; you may or may not speak it. Somewhere along the way, you learned the language, either relatively painlessly as a child or with significant exertion later (how significant depends a lot on who you are, how you were taught, and what other languages you already happened to know). This book is for both kinds of English-users, but it is not fundamentally a book about English. It is a history of scientific languages, the set of languages by means of which scientific knowledge has been produced and communicated. Whether

*"The scholars of other nations, to whom we have provided an example, believed with reason that they would write even better in their language than in ours. England has thus imitated us; Germany, where Latin seems to have taken refuge, begins insensibly to lose the use of it: I do not doubt that it will soon be followed by the Swedes, the Danes, and the Russians. Thus, before the end of the 18th century, a philosopher who would like to instruct himself about his predecessors' discoveries will be required to load his memory with seven to eight different languages; and after having consumed the most precious time of his life in acquiring them, he will die before having begun to instruct himself."

you are a scientist or have studiously avoided the sciences throughout your life (so far), the history of scientific languages is a constitutive part of your world. The story ends with the most resolutely monoglot international community the world has ever seen—we call them *scientists*—and the exclusive language they use to communicate today to their international peers is English. The collapse into monolingualism is, historically speaking, a very strange outcome, since most of humanity for most of its existence has been to a greater or lesser degree multilingual. The goals of this book are not only to show how we came to this point, but also to illustrate how deeply anomalous our current state of affairs would have seemed in the past.

For both ends, I have introduced what may seem the book's oddest feature: the footnotes. Every quotation in the text, except the epigraphs, appears in English. (The epigraphs, as you can see right here, always appear in their original language, and are translated in the first footnote.) For any quotation that was originally composed in a language other than English I have, where possible, tracked down the original and reproduced it on the bottom of the page, in its original orthography, with my own translation in the text. (When I have been unable to do so, I explicitly credit the translator.) I do this not because I am a perfect translator, but rather because I am a flawed one. You may indeed find mistakes in some of the renderings, and that is precisely the point. Every history has those flaws, but I want to expose the reader to the friction caused by languages one knows imperfectly, the alienating quality of other people's words, to make the active translation *visible*.[2] The past did not happen exclusively in English, though many histories make it seem as though it did. The footnotes also make the historical trajectory evident: as the book progresses, fewer and fewer footnotes appear; that's because the conversation in science has transitioned to English. (The footnotes can also be fun. Try reading Esperanto—you might like it!) Likewise, many of my sources wrote in foreign languages *poorly*. I have left their bad spelling and grammatical infelicities unadorned by the scholarly "*sic*," except in cases of typographical error. You are also, of course, free to ignore the footnotes and read the text through entirely in English. That is, in truth, how most of science is done today.

But it wasn't always that way: the languages of science used to be multiple. This is a book about *scientific languages*, and I use both terms with their most straightforward meanings. I certainly do not mean that some languages are intrinsically "more scientific" than others (although many have made such claims in the past and still do today, as we will see). I define *science* rather narrowly, consistent with modern Anglo-American usage, to refer to what are often further specified as the natural sciences. To be even more precise, I focus on the comparatively small community of elite, professional scientists, a community that has engaged in international communication for centuries and maintains to the present the highest prestige among investigators of nature. (I exclude here medicine and certain applied sciences, such as agronomy, in part because those practitioners' need to communicate with a nonscientist client base introduces significant complicating issues of popularization that are ancillary to the main issues in this book.[3]) The narrowness of *science* in English is distinctive. Other languages, such as French (*science*), German (*Wissenschaft*), or Russian (наука, *nauka*), use the term to encompass scholarship in a broad sense, including the social sciences and often also the humanities.[4] I follow English usage simply out of conceptual economy, although the ways languages have shifted in those disciplines are interesting and they exhibit a similar linguistic narrowing as the "natural sciences." The natural sciences (physics, biology, chemistry) display the phenomenon I am tracking more vividly.[5] I emphasize these sciences because they are at present almost exclusively in English, and they have been so for decades. If you are interested in what it would be like to live in a world with one language of communication, a world with no Babel, you should look to the natural scientists. They come from there.

At one level, the history of scientific languages is recorded in academic publications, as different scholars investigate nature and then try to persuade their colleagues of the detailed organization of the universe. But it is also a story of informal correspondence, friendly banter at conferences, government reports about the transformation of the scientific infrastructure, press releases, anti-Semitic diatribes, and muttering to oneself during a lonely night in the laboratory. This book ranges from the poetry of ancient Rome to attempts to communicate with alien civilizations, from the nationalist conflicts of the nineteenth century to the dawn of computerized machine translation, with a cast of characters including the greatest scientists of their day as well as (almost) anonymous librarians, politicians alongside linguists, frenzied debaters over the merits of artificial languages spoken by only a few dozen contrasted with attempts to standardize a language across the largest land empire the world has ever seen. It is an intimate and a public history, as befits language—something we all feel intensely about, while at the same time sharing it with communities of strangers.

Here is a truism: scientific activity is communicated in a language. I do not simply mean "in words"; I mean in a particular, specific language, shared by a community of speakers. People can have scientific thoughts, do scientific experiments, have scientific conversations, in whichever language they wish to use—in theory. But in practice, science has not been so conducted. Scientific findings are not usually communicated in Ibo, Bengali, or Polish, at least not at the dawn of the twenty-first century, and not at the dawn of the nineteenth, either. Science, as a lived human activity, has always traveled within a highly constrained set of languages. If we adopt the narrow stratum of elite science and look at the dominant languages in which it has been communicated to the international community of researchers from the beginning of recorded history to now, we end up with a rather limited list. Taking languages that register a statistically significant proportion of the world production of something we might now call *science*, we find (in alphabetical order): Arabic, Chinese (classical), Danish, Dutch, English, French, German, Greek (ancient), Italian, Japanese, Latin, Persian, Russian, Sanskrit, Swedish, Syriac, and Turkish (Ottoman). (I apologize for those I have excluded at the edges; even if you include them, the list does not grow significantly.) There is no other sphere of human cultural activity—trade, poetry, politics, what have you—that takes place in such a limited set of tongues.[6] Behind the truism, therefore, is a fact of tremendous importance. This book is about life in Scientific Babel: how scientists managed to work among this (limited) profusion of tongues, how they hoped to conquer it, and how it came about that the Babel was no more.

Every time you utter something, you need to balance between two competing demands. On the one hand, you would like to express your internal notions, to say exactly what you are thinking or feeling. Of course, this is an ideal; we have all experienced the disconnect between what's in our minds and the clumsiness by which we can formulate it.[7] Yet, for most of us, we get closest to this ideal in our native language or in the language we use most fluently; it is, fundamentally, a speaker-centric choice. I call this *identity*, and it is surely possible for a particular speaker to have multiple distinct identities, speaking to children in her role as a parent most easily in one language, to a spouse in her role as a wife in another, at work as a lawyer in a third. Nonetheless, in this kind of speech, the speaker focuses on the capacity to express herself or himself in that particular role. But what about the audience? With most utterances, you have some particular recipients in mind, real or imagined, present or absent. You want your interlocutor to understand what you say, and this is easiest to achieve by using the language your listener (or reader) understands best, or at least the strongest language you have in common—that is, using what is called by linguists a *vehicular language*. This choice is audience-centric, and I describe it as *communication*. Irreducibly, all utterances occupy a spot on the continuum, trying to express oneself as accurately as possible while at the same time making efforts to be understood correctly.[8] The tension exists within a single language—I am not certain that even now I am presenting my thoughts accurately in what is both my native language and a vehicular language we have already established you understand—but the challenge is magnified significantly when you add language barriers to the mix.

Scientific utterances are no different from ordinary utterances in this regard. Today's overwhelming dominance of one vehicular language may give the impression that science naturally trends toward communication and away from identity, since one's scientific peers need to vouchsafe the validity of one's claims—and, indeed, today science works this way, which helps explain the pressure toward fewer languages. But not necessarily to a single one, for there was a moment when European naturalists *had* a single language—it was called Latin—and they deliberately, consciously chose to give it up. Latin remained a language of communication, but it was joined by Dutch, English, Swedish, Italian, and some others. Identity was allowed in, to a certain extent, for a particular range of tongues. (One might also understand this as communication with a different, more local audience, as we will see.) Where communities fall on the spectrum between identity and communication is historically contingent; different tensions are tolerated differently at different times, but they have not gone away, even if scientific communication happens in a single language. It is, in fact, an omnipresent feature of all interchange, strongly dramatized in the case of science by its prominent intellectual creativity (identity) and its social organization (communication), and that allows us to see how creativity and social organization interact within the spheres of language and language choice. Yet the dilemma is not symmetric. If you are a native speaker of English, your language of identity equals your language of communication; your burden is reduced to the irreducible problem of saying what you mean, shared by all speakers everywhere, without the additional load of strug-

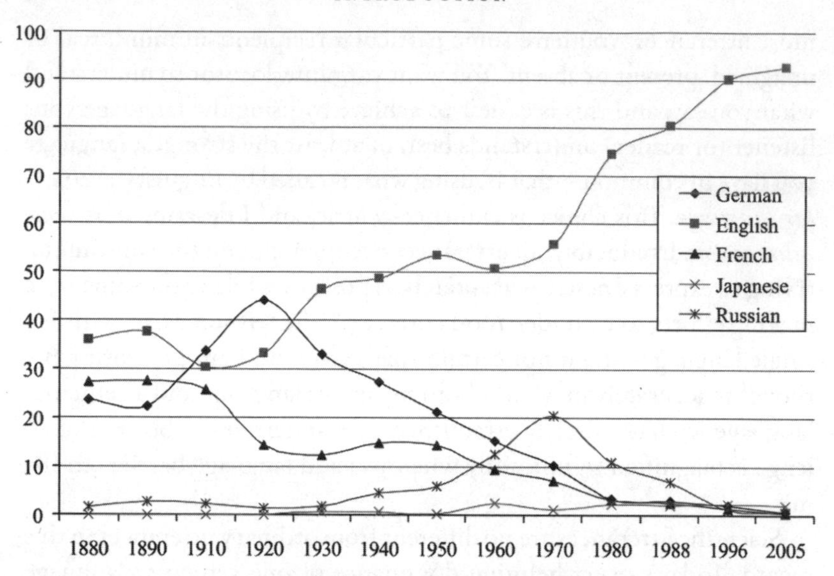

FIGURE 0.1. Graph of the languages in which science has been published from 1880 to 2005, plotted as a percentage of the global scientific literature. Ulrich Ammon, "Linguistic Inequality and Its Effects on Participation in Scientific Discourse and on Global Knowledge Accumulation—With a Closer Look at the Problems of the Second-Rank Language Communities," *Applied Linguistics Review* 3, no. 2 (2012): 333–355, on 338.

gling with a foreign tongue.[9] That is an enormous privilege, but it is a privilege that Anglophones are largely blind to. One goal of this book is to make visible this asymmetry and its consequences.

English is dominant in science today, and we can even say roughly how much. Sociolinguists have been collecting data for the past several decades on the proportions of the world scientific literature that are published in various tongues, which reveal a consistent pattern. Fig. 0.1 exhibits several striking features, and most of the chapters of this book—after an introductory chapter about Latin—move across the same years that are plotted here. In each chapter, I focus on a language or set of languages in order to highlight the lived experience of scientists, and those features are sometimes obscured as well as revealed by these curves. Starting from the most recent end of this figure and walking back, we can begin to uncover elements of this largely invisible story. The most obvious and startling aspect of this graph is the dramatic rise of English beginning from a low point at 1910. The situation is actually even more dramatic than it appears from this graph, for these are percentages of scientific publication—slices of a pie, if you will—and that pie is not static. On the contrary, scientific publication exploded across this period, which means that even in the period from 1940 to 1970 when English seems mostly flat, it is actually a constant percentage of an exponentially growing baseline.[10] By the 1990s, we witness a significant ramp-up on top of an increasingly massive foundation: waves on top of deluges on top of tsunamis of scientific English. This is, in my view, the broadest single transformation in the history of modern science, and we have no history of it. That is where the book will end, with a cluster of chapters focusing on the phenomenon of global scientific English, the way speakers of other once dominant languages (principally, French and German) adjusted to the change, preceded by how Anglophones in the Cold War confronted another prominent feature of the midpoint of the graph (1935–1965): the dramatic growth of scientific Russian.

But, on second glance, one of the most interesting aspects of this figure is how much of it is *not* about English, how the story of scientific language correlates with, but does not slavishly follow, the trajectory of globalization. Knowledge and power are bedfellows; they are not twins. Simply swinging our gaze leftward across the graph sets aside the juggernaut of English and allows other, overshadowed aspects of these curves (such as the rise of Russian) to come to the fore. Before Russian, in the period 1910 to 1945, the central feature of the graph is no longer English but the prominent rise and decline of German as a scientific language. German, according to this figure, was the only language ever to overtake English since 1880, and during that era a scientist would have had excellent grounds to conclude that German was well poised to dominate scientific communication. The story of the twentieth century, which from the point of view of the history of globalization is ever-rising English, from the perspective of scientific languages might be better reformulated as the decline of German. That decline started, one can see, before the advent of the Nazi regime in 1933, and one of the main arguments in this book is that the aftermath of World War I was central in cementing both the collapse of scientific German and the ballistic ascent of English. We can move further left still, and in the period from 1880 to 1910 we see an almost equal partition of publications, hovering around 30% apiece for English, French, and German, a set I will call the "triumvirate." (The existence of the triumvirate is simply observed as a fact in this book; I do not propose to trace the history of its emergence.) French underwent a monotonic decline throughout the twentieth century; one gets the impression (although the data is lacking) that this decline began before our curve does, but to participants in the scientific community at the beginning of our modern story, it appeared stable. My narrative for this earlier period comes in two forms: the emergence of Russian, with a minor peak in the late nineteenth century, as the first new language to threaten to seriously destabilize the triumvirate; and the countervailing alternative (never broadly popular but still quite revealing in microcosm) to replace the multilingual scientific communication system with one conducted in a constructed language such as Esperanto. Long before all of this data, all of these transformations, there was Latin, and that is where the book properly begins.

For all the visual power of the graph, most of this book pushes against its most straightforward reading: the seemingly inexorable rise of English. Behind the graph lie a million stories, and it is history's task to uncover them. There are other reasons for caution. For starters, we must be careful not to take its *quantitative* proclamations as gospel truth. The data comes from abstract journals: periodicals that supply an index of abstracts of scientific publications every year, an index to assist in taming the avalanche of information. (The history of these objects is an important subplot in this book.) A Japanese bibliographer named Minoru Tsunoda gathered a list of percentages of publications from numerous abstract journals (which he chose to publish in French, but in a Japanese journal), and then German sociolinguist Ulrich Ammon—the leading researcher today on the question of scientific languages—plotted the information in graph form, updating it as new information came in.[11] Abstract journals are, however, already a simplification of global production, and what we see here is therefore a selection of which periodicals abstract journals have chosen to include, and this culling obviously biases the results toward the dominant languages. For example, 5,986 scientific and technical journals were published in Brazil in 2007, but only 17 were registered in the Science Citation Index, and therefore the majority do not show up in this kind of data.[12] That obviously hurts the statistics for Portuguese (although quite a few of those journals might publish in several languages, or exclusively in English). Abstract journals, although they do reflect how elite scientists encounter the cutting-edge literature in their fields, do nonetheless generate some distortion, and we should view this curve more to gain a *qualitative* impression rather than a rigorous result. And that impression is extremely difficult to ignore.

As is evident from the above, I use the word *language* in a specific, but rather everyday, manner. I have not written a technical linguistic study, but neither do I use *language* in a literary fashion. There is a sense in which we can talk about "scientific languages" metaphorically: that scientists use a jargon that is not the same as ordinary language; or that biologists and geologists "speak different languages"; or that each individual laboratory has its own particular idiolect that outsiders have a hard time penetrating. Much of the scholarship on the history of science and language concerns this metaphorical sense, and a good deal of it is of the highest intellectual rigor and utterly fascinating. However, precisely this sense, which I will refer to as *discourse*, is not my quarry here.[13] I mean *language* in the brute forms of English, Swahili, Korean, or Russian. That is, I am interested in which languages people choose to use—and *not* use—in various contexts, at different times, in assorted places. I explore the history of these scientific languages mostly from 1850 to the present (although with a necessary excursus into Latin at the beginning), and with a principal focus on Europe and North America, with occasional visits to other parts of the globe. The comprehensive story is obviously bigger than that and could include all of the world over all of recorded history. I restrict myself to this narrower swath for two reasons: one intellectual and one practical. The first is that the phenomenon of global English started there, as did the basic institutions of modern science that were exported (sometimes forcibly, sometimes not) to other parts of the world. That is one significant justification for limiting this first pass, leaving you with a book of manageable size you can hold in your hands.

The second reason is no less important: the languages I happen to know are a subset of these languages of European origin, and I cannot write a history from sources I cannot read and understand. That is a frank admission of ignorance, and you don't come across such things very often in books like this one, but without it you will lack a crucial piece for understanding not just this specific book, but *any* book on the question of scientific languages. To write this book I have used sources in English, French, German, Russian, Latin, Esperanto, and the latter's offshoot, Ido.[14] I hesitate to say that I "know" these languages, because competence in a tongue is always a relative matter, and I am more fluent and subtle in some of these languages (my native English and also Russian) than in others (French and German), and some, such as the Latin I learned in order to write this book, are still very much works in progress. I pen this confessional paragraph to illustrate several points that condition the following historical narrative.

The first is that *knowing* a language is measured by a standard that changes over historical time. Many of the scientists I discuss read and published science in three or four languages as a matter of course. Was this a vanished race of polyglot naturalists? Of course not. Some of them were more linguistically gifted than others, to be sure, but most of them managed with a dictionary and consultation with those more adept (as I often did). Today, scientists expect their peers to be relatively fluent not just in reading and writing English, but also speaking it.[15] The standard of fluency has gone up; the standard of quantity has gone down. The second point is that I happen to read these languages *and not others*. I chose to learn Latin to write this book. I wish I had the time and energy to learn Japanese, which has an important role to play in the history of scientific languages in the twentieth century, or Dutch, which was central in the seventeenth and eighteenth, or Italian, which continued its salience into the early nineteenth century. If I had, the story you read would be different. (I particularly regret the comparative neglect of East Asia in this account.) The few extant studies of scientific languages are written by those who do not know Russian, and those renditions look rather different than mine, which emphasizes that language quite a bit. I hope that those with different linguistic capacities—or even the same ones, calibrated to different degrees—will take the question of languages in communicating knowledge and run with it. We need more, and more diverse, accounts.

It is necessary to state all this up front because of the seemingly universal phenomenon of linguistic citation bias. Scholars disproportionately cite literature in the languages they feel most comfortable with, which are often their native languages. According to results cited in one 1981 study, American and Indian journals offer citations that are 90% to English-language literature, which was greater than the proportion (roughly 75%) of English material in the scientific literature in that day. Quality and even relative quantity, therefore, is not a full explanation. Likewise, the French cited 29% French, Germans 22% German, Japanese 25% Japanese, Soviet researchers 67% Russian—all in greater proportion than the baseline literature would suggest. (Articles in Chinese were cited only in China, for example.)[16] I doubt I am an exception to this rule—many of my citations are to Anglophone literature, and I include almost no citations outside of my dominant linguistic core set. The scholarship you read is always biased by the linguistic capacities of the scholar. It's only honest to admit it.

This is all the more important because of a very widespread notion that translation is trivial with respect to science, such that some studies neglect to mention a language barrier at all, or recognize that "[a]l-though language of publication is an inescapable feature of scientific communication, it is most often treated as background noise."[17] Or, in what amounts to the same thing, that science has uniform content and is therefore beyond translation: "Scientific prose has in fact a valuable and a not uninteresting characteristic—almost alone among all the different categories of prose it can be translated into languages other than the language in which it was first written, not merely satisfactorily but perfectly."[18] Such statements are based on a philosophical assumption that scientific claims represent the world unfiltered, and therefore scientific utterances are a kind of "metalanguage" that are only partially expressed in any individual tongue but are equally true in all of them. This belief is a central one to many of the scientists we will encounter in this book, but it is a view that is complicated by the experiences of those individuals who daily have to translate between and among various scientific languages. For them, translation has been a source of frustration, and often conceptual confusion.

The power of this notion of a metalanguage stems from the unquestionable success of mathematization of the sciences.[19] When I have discussed this project with both scientists and humanists, I have often been told that there is no need to pay attention to the languages in which science is written because scientists can simply read the equations and figure out what is going on. This might be true in certain cases, but it is hardly true generally. Even for an ostensibly "hard" science like chemistry, papers contain more than isolated chemical formulae and mathematical equations. You read descriptions of the reaction, analyses of colors and odors, detailed explanations of method. This verbosity is one of the reasons why the emphasis in this book will be upon chemistry, which shares both in mathematical formulations and in more descriptive scientific traditions, and therefore exposes the capacities and limits of each. Even in cases of strongly mathematized sciences, like classical mechanics, a bare equation never tells you all you need to know. Consider this simple one:

$$M \propto WgT^2/l$$

What does it say? Without further context, you can tell me that M is directly proportional to the square of T, and inversely proportion to l, but what does it *mean*? Mathematical equations are incredibly powerful tools, economically expressing detailed relationships and enabling stupendous manipulations that seem impossible without the formalism. But they are also parasitic on the human languages that surround them, the words that tell you what the variables represent.[20] Without the context, an equation like the above is neither true nor false as a scientific claim.

So what does that expression say? This particular formula is Indian-American astrophysicist Subrahmanyan Chandrasekhar's rendition of Proposition XXIV, Theorem XIX, in Book II of Isaac Newton's *Philosophiae naturalis principia mathematica*, usually abbreviated as the *Principia*, of 1687.[21] Here is what the text says:

> *In simple pendulums whose centers of oscillation are equally distant from the center of suspension, the quantities of matter are in a ratio compounded of the ratio of the weights and the squared ratio of the times of oscillation in a vacuum.*[22]

The formula above transcribes this prose, and describes how a pendulum moves. However, the above English is *not* what Newton wrote, but instead is I. Bernard Cohen and Anne Whitman's 1999 translation of the *Principia*. (Chandrasekhar used an earlier translation.) What Newton *actually* wrote was:

> Quantitates materiae in corporibus funependulis, quorum centra oscillationum a centro suspensionis aequaliter distant, sunt in ratione composita ex ratione ponderum & ratione duplicata temporum oscillationum in vacuo.[23]

Is that the same thing as the formula? Well, it is and it isn't. My point is that calling mathematics a language is a move in the direction of discourse, and does not, in any event, overcome the problem of the language barrier.

Mathematicians experienced the same tension between identity and communication in their professional lives; the language barrier and the difficulty of translation have historically been neither incidental nor irrelevant to mathematics, washed away by the balm of the formalism. The mathematical community today, like other scientific communities, has also been strongly squeezed (by publishers, by international conferences, by the exigencies of communication) into English, but formalism does indeed help, for the transition has been less total and less rapid than in the more descriptive sciences, and mathematicians are often justly proud of their ability to read papers published in other (usually

European) languages. In the late nineteenth century, German dominated mathematical publication, but not exclusively, and mathematicians were expected to keep track of developments in several tongues—not just through reading, but also through lecturing and conversing with their international peers.[24] But even linguistically gifted mathematicians recognized that the formalism was a vital tool in bridging Scientific Babel. In 1909, French mathematician Henri Poincaré gave a series of lectures at the German university town of Göttingen, then the epicenter of world mathematics. For his final lecture, he chose to abandon German:

> Today I have to speak French, and I must apologize for it. It is true that in my earlier lectures I expressed myself in German, in very bad German: to speak foreign languages, you see, is to want to walk while one is lame; it is necessary to have crutches; my crutches were until now mathematical formulas, and you could not imagine what a support they are for an orator who does not feel himself very firm.—In this evening's lecture, I do not want to use formulas, I am without crutches, and that is why I must speak French.*[25]

I assume most of his audience understood it: languages had been built into their scientific training. Learning how to handle yourself in several languages, even only passively—being able to listen and read but not speak or write—was part of the scientific life. Both the (comparative) equality of the burden and the degree of fluency have changed; the problem has not.

Today's situation raises obvious issues of fairness, whereby non-Anglophones have to study English intensively and deploy it with some high level of fluency, while native speakers of English can conduct their science without that educational burden. Questions of equity will come up often in our story. But aside from those, does this almost total dominance by a single language—or, earlier, a smaller set of languages (for

*"Aujourd'hui, je suis obligé de parler français, et il faut que je m'en excuse. Il est vrai que dans mes précédentes conférences je me suis exprimé en allemand, en un très mauvais allemand: parler les langues étrangères, voyez-vous, c'est vouloir marcher lorsqu'on est boiteux; il est nécessaire d'avoir des béquilles; mes béquilles, c'étaient jusqu'ici les formules mathématiques et vous ne sauriez vous imaginer quel appui elles sont pour un orateur qui ne se sent pas très solide.—Dans la conférence de ce soir, je ne veux pas user de formules, je suis sans béquilles, et c'est pourquoi je dois parler français."

Albanian and Zulu were never even "minor languages" of science)—have implications for the *content* of science? That is, does it *matter* that science has a particular linguistic structure? There are two ways of understanding that latter query, one philosophical and the other pragmatic.

Taking the first tack, we come to the Whorfian hypothesis, named after Benjamin Lee Whorf, a part-time linguist (and full-time Connecticut fire inspector) who argued for a strong form of linguistic relativism that posited that the languages in which we think not only shape our perceptions of reality, but in some way determine them.[26] Whorf formulated his basic principle of linguistic "relativity" (a nod to Albert Einstein's principle of relativity from physics)—namely, that "all observers are not led by the same physical evidence to the same picture of the universe, unless their linguistic backgrounds are similar, or can in some way be calibrated"—in a series of articles published in 1940 in *Technology Review*, the house journal of his alma mater, MIT. He explicitly situated this enormously influential idea in the context of scientists, arguing that we should not be surprised that there was considerable agreement about the laws of nature, since those were developed by individuals speaking closely related languages: French, English, and German, all members of the Indo-European language family, just like the ur-scientific language, Latin. Whorf contended that a person's native language generated the categories through which she viewed the world, so that speakers of languages with very different notions of, say, time—like Latin and Hopi—would come to different physical conceptions. Tell me what you speak, and I will tell you what you think. What then should we make of the agreement in the sciences, given that not all scientists, even at the time of Whorf's writing, were native speakers of Indo-European tongues? No worries, for Whorf: "That modern Chinese or Turkish scientists describe the world in the same terms as Western scientists means, of course, only that they have taken over bodily the entire Western system of rationalizations, not that they have corroborated that system from their native posts of observation."[27] Whorf's notion has been tremendously controversial, and the evidence for it (for example, different ways of parsing colors) is strongly contested.[28] Nonetheless, if it were true, even in a limited degree, then one might worry that the reduction in scientific languages has produced a concomitant reduction in conceptual breadth. I am agnostic on the outcome of this debate; I only note that the debate itself is an emergent part of our history, and motivated many of the scientists and intellectuals we will meet in later chapters.

Repeatedly in the pages that follow, we will find instances of scientists arguing that the choice of language of publication makes an active difference; whether that claim seems credible very much depends on the situation, and this brings us to the second way in which the choice of a scientific language matters. Until the almost universal dominance of English, choosing to publish in a particular language always carried the possibility that you would not be understood, simply because your peers could not (or would not) read the work. The language barrier can be understood as a kind of *friction*, and regardless of whether it changes the content of the science (as Whorfians would have it), there is no question that language friction has shaped the manner in which scientists have operated in the real world. Before beginning the story from the dusty conjugations of Latin, it would be helpful to get a sense of how such a phenomenon worked in the historical past, thereby illustrating how many well-worn episodes in the history of science take on a different tenor if viewed through the lens of scientific languages. In that spirit, allow me to offer here a brief account of one of the most archetypal set-pieces: the Chemical Revolution of the late eighteenth century. There are few topics in the history of science which have been so often addressed in terms of language than the development in the final decades of the eighteenth century of the oxygen theory of chemistry by Frenchman Antoine Lavoisier and the overthrow of Englishman Joseph Priestley's phlogiston theory of combustion. This scholarship uses to the fullest the notion of language in the metaphorical sense.[29]

The basic events of the Chemical Revolution lend themselves well to this kind of analysis. Beginning around 1770, both Priestley and Lavoisier came to be dissatisfied with the regnant theory of burning, which posited the existence of a principle of combustion called phlogiston. For decades, *combustion* had been defined as the exit of phlogiston from a substance: wood stopped burning when all the phlogiston had left; certain gases, notably "fixed air" (we now call it carbon dioxide), snuffed out flames because these gases could absorb no more phlogiston. It was a wonderful qualitative theory, providing a theory of acidity and color to boot, but it had problems of quantity—burned substances seemed to be heavier than their source materials, meaning phlogiston might have "negative weight"—and other difficulties associated with the proliferation of new "airs" (Lavoisier called them *gases*) released in chemical manipulations. Priestley sought to reform phlogiston theory to account for the objections; Lavoisier threw the whole notion overboard. For him, burning was not the *release* of phlogiston, but *combination* with a new gas, that he dubbed "oxygen," from the surrounding air. Given that everyone has heard of oxygen, and phlogiston survives only in the anecdotes of historians of science, you can surmise who won. For Lavoisier, it was a victory of method, which was nothing more than discourse: "Thus an analytic method is a language, a language is an analytic method, and the two expressions are, in a certain sense, synonymous."* [30] What might we learn if we examine this same dispute by tracing the languages in which it was conducted? That is, as a disagreement between a man who functioned primarily in French, and one who wrote his important works in English?

French has often been proclaimed the central language of intellectual life, including natural philosophy, in the eighteenth century.[31] Intellectuals across Europe either read the French language, or devoured vernacular translations of French texts, or, failing that, translations into Latin from the French. The rise of the French vernacular was a long time in coming, dating plausibly back to before the ninth century, though Old French began to stabilize into a modern standard, centered on the Parisian dialect, only in the twelfth century, and by the seventeenth—and the reign of Louis XIV, *le roi soleil*—a powerful myth of an unchangeable, perfect French had already materialized.[32] When Louis revoked the Edict of Nantes in 1685, newly persecuted Protestant Huguenots fanned out across Europe, carrying the prestigious French language with them.[33] The Treaty of Rastatt in 1714 began the enshrining of French as the leading language of international diplomacy, even among the Germanophone principalities of the Holy Roman Empire, a transformation complete by the Treaty of Hubertusburg (1763), when the French text acquired priority over the Latin.[34] These well-known milestones demonstrate the salience and prestige of the Parisian language.

French was so ubiquitous in intellectual life that the Prussian Academy of Sciences in Berlin operated in the language and played a central role in propagating it as the only fitting tongue for scholarly interchange, most notably in its 1783 prize question. Essays were solicited to answer the following question (posed, ironically, originally in Latin, but most commonly rendered in French):

*"Ainsi une méthode analytique est une langue; une langue est une méthode analytique, et ces deux expressions sont, dans un certain sens, synonymes."

What has rendered the French language universal?
—Why does it merit this prerogative?
—May one presume that it will maintain it?* [35]

The prize was jointly awarded to the Comte de Rivarol, for an essay now lauded as a monument to French prose style, and Johann Christoph Schwab, whose response was submitted in German but was widely circulated in an 1803 French translation by Denis Robelot. Both of them echoed long-standing Enlightenment notions that the dominance of French was not merely a consequence of Parisian political power, but a logical entailment of the clarity of the language.[36] For Rivarol,

> [w]hat distinguishes our language from ancient and modern languages is the order and the construction of the sentence. This order must always be direct and necessarily clear. French names at first the *subject* of the discourse, then the *verb* which is the action, and finally the *object* of this action: this is what comprises common sense.[. . .] French syntax is incorruptible. From this that admirable clarity results, the eternal foundation of our language. *That which is not clear is not French*; that which is not clear is still English, Italian, Greek, or Latin.† [37]

And for Schwab, addressing his German compatriots, the state of affairs was similar:

> I thus say: not only should we not be jealous of the empire of the French language, but we should join our wishes and our efforts so that it becomes universal. The extensive connections which are formed on all sides among Europeans provides them with an absolutely necessary universal instrument of communication. Latin is

* "Qu'est-ce qui a rendu la langue française universelle?
—Pourquoi mérite-t-elle cette prérogative?
—Est-il à présumer qu'elle la conserve?"

† "Ce qui distingue notre langue des langues anciennes et modernes, c'est l'ordre et la construction de la phrase. Cet ordre doit toujours être direct et nécessairement clair. Le français nomme d'abord le *sujet* du discours, ensuite le *verbe* qui est l'action, et enfin l'*objet* de cette action: voilà ce qui constitue le sens commun.[. . .] [L]a syntaxe française est incorruptible. C'est de là que résulte cette admirable clarté, base éternelle de notre langue. *Ce qui n'est pas clair n'est pas français*; ce qui n'est pas clair est encore anglais, italien, grec ou latin."

> dead, it cannot be this universal instrument. The language of the Frenchman has become this [instrument] because of its merit; let it therefore retain its universality.* [38]

These quotations resemble rather strikingly comments about German in the early twentieth century, and English in the early twenty-first. Nothing seems clearer to native speakers than the limpidity of their own tongue. You would scarcely guess from these paeans to French's universality that in the midst of the French Revolution in the 1790s, Henri Grégoire estimated that French was dominant in only 15 of the country's 89 departments, sharing the stage with German, Basque, Breton, Occitan, Provençal, and other patois. One of the great crusades of the early Revolution was, in fact, to make French universal *in France*.[39]

At the moment of the other revolution, then—the Chemical one—French was simultaneously touted as a universal scholarly language and yet not quite one. We know that Priestley understood French fairly well, keeping abreast of publications that came out of the anti-phlogistonist group around Paris in that language and in Latin. (Priestley also taught Latin and possibly Greek, and studied "High Dutch"—what we now call German—in order to follow the scientific literature from Central Europe.[40]) The new chemical journal, the *Annales de Chimie et de Physique*, accepted submissions only in French, and the British simply read it that way. (An early effort at translation foundered due to lack of interest.[41]) We also know that, although Lavoisier had a good grounding in classical languages, he understood no English.[42] As a native speaker of what was touted as the universal language, he saw no need to learn the awkward speech from across the Channel, for the English would necessarily read his own work without his publishing in their language. This insularity concomitant with a language's dominance is a common historical pattern, and we will encounter it many times.

Nonetheless, Lavoisier knew that the English "pneumatic chemists"

*"Je dis donc: non-seulement nous ne devons pas être jaloux de l'empire de la langue françoise, mais nous devons réunir nos vœux et nos efforts, pour qu'elle devienne universelle. Les liaisons étendues qui se sont formées de tous côtés, entre les Européens, leur rendent un instrument universel de communication absolument nécessaire. La langue latine est une langue morte, elle ne peut être cet instrument universel. C'est par son mérite que celle des François l'est devenu; qu'elle conserve donc son universalité."

were discovering new airs and that their modifications of the phlogiston concept could be central to his own theory of combustion. How did Lavoisier learn what the Britons were up to? The same way language barriers in science were always transcended before they ceased to exist: through translations and polyglot collaborators. Lavoisier read about the pneumatic chemistry experiments of Stephen Hales, published in English in 1727, through the German Johann Theodor Eller's thesis on the elements, published in French in the 1746 *Mémoires* of the Berlin Academy. Lavoisier came across it twenty years later.[43] He eventually read Hales in the 1735 translation of the distinguished French naturalist, the Comte de Buffon. Translations slow things down. But Priestley's work he discovered more rapidly. He heard of it through itinerant factotums who called themselves *intelligencers*. Jean Hyacinthe de Magellan came across Priestley's English-language publications on the release of "dephlogisticated air"—an invigorating gas that supported combustion brilliantly, and that we now call oxygen—and quickly produced a long summary in French, forwarding it along with the original to Trudaine de Montigny, the Director of the Royal Bureau of Commerce. The latter was an amateur scientist and rewarded Magellan's tips, while at the same time passing the note along to Lavoisier, who read it to the Académie de Sciences on 18 July 1772. A translation of Priestley was published the following year.[44] In October 1774, Priestley himself traveled to France, and his *Experiments and Observations*, including the crucial experiments on dephlogisticated air, came out the following year, translated by Jacques Gibelin.[45] Lavoisier never had to budge from his native language.

Priestley's Irish colleague Richard Kirwan stepped up to defend phlogiston against the oxygen chemists. Kirwan learned about the French views the old-fashioned way: by reading them in the original. He published his *Essay on Phlogiston* in 1787, one of the most sophisticated chemical treatises of the decade and the last major defense of Priestley's reformed theory. Lavoisier and his peers had to respond to it—but how? They couldn't read it. Lavoisier turned to his wife, Marie-Anne Pierrette Paulze, who had learned English precisely for such purposes. Her French translation came out only a year later, complete with extensive footnotes and interstitial essays by her husband and his colleagues, dissecting and refuting Kirwan's arguments. (She was aided by Madame Picardet, the assistant and mistress of Lavoisier's fellow chemist Louis-Bernard Guyton de Morveau.[46]) The challenge, she noted in an unsigned preface, was formidable:

> If the French Chemists whom [Kirwan] has battled destroy his objections, perhaps one would be right to conclude that there was nothing solid in making them? It is principally upon this last consideration that one is determined to undertake the Translation of the Essay on Phlogiston: one has tried to render it as literally as the difference of the languages can allow, & to express, in the clearest and most precise manner, Mr. Kirwan's ideas: the extreme exactitude which scientific matters demand requires the greatest severity in the choice of expressions.* [47]

This French edition was then translated *back* into English in 1789, complete with the anti-phlogistonist commentaries.[48] Kirwan was not convinced, but he retired from the fray. After that, Priestley was the lone defender of his revised phlogiston chemistry, while Lavoisier's theory and the accompanying nomenclature was translated and distributed across Europe (although not without significant linguistic obstacles).[49]

Of course, the Chemical Revolution was centrally about concepts of combustion, and nothing in the above brief story disputes that. But if we pay attention to the frictions and asymmetries imposed by Scientific Babel, by the need to translate from English into French, and *not* from French into English—all at a moment when the cultural status of French as a universal language was both taken for granted and not at all secure, not even in France—we are forced to pay attention to things like timing, social status, the labor of bibliographic searching, and cultural miscommunications. Such hiccups, backtracks, and rethinkings are at the heart of this book, which focuses less on grand demographic and geopolitical transformations than on the careers and perceptions of individual scientists, struggling to understand and make themselves understood in a polyglot world.

Reading the story of the Chemical Revolution in miniature also introduces the major science that will dominate the account that fol-

*"Si les Chimistes François qu'il a combattus détruisent ses objections, peut-être sera-t-on en droit de conclurre qu'il n'y en a pas de solides à leur faire? C'est principalement d'après cette dernière considération, qu'on s'est déterminé à entreprendre la Traduction de l'Essai sur le Phlogistique: on s'est appliqué à la rendre aussi littérale que la différence des langues a pu le permettre, & à exprimer, de la manière la plus claire & la plus précise, les idées de M. Kirwan: l'extrême exactitude qu'exigent les matières scientifiques, oblige à la plus grande sévérité dans le choix des expressions."

lows. In order to rein in the proliferating cases of language, translation, and counter-translation culled from the history of science of the past several centuries, I have emphasized the science of chemistry, although I have not been slavish about this and have gladly appropriated stories from mathematics to botany to physics when they serve to illustrate a point. There are three reasons why chemistry provides a fitting entrée into the world of scientific languages. The first, as we have just seen, is that chemistry and language have been explicitly entwined from its modern beginnings in the eighteenth century, for chemistry is a science of description, taxonomy, and nomenclature as much as it is about test tubes, pipettes, and Bunsen burners. Chemists worry about what to name things and how to make those names correspond across human languages. Second, chemistry has its own formulae, established in the early nineteenth century and serving as another foundation that highlights the tension between the universality of the symbolism and the diversity of individual chemists' tongues.[50] And, third, sheer numbers: chemistry was, in the nineteenth and twentieth centuries, simply the largest science, spanning the gamut from pharmaceuticals to dyestuffs to weaponry to quantum theory. The larger the science, the more global its span, the more Scientific Babel becomes visible. There are of course other histories on these questions that could be written. For now, I will begin here.

Where, precisely, is that? The core of this book traces the story from the consolidation of the triumvirate of English, French, and German around 1850—that consolidation itself is skipped over, and must be left to another history—and then follows the graph of scientific languages forward, moving through the decades up to the present. Each chapter focuses on a principal language (Russian, Esperanto, Ido, German, English) but not exclusively so, because we cannot understand the history of any individual language without seeing how its users deploy it in dialogue with its competitors, shaping it to the lacunae of others. Certain characters trace through several chapters, others blaze across the storyline briefly before fading away. Some episodes or themes could have found a home in any of several chapters, and I have addressed each where it seemed most fitting. Along the way, each chapter presents a different central question in thinking about scientific languages: from translation to publishing, from computerization to emigration, from standardizing a new scientific language to attempting to preserve a venerable one from extinction. It is perhaps difficult to see these various experiments at transcending the dilemmas of Scientific Babel as hopefully as their advocates did, given that we know the outcome. It is important to underscore, however, that the actors in these pages did not know how things were going to turn out—they knew only where things had begun, in their Western, European scientific tradition. And so this book begins where they thought it should: with the dream of universal Latin.

CHAPTER 1

The Perfect Past That Almost Was

> Nec me animi fallit Graiorum obscura reperta
> difficile inlustrare Latinis versibus esse,
> multa novis verbis praesertim cum sit agendum
> propter egestatem linguae et rerum novitatem[. . .].*
>
> LUCRETIUS[1]

All languages are, in an important sense, imagined. This might sound absurd: you use language every day; I am using it right now to put these words in sequence in order to convey meaning. What's imaginary about that? But I did not say "imaginary." I said "imagined." The things that we refer to as languages—Swahili, Mongolian, Thai, English—are not objects sitting out there in the world, like a peculiar rock or a specific yellow clapboard house. All around us, words flow (spoken, written, gestured), and we use those words to communicate with other people. Sometimes, communication fails. If you don't know Telugu and your neighbor addresses you in it, then mutual intelligibility is zero. If you know Russian and your neighbor knows Ukrainian, then mutual intelligibility can be quite sizable. You are communicating even though you are not speaking the same language. If you both use English, then mutual intelligibility is almost total. Almost, but not quite—and that is the essence of what I mean by "imagined." We each speak our own idiolect, our own storehouses of words put together by our own grammars. When our own specific set of language rules meshes with someone else's, we call that speaking the same language. It is an imagined convergence.

Imagined in precisely the same sense that Holland or Canada is imagined. There are borders to nations, which are sometimes natural barriers

* "Nor does it escape my mind that the dark discoveries of the Greeks / Are difficult to illuminate in Latin verses, / Principally since one must make many new words / Because of the poverty of the language and the novelty of things[. . .]."

(a chain of mountains, a deep river) and sometimes merely conventions, lines drawn by explicit agreement or simply by habit. But on the edge between Canada and Minnesota, it is not obvious which side you are on unless someone with the imprimatur of officialdom tells you: Manitoba. Likewise, on the conceptual border between Dutch and German sits a range of language mixtures, blends that are purged through formal education. What we do—routinely, habitually, necessarily—is draw artificial lines around tongues and designate them as separate. *That* person is speaking English; *this* one speaks Welsh. The woman there is speaking English too, she's just from Glasgow. (We imagine that as English.) Languages are no less real for being imagined, and it matters how they are imagined, and by whom. The entity that Chaucer would have called "English" is not the same as Shakespeare's, or Hemingway's, or yours.

"Scientific languages" are either specific forms of a given language that are used in conducting science, or they are the set of distinct languages in which science is done. In either event, we are talking about imagined constructs, and the goal of this book is to trace out the historical variability and specificity of both meanings over time: how they relate to each other, how they diverge, how the set of languages that can participate grows or shrinks. Since this is a history of Western science, we must begin with the most persistent archetype of a scientific language: Latin. Almost every time a person makes an assertion about scientific languages, their imagined yardstick is the native language of a Mediterranean city-state that flourished over two thousand years ago.

Latin has been imagined in two primary ways in the history of Western science from the early modern period (roughly, fifteenth through eighteenth centuries) to the present. Those living in a world surrounded by various learned languages—French, Dutch, German, Italian, and so on—tended to imagine Latin either as a Paradise lost, a moment of universal comity before the descent of Babel, or as an artificial straightjacket that Europe is better off without. Readers of this book, however, do not live in such a multilingual universe; for you, science is performed almost universally in English. The contemporary status of English changes the way we view Latin. If you think that one language for science improves efficiency and understanding, the rejection of Latin appears as a monument to human folly; if you lament the loss of individuality and heterogeneity, then we are back to Paradise lost, but this time our Eden is polyglot.[2] English has sometimes been called a *lingua franca*—a problematic category named after a complex trading pidgin of the Renaissance Mediterranean—but that is not quite right. English is not a pidgin, it is not low status, and it is not (in its scientific form) variable. English is not today's lingua franca; it is our Latin.[3]

This chapter has a double task. On the one hand, we will follow Latin from ancient Rome (Republic and Empire), exploring its detailed history with the assistance of a learned army that has mapped the ins and outs of this storied tongue. (Among the panoply of scientific languages, none has been more thoroughly and well researched than Latin, and I gratefully acknowledge my debt to these scholars.) We will see that while Latin did function for a period as a universal language of scholarship and natural philosophy—the predecessor to what Anglophones have come to call, since the early nineteenth century, *science*—it served in this role for a relatively short span of its long history. The dominance of Latin started almost a thousand years after the fall of Rome, and fell into decline (but not extinction) three centuries later. "Scientific Latin" both started later and lived longer than you might expect. Our second story runs alongside this in counterpoint: how people have imagined, lauded, and berated Latin throughout this long history, and especially how they understood the eclipse of universality in scientific communication. This chapter is about the birth of Scientific Babel: not just the origin of the profusion of tongues for research, but the emergence of the idea that multilingual scientific communication *was* a Babel, a curse afflicted upon the scholarly community. We begin to imagine scientific languages by imagining Latin.

The Roman Language of Science

In our modern Anglophone world, one cannot assume readers know Latin, even a smidgen redolent of the dust of forgotten schoolbooks and diligent turns copying declensions on the blackboard. So our story must begin with a somewhat abstract tour of Latin's linguistic features, enough to understand both the charms its enthusiasts saw in it as well as the torment and frustration that afflicted two millennia of schoolchildren. Chances are, you come across a healthy dose of Latin in your casual readings and meanderings, and you know enough—even if you don't "know Latin"—to identify ipso facto, cogito ergo sum, ecce homo, and carpe diem as Latin, and perhaps even what these phrases mean. It is a language you can imagine without study, for the Western tradition is saturated with it.

Latin is an Indo-European language, sharing a common ancestor with every other major language of science in the modern period with the exception of Japanese. It is a case language, which means that nouns and adjectives indicate their grammatical function in their form (by inflections, exhibited in the case of Latin through suffixes), enabling a much freer word order than we are accustomed to in English (although there are general regularities, such as a preference for the verb at the end of a sentence or clause). Sometimes, a noun is the subject of the sentence, the doer of action: *The animal* eats the apple. This is the nominative case. Sometimes, it is the direct object of action: The boy eats *the animal*—the accusative. Other relations are possible, marked off in different Latin cases: genitive (the *animal's* apple), dative (the boy gives the apple *to the animal*), and ablative (a hodge-podge of possibilities: the boy walks *with the animal*; the boy steps *away from the animal*; the boy rides *the animal* to the store). Occasionally you even see a vocative (*Animal*, get over here!). That all seems relatively straightforward, requiring only that you memorize the pattern for five different cases which govern the inflections. Except that there are five separate *patterns* (called declensions) across which the Latin vocabulary is strewn, with different inflections for the three genders (masculine, feminine, neuter), and for the plurals of each of these. Adjectives agree with nouns in number, gender, and case, but have their own declensions. And then there is the verbal system: four (or five, depending how you count) different categories of *regular* verbs, each with six basic tenses, completely different endings for the passive voice—English does the passive and many tenses with helping verbs—participles, gerunds, a rich subjunctive mood, and more. There are radically different grammatical forms for reporting the speech of others, depending on whether what you are reporting is a command, a question, or (worst of all) statements. It's fiendishly complicated and entrancingly beautiful, all at once.

It was also spoken, as a matter of course, by senators, slaves, four-year-old children, and village idiots for hundreds of years as the language of one city, and then across the sprawling Roman Empire—encompassing what is today France, parts of Britain (for a while), Spain and Portugal, North Africa, Egypt, much of the Middle East, Turkey, and the Balkans. What strikes the student as an immensely complicated structure was ordinary, everyday language, no more difficult to grasp than the native Anglophone's easy choice of *a*, *the*, or nothing to preface nouns. (This is not a trivial matter, as you will easily find if you try to enumerate rules for definite and indefinite articles. Those who do not speak languages with articles, such as Russian, will thank you. Latin is also article-free.) Appreciating Latin's past *ordinariness* is essential for grasping its position as a language of science in ancient Rome.

Latin started as a local language of the region around the city of Rome on the Italian peninsula, one member of the Italic language family that cohered into a sophisticated and flexible tongue as various dialects from surrounding Latium congregated in the new metropolis.[4] By the end of the first century AD, Latin had eradicated every other native language in Italy, except for Greek, spoken by the descendants of colonists from the Greek city states who populated towns in the south and on the nearby island of Sicily. As the Roman Republic conquered new territories and they were incorporated into an eventual Empire, the language spread. Native Celtic languages were extinguished in Iberia and Gaul, and eventually even the Punic and Berber languages of North Africa were displaced by Latin. "Latin," of course, is imagined. As a lived language, ancient Latin exhibited extensive regional diversity, as you would expect over such a broad geographic area. African Latin was the most distinctive, but all forms had shades of vocabulary and even syntax, variability traceable today in the descendants of Latin's fragmentation, the Romance languages (Catalan, French, Italian, Portuguese, Provençal, Romanian, Sardinian, Spanish, and others).[5] Most of the languages treated in this book cleave doggedly to a written standard, and that is because this is a history of scholars, who like such things. The standard most commonly hoisted for Latin is that of Marcus Tullius Cicero, about whom more in a moment.[6]

First, however, something about the spread of Latin bears a second look: Latin eradicated all the languages of Italy *except Greek*. Greek was special. Throughout the Eastern Mediterranean, Latin shared space with the language of the Greeks, which functioned as a vehicular language for centuries for everything but official Roman administration. And not just in the East, but in Rome itself. From the height of the Republic until the collapse of the Empire, the Roman elite were bilingual, an indication of the immense admiration the upper classes possessed for the art and learning of the ancient Greek city-states and for the Macedonian conqueror Alexander the Great, who in the fourth century BC spread Greek to Egypt in the South and the borders of Persia in the East. The Emperor Claudius spoke Greek in the Senate to Greek-speaking ambassadors, foreshadowing the philhellenism of Emperors

Hadrian, Antoninus Pius, and Marcus Aurelius, and the children of the elite learned Greek from slaves, private tutors, and grand tours to Hellas. Greek was the only foreign language so esteemed.[7]

Greek's privileged status in the East of the Roman Empire represented a particular stage in the history of the longest continuously attested language in the European sphere. It was never seriously threatened as a vehicular language, although we should keep in mind that most of the population of these regions spoke neither Greek nor Latin as a first language (Aramaic, Coptic, and Armenian come to mind, among many other languages, now lost).[8] A particular variant of Greek, called *Koine*, was essential for administration and learning.[9] Ancient Greek had been a cluster of different dialects, such as the Ionian of Homer or the Attic of Sophocles's Athens, but by the time of Plato and Aristotle Attic had emerged as dominant, and it later evolved into the transregional *Koine*. This was the language of the Eastern Roman Empire, formalized in 212 AD, and it evolved into the medieval Greek of Byzantium. Although Latin was far from absent in the East, it is no exaggeration to say that the impact of Greek on Latin was substantially greater than the reverse, and became more so after a worsening of relations in the second century symbolized by the Roman sack of Corinth in 146.[10] *Koine* would become the language of early Christianity, but before then it had long served as the language of intellectual intercourse in the late Roman Republic and early Empire.[11]

That meant it was the language of science, too. Greek was the language of philosophical speculation (Aristotle), mathematics (Euclid and Archimedes), astronomy (Ptolemy), and medicine (the ubiquitous Galen, as well as the collection of authors conventionally blended together as "Hippocrates"). That these scholars wrote in Greek is undeniable. It is equally irrefutable that many of these Hellenophone scholars were, in most meaningful senses of the term, Roman. Consider that the first names of both Ptolemy and Galen, second-century authors, was "Claudius," the most Roman of monikers. While Latin speakers would constantly lament the superiority of Greek for natural science—as Lucretius does in the epigraph to this chapter—and composed most of their natural philosophy in Greek, it is simply incorrect to declare, as some have, that "[i]t is a universally recognized fact that imperial Rome was utterly uninterested in pure scientific speculation."[12] Lucretius shows us the origins of the error: Latin was often a language used to popularize Greek work, to take cutting-edge natural philosophy and expose it to a broader, less elite (and therefore not bilingual) Roman audience. Declaring Latin not a language of science excludes popularization from the realm of scientific activity, and (what is worse) denies that "engineering"—at which the Romans excelled—is not an important aspect of natural knowledge.[13]

It also ignores the obstacles to becoming a scientific language. By the time Romans began to speculate about the nature of matter or the motions of the cosmos, writers had been expounding on such topics for centuries in Greek. Scientific languages are not born, they are made, and made with a good deal of effort. We will later see how hard it was to make German or Russian capable of "holding" science, and it is therefore noteworthy that even the iconic scientific language, Latin, faced this same hurdle in the face of Greek. Cicero, paragon of Latin eloquence and masterful reader of Greek learning, saw the conundrum clearly. "I thought to illustrate this [philosophical question about immortality] by writing in Latin, not because philosophy cannot be grasped by the Greek language and through Greek instructors," he noted, "but my judgment has always been that our people have found all things more wisely than the Greeks, or have improved upon those things which they accepted from the Greeks, when they thought it worth the effort."* [14]

Improvement is an act of labor. For example, Cicero confronted a Latin that did not have a word for the abstract notion of "muchness." In analogy to Greek, he performed some grammatical manipulation to the ubiquitous question word "how much," *quantus*, producing *quantitas*, the root word for our own "quantity." Someone had to create the word "quantity." It is not an obvious concept, certainly less so than "eye" or "tree," and yet it is difficult to imagine science without it. This particular bit of linguistic alchemy takes place in Cicero's *Academics*, a staged dialogue that represents a conversation at Cicero's country estate with his good friend Atticus and their neighbor Marcus Varro. The encounter begins with Cicero and Atticus inducing Varro to talk about his philosophical work, but to do so *in Latin*. Varro confesses that he had never thought of doing this before, invoking a simple Catch-22 that can be found across the history of scientific languages:

*"hoc mihi Latinis litteris illustrandum putavi, non quia philosophia Graecis et litteris et doctoribus percipi non posset, sed meum semper iudicium fuit omnia nostros aut invenisse per se sapientius quam Graecos aut accepta ab illis fecisse meliora, quae quidem digna statuissent in quibus elaborarent."

> For when I saw that philosophy was explicated most diligently in Greek, I reckoned that those of us who were gripped by its study, if they were learned in Greek doctrines, would rather read Greek than our language; if they shun the skills and disciplines of the Greeks, they would certainly not care for these topics, which cannot be understood without the erudition of the Greeks: thus I did not want to write what the unschooled could not understand and the learned would not bother to read.* [15]

Cicero conceded the justice of this claim, up to a point. "Although you put forward a probable case—either that those who are learned will indeed prefer to read Greek, or that those who don't know such matters will not read these—but grant me now: have you sufficiently proved your point?," he countered. "On the contrary, it is true that who cannot read Greek will read these, and those who can won't scorn their own language. What possible reason is there why those literate in Greek should read Latin poets but not read Latin philosophers?"† [16] Varro agrees to play along with this linguistic experiment. Atticus was even willing to spot Varro a mulligan or two: "[Y]ou will be allowed to use Greek words when you want, if Latin ones happen to desert you."‡ Varro hoped he would not need them: "Truly you are kind; but I will make an effort to speak in Latin, except in these instances of certain words, such as naming 'philosophy' or 'rhetoric' or 'physics' or 'dialectics,' which along with many others are now customarily used in Latin."§ [17]

By the end of the Republican period (which died approximately when Cicero did), then, we see the formation of Latin as a language

*"Nam cum philosophiam viderem diligentissime Graecis litteris explicatam, existimavi si qui de nostris eius studio tenerentur, si essent Graecis doctrinis eruditi, Graeca potius quam nostra lecturos; sin a Graecorum artibus et disciplinis abhorrerent, ne haec quidem curaturos quae sine eruditione Graeca intelligi non possunt: itaque ea nolui scribere quae nec indocti intellegere possent nec docti legere curarent."

†"Causam autem probabilem tu quidem adfers, aut enim Graeca legere malent qui erunt eruditi, aut ne haec quidem qui illa nesciunt; sed da mihi nunc—satisne probas? Immo vero et haec qui illa non poterunt et qui Graeca poterunt non contemnent sua. Quid enim causae est cur poëtas Latinos Graecis litteris eruditi legant philosophos non legant?"

‡"quin etiam Graecis licebit utare cum voles, si te Latina forte deficient."

§"Bene sane facis; sed enitar ut Latine loquar, nisi in huiusce modi verbis, ut philosophiam aut rhetoricam aut physicam aut dialecticam appellem, quibus ut aliis multis consuetudo iam utitur pro Latinis."

of science. This was accomplished through direct translation of foreign concepts into a vernacular idiom—in this case, the vernacular was Latin. The operation was successful, insofar as Aulus Cornelius Celsus composed medical texts in Latin in the first century AD without (so far as we know from surviving texts) any of the explicit apologetics that Cicero used just over a century earlier. Galen, likewise, apparently read some Latin texts, even if his entire oeuvre was written in Greek.[18] But some scientific writings are hardly sufficient to generate an entire language of science. Latin was not widely used as a scientific language by any of the major schools of natural philosophy for at least half a millennium after the fall of Rome. There were other scientific languages that dwarfed Latin, for the Romans and for their successors.

How Latin Got Its Groove

A language of science needs two features: it must have the requisite flexibility to adapt to changing discoveries and theories; and scientists (or, before that English term was coined in 1833, natural philosophers) have to actually *use* it. In late antiquity and the early Middle Ages—up to, in fact, the twelfth century—neither of these held true for Latin.

They had even ceased to hold for Greek. When the city of Constantinople (today's Istanbul) was founded in 330 AD, Latinity in the East briefly revived in this center for administration and propagation of the language of the West. Yet after the Emperor Theodosius's death in 395 and the complete partition of the Empire, the limited Latin that had percolated into this region dried up completely.[19] The ensuing Byzantine Empire lasted for a millennium, governed through a medieval mutation of the dominant *Koine* of the Roman East—with healthy doses of bilingualism in Armenian, Arabic, Slavic languages, and others—but without developing a significant research tradition in the sciences.[20] So while the medieval East never had a significant language barrier to reading *Koine* texts, we nonetheless do not find much engagement with this material.[21]

In the Latin-speaking medieval West, on the other hand, the language barrier proved extremely important, as knowledge of Greek—traditionally dependent on some initial childhood exposure to native speakers—became as scarce in the West as Latin had become in the East.[22] (Contact with Greek was never fully sundered in Sicily, but this was a highly localized phenomenon.[23]) The difficulty of acquiring Greek in Western Europe meant that the scholarship that was encouraged by the likes of Charlemagne, who brought Alcuin of York to his court to overhaul education and revive the status of Latin as a clerical and liturgical language, was largely confined to a set of texts either originally written in Latin or translated from the Greek in late antiquity. The canon is small, and bespeaks the limited interest in natural philosophy in the Latin West: Pliny's *Natural History*, Aulus Gellius's *Attic Nights*, Solinus's *Collection of Remarkable Facts*, Macrobius's *Commentary on the Dream of Scipio* (a Cicero text), Martianus Capella's *The Marriage of Mercury and Philology*, the encyclopedic gleanings of Isidore of Seville, Chalcidius's Latin commentary and partial translation of Plato's *Timaeus*, and the translations of Aristotle's and Porphyry's logical works and Euclid's geometry by the venerated polymath Boethius.[24] The rest of classical and contemporary learning was locked out of Latin, inaccessible to the few who cared about how nature operates.

The hangup was with natural philosophy, not with Latin, which was *the* crucial language of medieval Europe by virtue of its role in the Catholic Church, the dominant institution across the entire region. Most literate individuals were connected to the Church and deployed their intellectual skills in its domains, which meant they knew (at least some) Latin to pray and work, but had little time or patience for the abstruseness of natural knowledge. The importance of the Roman adoption of Christianity and the Latin West's retention of the language of the Empire for the religion is of absolutely pivotal importance, and is such a dominating fact that I will take it for granted and say little more about it explicitly—though it perpetually hovers in the background.

Latin was a strange beast in the Middle Ages, neither living nor dead. After the collapse of Rome, no one—with idiosyncratic exceptions in the Renaissance—learned Latin as their native language, the classic definition of a "dead" language, such as Hittite or Biblical Hebrew. On the other hand, children of the elite or those headed for a career in the Church studied Latin from a very early age, and it functioned as the spoken and especially written language of choice in most instances. Medieval Latin went through substantial variation, development, and enrichment as it adapted to a whole host of innovations, although many of these changes in the tongue came to be demonized by later purists who opted for a more Ciceronian Latin (more on that soon).[25]

Eventually, Latin did become a widely used language of science, employed by countless users who manipulated its strikingly varied vocabulary of philosophical terms. How did this happen? The same way that Cicero had envisioned: through translation. This time, however, the translation was not from Greek but from what has been called "the second classical language, even before Latin": Arabic.[26] Arabic was, far and away, the leading scientific language of the medieval period, serving for longer than any other language in the West for the codification, elaboration, and expansion of natural philosophy. Beginning in the eighth century, the Abbasid dynasty, based in its new capital of Baghdad, endorsed and patronized a gigantic translation enterprise, rendering most of the philosophical and scientific texts of Greek learning into Arabic (often through the bridge of Syriac, another Semitic language), in the process transforming the target language into a supple resource for continued development of these sciences. Two centuries of expensive translation do not happen unless the translators want to *use* the knowledge for something—say, for astronomical research—and the storied commentaries and revisions of Greek science constituted the high-water mark for natural philosophy for centuries. Abetted by the contemporary arrival of Chinese paper making in Baghdad, which made the production of texts much cheaper, the Islamic Empire that spanned from Persia to Spain transformed natural philosophy from a Hellenophone to an Arabophone enterprise.[27] Greek ceased to be a major language of science not because it ceased to be a language, but because people had stopped doing science in it; Latin became a scientific language through its encounter with Arabic.

That encounter is now often called "the Renaissance of the twelfth century," a full two centuries before the storied Italian Renaissance. Latin scholarship had already begun a slow process of reemergence with the revival of monasteries such as Montecassino, whose libraries offered to itinerant researchers access to musty tomes of older learning.[28] Of course, the common language of Latin (despite strikingly varied pronunciation) enabled this scholarly mobility, meaning that once a text was rendered into Latin it became accessible across a broad region where the language fulfilled ecclesiastical and administrative functions.[29] This early sparking exploded into the fervor of scholarship after a few curious researchers began searching for reports that ancient Greek wisdom—especially Ptolemy's astronomical masterpiece, the *Almagest*—might be read if only one could find the Arabic books.[30] (The common title of Ptolemy's work, composed in Greek, bears traces of its active life in Arabic.)

The translation movement from Arabic into Latin also followed the migration of paper-making technology, which arrived at Europe in the tenth and eleventh centuries from the Arab world. There were numerous sites where the intellectual encounter via translation transpired; suffice it to point here to the Spanish city of Toledo, conquered from the Moors in 1085 by Alfonso VI, king of Castile and Léon. Latinate scholars looking for scientific texts, such as the Herculean translator Gerard of Cremona, converged on the city and set up collaborative translation partnerships with bilinguals in Arabic and Mozarabic (the Spanish vernacular), typically Jews, until they had mastered the language for themselves.[31] Latin renditions—at times of iffy quality—of ancient wisdom (multiple versions of Euclid, tomes on alchemy, Aristotle's surviving corpus) poured out of Spain, accompanied by the cutting-edge commentaries of the Arabic philosophers.[32] Other translators worked in Sicily, producing Latin versions of these canonical works directly from the Greek, but these circulated far less widely. "The Latin world could have got its Aristotle and its Galen, its Ptolemy and Euclid, largely through these Graeco-Latin versions," noted renowned medievalist Charles Homer Haskins. "It *could* have got much Greek science in this way, but for the most part it *did* not. The current language of science was by this time Arabic."[33]

Now that ancient Greek and medieval Arabic natural philosophy was available in Latin, the metamorphosis this translation effort wrought on the language quickly elevated Latin to the universal language of science for Western Europe. Yet scientific Latin in the twelfth century and until the eighteenth was a complicated and varied organism. Today, we call it "Scholastic Latin," after the "schools" (i.e., universities) where it was studied. At the time, it was understood as a specialized subset of Latin for technical discussions of the kind natural philosophers were wont to have. Real linguistic adjustments to Latin had to be made in the process of translation, because transferring the highly inflected technical lexicon of Arabic and Greek generated challenges beyond vocabulary. For example, classical Latin has no present or past participle for the verb "to be" (*sum/esse*), which makes rendering medieval metaphysics rather dicey. The absence of definite articles—present in both Greek and Arabic—was another obstacle, circumvented by thirteenth-century translator Willem van Moerbeke by simply inserting the French "le."[34]

With the emergence of humanism in the fourteenth century, driven by a reverence for Ciceronian Latin, precisely such tinkering was decried as "barbarism." If there ever were a moment when Latin was placed on a pedestal and ascribed exalted status as *the* universal language of scholarship—because it was perceived as uniquely suited *linguistically* for the purpose—it was the Renaissance, beginning from Italy and radiating outwards in multiple and varied incarnations. On the surface, much looks the same. Latin was still a language of bilinguals, with certain (mostly male) children taught the language so they could engage a continent-wide community of elite scholars and ecclesiastics.[35] But the differences were sensible at all levels. Many of the countries of Western Europe had begun to use their local vernaculars for administrative purposes. Latin as a language of statecraft persisted in international diplomacy and in the governance of polyglot regions in the East, such as the Polish-Lithuanian Commonwealth and the sprawling Habsburg lands.[36] Vernacular had seeped into many of the old functions of Latin, but not in the disciplines surrounding the sciences.

That difference is essential, because it came to define Latin as the European scientific language. That is, Latin's status as a language of science rested on the *contrast* it made with the use of the vernacular in other contexts. The bilingualism of scholars underscored scholarship as a distinctive activity.[37] The very artificiality of using a classical language added to this logic, feeding back into a purist quest for the most pristine Latin style. The rules for indirect speech were reintroduced, eliminating the *dixit quod* and *dixit quoniam* which had substituted for it in Medieval Latin, and Cicero's intricate periodic sentence returned in full force.[38] The more specialized scholarship became, the more unvernacular Latin had to be, enabling certain scholars like the Dutchman Erasmus to have a continental reputation while, according to one biographer, producing an "estrangement" that prevented him from ever being "thoroughly at home" among his countrymen.[39] Whether or not that sense of alienation obtained, the quest for Latinity manifestly helped give life to what was becoming a European community of learning. Humanists undertook a search for more perfect editions of the Greek and sharper translations into Latin, in the process changing the classical tradition and extending their research into the natural world around them.[40] Latin had arrived as a language of science.

Universal But Not Global

Latin was not, as we have seen, a dominant language of science until the high Middle Ages and the Renaissance, having been subordinate in the continental and Mediterranean regions to Greek and Arabic for over a millennium. Even when it became the obvious vehicular language among scholars, its reach was relatively limited. Leave aside the New World, which was only just at this moment being colonized by Western

European powers; Latin had no purchase among the indigenous who lived there or the adventurers who arrived (although natural historical findings would eventually come back to Europe in Latin guise). Even in the Eastern half of Europe, scholarship often still bore Greek dress. Head farther East, as Europeans did in greater numbers during the Renaissance, and two other languages of scholarship and science hove into view. Each commanded a much stricter monopoly over knowledge than Latin had (or did) in Europe and lasted for longer over a wider area. Our imagination of what it means to be a scientific language is strongly shaped by the historical experience of Latin, but the image of that language, at the moment of its eclipse, was in turn conditioned by contemporary understandings of two vehicular scholarly languages from other parts of the world.

Heading toward the sunrise, we meet Sanskrit. The encounter with this ancient tongue and the recognition of its kinship with Latin and Greek would in the late eighteenth century generate the category of "Indo-European" and spark modern historical linguistics, but in the seventeenth century it played essentially no role in the European debates about languages of science. That said, a cursory examination of Sanskrit's reach and role is instructive about the parochial character of universal Latin. After the dawn of the first millennium, Sanskrit ceased to be—as its name states—"holy writing," reserved for ritual purposes in the custody of a specific group, and broadened out, becoming a vehicular language of scholarship, correspondence, and literature across a staggeringly broad region: from Afghanistan to Southeast Asia, from Sri Lanka in the South to the steppes of Central Asia in the North. Viewed through the distorting lenses of Latin's history, Sanskrit represents a series of absences: the absence of military conquest, the absence of a scriptural religion, the absence of a recovery narrative of a lost classical tongue. This is, of course, the wrong way to look at it. From the perspective of the Sanskrit-linked world, it was simply the mode of vehicular communication in use, accompanied by an astonishingly rich body of learning. And it lasted until the end of the eighteenth century, when European languages—both the consequence and the medium of British, Portuguese, and French colonization—began to erode its scholarly functions.[41]

This is an important story, not least because it demonstrates the durability of vehicular languages and writing systems. We should pause for a moment longer on the role of Sanskrit as a language of science, principally in mathematics and astronomy. The range of Sanskrit's penetration (both horizontally across space and vertically within social groups) enabled a particular efflorescence in the sciences.[42] In addition, Sanskrit was also the first language—as far as I can find in the scholarly literature—to be specifically examined for the linguistic consequences generated by being a language of science. In 1903, the distinguished German scholar Hermann Jacobi published a ground-breaking article on "The Nominal Style of Scientific Sanskrit," revealing the tendency of the language to shift toward ever more complex noun forms to encompass the kinds of abstractions demanded by scientific and mathematical thinking.[43] Jacobi's insights later reverberated through other linguists' efforts in the early twentieth century to make sense of what was happening to their own languages. Hence Otto Jespersen, a figure we will encounter more than once, in 1924: "German scientific prose sometimes approaches the Sanskrit style described by Jacobi. When we express by means of nouns what is generally expressed by finite verbs, our language becomes not only more abstract, but more abstruse[. . .]."[44] Although the history of Sanskrit happened largely detached from that of the European languages, it has structured our understanding of what it means to be a scientific language, and how those dialects are transformed through the act of exploring nature.

Sanskrit's spread across the Eurasian landmass was stopped by its encounter with another venerable language of scholarship that cloaked the Eastern Pacific Rim, from the islands of Japan and the peninsula of Korea down to Vietnam: Classical Chinese. This book is not the place to summarize the vibrant scholarship about Chinese science across two millennia; my goal is much more narrow: to show how a certain kind of imagined Chinese forced Europeans to rethink the purposes and potential of scientific communication.[45] Classical Chinese was no less imagined for those who used it in East Asia than for the Europeans who encountered it at the close of the Renaissance. The reality is difficult to pin down, precisely because we have thousands of years of writing, but no full sense of how Classical Chinese was used orally. (This is also a problem for medieval Europe, where vernacular utterances are hidden in Latin transcriptions.)

Chinese both was and wasn't the "Latin" of East Asia. Chinese script, as is well known, extended across the entire region. It is not, despite popular misconceptions, an essentially "ideographic" system, with characters depicting images or concepts, but rather contains significant phonological cues. That said, the degree of abstraction allows for people speaking very different kinds of Chinese—in Shanghai, Beijing, and Hong Kong—to use the same script. The immense prestige and power of Chinese philosophy, Buddhism (itself translated from South Asia), medicine, and natural science promoted the study of the language in Korea and Japan, and scholars there read it with avidity and sometimes composed texts in it. As a rule, however, those texts did not circulate back to China or even among these various peripheries, and Chinese was almost never learned as a spoken language by Koreans, Japanese, and Vietnamese. Communication between foreigners, even in person, sometimes relied on writing out Chinese characters. The script was so ubiquitous among the literate that it was eventually adapted to local vernaculars, even though the fit was poor with, say, Japanese, and demanded significant adjustments. Some of this resembles Latin quite closely: most international communication in it was written, and pronunciation (especially between the English and the continent) was widely divergent. The major contrast is also highly significant: the Chinese language belonged to the Chinese, the residents of this immensely powerful Empire around whose periphery the other Asians lived. Vernacular works in Japanese or Vietnamese were never translated into Chinese for circulation abroad, and every use of the language was marked, not neutral.[46]

The European literati who returned from visits to China saw the universality of the script but missed out on its freighted quality. In fact, they took their impression of Chinese writing as ideographic—based, as it happens, on a mix of misunderstanding, guesswork, and absorption of mistaken popular conceptions published by the Chinese—and used it to generate a conception of a truly universal scientific language.[47] After a good two centuries of esteem and adoration, the humanistic bloom was off the Latin rose, and influential natural philosophers in several different lands began to think of a replacement, unencumbered by historical oddities and adapted to the innovations of the transformation of science underway in the seventeenth century. In order to get such a "philosophical language" that was utterly natural, they decided to invent it.

The idea of creating a language for an express purpose dates back at least to the twelfth-century mystic Hildegard of Bingen, who created what she called *lingua ignota* for her own use, but applications of the notion to natural philosophy emerged only in the seventeenth century. The canonical first mention of such an idealized philosophical language is usually located in a letter from philosopher and mathematician René Descartes to frenetic correspondent Marin Mersenne, dated 20

November 1629. In that same letter, however, Descartes lamented about the possibility that such a language would ever come about. "But do not hope to ever see it in use," he wrote; "that presupposes such great changes in the order of things, and it would require that the entire world become a terrestrial paradise, that it is only worth proposing in the land of novels."* [48] Hope did not wither on the vine; on the contrary, it flourished across the seventeenth and eighteenth centuries, producing at least a dozen serious projects, before temporarily disappearing in the early nineteenth century.[49]

The idea that extant human languages were somehow inferior to a potential philosophical language stemmed from three sources. The first was a lingering dissatisfaction with Latin, in itself a consequence of the ambition and vigor of the vernacular languages (especially French), which led philosophers to question the adequacy of their current implement.[50] A logical consequence of expanding the vernacular languages was the creation of a Scientific Babel, a confusion of tongues whereby the Dutch would be unable to communicate as easily with the Italians as they had in Erasmus's day. The second source provided an answer for that: the apprehension of Chinese script as a universal writing system in which you could read what a Pole wrote down and interpret it as German, in the manner they imagined the Japanese simply read Chinese *as* Japanese.[51] (This was not in fact so. The Japanese developed an elaborate system of annotations to Chinese script to enable legibility.) The third factor bolstered the plausibility of constructing such a language: the seventeenth-century innovations in mathematical formalism and to a lesser extent in musical notation provided an analogy to universal writing that inspired German polymath Gottfried Wilhelm von Leibniz.[52]

The most prominent of these experiments in philosophical languages was that of John Wilkins, Oliver Cromwell's brother-in-law and—after the Restoration of the monarchy following the English Revolution—a founder of the Royal Society of London, one of the world's first scientific societies. Wilkins had long been interested in generalizing knowledge, both in the sense of making what was known accessible to more people, and of encompassing more of the world into what was known, an ambition already visible in his 1641 book on codes and ciphers, *Mercury, or the Secret and Swift Messenger*.[53] His project for a full-fledged

*"Mais n'esperez pas de la voir jamais en usage; cela présupose de grans changemens en l'ordre des choses, et il faudroit que tout le monde ne fust qu'un paradis terrestre, ce qui n'est bon à proposer que dans le pays des romans."

philosophical language, *An Essay towards a Real Character and a Philosophical Language* (1668), had to wait for over two decades, delayed not least by the incineration of the first manuscript in the Fire of London of 1666. Wilkins's idea was simple: things in the world have relations among each other, and so should the words that pick them out. A sparrow, an eagle, and a penguin are all birds, but nothing in their English names would lead you to suspect that, or even to tell you they were animals and thus more closely related to pigs than to daffodils. Wilkins assigned letters to concepts, building up a quasi-mathematical representation that would encode the map of the universe. You can see the difficulty right away: do we class peanuts with peas or nuts? On the one hand, they are legumes, on the other, they are, well, nutty. If we do not understand the universal map of nature—or, worse, if there simply is no single map—then the project cannot work, however noble its inspiration. Wilkins's grand scheme has received detailed treatment by scholars, and I recommend those who are interested to peruse those works, or just to read the original, which is in English.[54]

That's an interesting fact: Wilkins's manifesto to repair world knowledge by embedding it in an analytical linguistic frame was composed in a vernacular that was generally understood only in one North Sea archipelago. In one sense, that is not surprising: although Wilkins knew Latin, he composed most of his works in English, given his desire to reach a broader audience. But Erasmus also wanted to reach a broader audience—one that spanned across Europe—and that is why he wrote in Latin instead of Dutch. Something had changed about the understanding of audiences between Erasmus and Wilkins. Wilkins thought about this issue intensely. Early in the volume, before he introduces his philosophical language, he narrates a history of the vehicular languages of the world (including Chinese). Wilkins thought of his philosophical language not only through mathematics, but also through actual human languages. As a case in point, he considered Malay, "which seems to be the *newest Language* in the World," as an argument that it is possible to build a language: "for the more facil *converse* with one another, they"—the Portuguese, the Dutch, and many fishermen of Southeast Asia—"agreed upon a distinct *Language*, which probably was made up by selecting the most soft and easy words belonging to each several Nation." This was an inspiration for Wilkins. "And this is the only *Language* (for ought I know) that hath ever been at once *invented*; if it may properly be styled a distinct Language, and not rather a *Medley* of many. But this being invented by rude Fishermen, it cannot be expected that it should have all those advantages, with which it might have been furnished by the rules of Philosophy."[55] A language could be built—naturally. The border between invented and natural is blurry.

We have, then, a book in English that uses Malay and Chinese to introduce an abstract creation. It was bound to restrict readers, precisely because Wilkins did not publish it in Latin, or even French, fast becoming the leading vehicular language of Western Europe (efforts to have the work translated into both fizzled).[56] We see in Wilkins a change in audience, one which characterizes the slipping hold of Latin on European science and the gradual drift toward the vernaculars. It was not that scientists ceased to wish to communicate and all shifted to expressing their identity; it was rather that their intended recipients had changed to readers and patrons closer to home, ones who might not be classically schooled in the humanists' Latin. You gain some audiences by switching to the vernacular instead of Latin, but you lose some, too.

How to Speak to Torbern Bergman

Find your nearest scientist and ask them how they feel about English being the universal language of learning, and odds are you will hear that it is fantastic that all science today is communicated in one language. This response immediately raises one of the most important questions in the history of scientific languages: "If an international language is desirable in science, why was Latin abandoned after it had been used for several centuries?"[57] Why indeed? A few answers spring to mind. First, it was difficult to adapt Latin, especially in the classicizing form beloved of Renaissance humanists, to the rapid changes in knowledge characteristic of the seventeenth and eighteenth centuries.[58] This can hardly be the whole story, for the way English deals with the same changeability and innovation in science today is to constantly coin new terms, largely by raiding . . . Latin. Another impulse, more directly significant to those living at the time, was Wilkins's: seek out the patrons closest to you, the people you want to convince who speak your own native language. This was Galileo's motivation when he shifted to Italian from Latin, and also Isaac Newton's, who turned to English for his 1704 *Opticks* but not for his 1687 *Principia*.[59] (Paracelsus, maverick of sixteenth-century chemistry, was ahead of the curve here as well, publishing mostly in German and relying on translators to spread his message.[60]) A host of other reasons charge in: the decline of power in the Catholic Church after the Protestant Reformation (although many Protestants were excellent

Latinists, and Protestant German scholars stayed Latinate far longer than their counterparts in Catholic France); a sense of self-conscious modernity; absence of classical learning, especially among the expanding circle of female readers.[61] There is some truth to each of these.

The effect is clearer than the causes. Across Western Europe, vernaculars came to be used, at different times and in different places, as competitors to or substitutes for Latin in all of its varied domains, not least natural science. As two scholars of the "Neo-Latin" of the postmedieval era put it, somewhat hyperbolically: "The main victim of the Scientific Revolution of the 16th and 17th centuries is without doubt the Latin language and its (quasi-)monopoly as the language for academic scientific teaching and publications."[62] Those who lamented the incipient Scientific Babel provided both the supply and the demand for the booming book trade of translations from vernaculars into Latin across the early modern period.[63] Latin also became a way of expressing identity, at least for the Germans, whose only other option would have been to communicate in French.[64] Decidedly *not* neutral. Medical texts had begun to appear in French in the sixteenth century (mostly between 1530 and 1597), but then vanished under the surface of Latin, only to reemerge to dominance around 1685.[65] The transition to the vernacular was so commonplace that by 1698 Jean-Baptiste Du Hamel, in his history of the French Académie des Sciences, had to defend his choice to publish in Latin:

> Neither do I think it deplorable that I began to write these things in Latin, not French: Of course it is demanded of me, so that it will be read not only by learned Frenchmen, but even by foreigners who do not know French. No matter how much indeed the Latin language is right now temporarily eroded and held by many in contempt, however we are allowed to usurp what Cicero once said of Greek concerning Latin: *Latin is read among almost all peoples, French is constrained by its borders, which are certainly narrow*. Notwithstanding that indeed these things are not entirely true, it nevertheless must be granted that the French language is not spread so broadly as Latin, which is the same among peoples everywhere, nor is it subject to so many changes as the vulgar languages. But enough of this.*[66]

* "Neque id reprehendendum puto quód Latine, non Gallice hæc scribere sim ingressus: Id quippe postulatum à me est, ut non ab eruditis modò Gallis, sed etiam ab exteris, qui Gallice non sciunt, legerentur. Quantumvis enim Latina lingua nunc tem-

His colleagues were displeased. By the second edition in 1701 this passage was gone. Latin would soon join it.

Living through the transition to Scientific Babel was complicated, and each naturalist experienced it differently. To give a sense of one particular path through the linguistic morass, I will follow the correspondence of Torbern Bergman. I know, not exactly a household name, but he once was, at least in certain circles. Bergman was one of the most important chemists of the eighteenth century, a contemporary of Lavoisier and Priestley who has been all but dropped from the history of chemistry. I choose him because this neglect is, at least in part, closely related to the fact that he was Swedish rather than French or English, or even German, and thus removed from the three scientific languages that would, by the mid-nineteenth century, consolidate in a triumvirate of chemical communication. Bergman was born in Catherineberg in West Gothland, Sweden, on 20 March 1735, the son of a tax collector.[67] He was sent to the University of Uppsala, where his father hoped that he would study theology or law, but young Bergman fell under the sway of natural philosophy, working at it so intensely that his health failed and he had to return home. His recuperation meant even more science—this time botanizing outdoors—and his father relented, returning him to the university with permission to supplement his legal studies with mathematics, physics, chemistry, botany, and entomology. His findings in the last drew the attention of Carl Linnaeus (of botanical nomenclature fame, and the leading light of contemporary Swedish science), who encouraged him to deepen his explorations. In 1758 Bergman earned a masters degree in pure mathematics and was appointed a *magister docens*—an assistantship peculiar to Uppsala—in natural philosophy. In a few years, he was appointed an adjunct in mathematics and physics, and in 1766 published, in Swedish, his first work to gain international attention, a comprehensive physical geography entitled *Physisk beskrifning öfver jord klotet* (Physical Description of the Earth). Neither the topic nor the language were that unusual, since Sweden in the eighteenth century was heir to a vibrant tradition of mining engineering,

---

poris deteratur, & a multis contemptui habeatur, id tamen quod olim de Græca dixit Tullius, de Latina nobis usurpare licet: *Latina leguntur in omnibus fere gentibus, Gallica suis finibus, exiguis sane continentur.* Tametsi enim hæc non sunt ex omni parte vera, id tamen fatendum est linguam Gallicam non esse tam late fusam, quàm Latinam, quæ ubique gentium eadem est, neque tot mutationibus obnoxia, quot linguæ vulgares. Sed de his satis."

typically conducted in Swedish in contrast to the more academic, and Latinate, chemistry of the universities.[68] Those who read it lauded Bergman's gifts, although most of them resorted to the German translation (it was also translated into Danish and, later, Russian).[69] The following year Bergman was appointed professor of chemistry and pharmacy at Uppsala as the successor of J. G. Wallerius, supposedly upon the intervention of the Crown Prince, the future Gustavus III.

It was as a chemist, not as a physical geographer, that Bergman solidified his European reputation as a thinker with a synoptic view of the entire science and a penchant for classifying and organizing its findings in a manner analogous to what his mentor Linnaeus did for flora. Carl von Linné, better known as Linnaeus, is probably the first name you think of when it comes to "Latin in the sciences," given his comprehensive program to categorize all plants with a binomial nomenclature that is still the most prominent use of Latin in today's science. Linnaeus's story would take us far afield, but it is worth noting that his choice of Latin was itself a feature of the looming Babel that he sensed (and, in his extensive Lutheran musings, contemplated through its rendition in the book of Genesis). Linnaeus wrote extensively in Swedish and in Latin, but was unable to read French despite the avowed Francomania of the Swedish elite; his own selection of the language of Rome was in no small part a declaration of relevance for the provincial kingdom in which he felt such enormous pride.[70] Bergman would follow Linnaeus's example in this as well.

Bergman's academic status was enabled by two factors: maintaining an avid correspondence, and publishing in Latin. Correspondence was central to the conduct of chemistry in this century when—with the exception of Paris and possibly Edinburgh—there were simply not enough chemists in one place to sustain a vibrant, self-contained community. The exceptionally well-informed Bergman never left Sweden; almost all of his information thus came from personal reports or the publications he received by the surprisingly good postal system of the eighteenth century. We know, sadly, much less about Bergman's half of these exchanges, but the letters to him have been collected in an invaluable published edition, and it offers one of the single best resources to understanding the Europe-wide conversation about the woes of phlogiston and gas chemistry discussed in the introduction to this book.

The very appearance of the letters is a revelation. There are, as one would expect, incoming letters in Swedish and also Danish (closely related to Bergman's native tongue), as well as French. So far, so humdrum. But Bergman also received, and obviously understood well enough, letters in English, German, and Latin. Several correspondents used different languages in sequential letters. Scientific Babel arrived at Bergman's doorstep routinely, and the letters convey a consciousness of language barriers. Johann Gottlob Georgi, a German chemist who later ended up teaching in St. Petersburg and an occasional translator from Swedish to German, wrote Bergman in 1768: "I apologize that I write to you not in your language but rather in mine. As dear to me as the former is, I do however write so poorly in it, [and] it was necessary that I express myself clearly about everything and Your Lordship reads the works of my country without offense."*[71] On the other hand, Richard Kirwan, the defender of phlogiston we have already met, never had the option of writing in Swedish and at first used Latin, but soon "was delighted to see by your letter to Mr. Magellan"—the same intelligencer who had communicated Priestley's work to Lavoisier—"that I received yesterday that you command the French language so well that I can write to you in this language, which is much more familiar to me than Latin."†[72] Occasionally, as with Anglophone German Franz Xaver Schwediauer, Bergman was even given his choice of languages: "If you shall be so obliging as to favour me with an answer, it may be written either in latin, french, english or German, as is most agreeable to you, and let me know which of these languages you would chuse that I should write to you in answer."[73]

Some of his correspondents even took measures to learn Swedish so they could follow the work of Bergman and his compatriots. Thus Fausto de Elhuyar, a Spanish mineralogist who with his brother Juan José de Elhuyar first isolated the element tungsten, writing to Bergman in eerily accentless French: "I would like to study the Swedish language a bit so that I can read the Memoires of the Stockholm Academy and several other excellent works which you have in your country."‡[74]

* "Ich entschuldige, dass ich nicht in Ihrer sondern in meiner Sprache geschrieben. So lieb mir auch die erste ist, so schreibe ich doch zu schlecht in derselben, es war nötig, dass ich mich über alles bestimt ausdrukte und Ewr Hochedelgeb. lesen die Werke meine Nation ohne Anstoss."

† "Je suis charmé de voir par votre lettre a Monsr Magellan que j'ai reçue hier que vous possedez la langue françoise si bien que je puis vous addresser dans cette langue qui m'est bien plus familiere que la Latine."

‡ "Je voudrois faire quelque etude de la langue Suedoise pour pouvoir lire les Memoires de l'Academie de Stockholme et plusieurs autres excellens Ouvrages que vous avez dans votre pais."

(He had read the physical geography in its German translation.) Bergman's most famous correspondents, Lavoisier's colleagues Pierre Macquer and Louis-Bernard Guyton de Morveau, popularized Bergman's work abroad, the first by translating him from Latin and the latter by actually learning Swedish. To do so, he followed de Elhuyar in asking Bergman to send Swedish grammars, dictionaries, and even novels. Guyton de Morveau induced his collaborators, including his mistress Madame de Picardet (who had helped Madame Lavoisier with her English), to assist in translating Bergman. (They did not get credit for it.)[75] Even Schwediauer expressed some interest in learning the language, but really wished he didn't have to: "I regret that you wish to unearth paleontology using the Swedish language, for very few chemists know Swedish, yet who nevertheless really ought to learn these new truths and promulgate them[. . .]."*[76] The obvious solution for these cases was to rely on vernacular translations, although sometimes the offer to vernacularize itself came in Latin: "It is fitting that this work be translated into several languages, so that the reputation of so illustrious and learned a man spreads to many peoples[. . .]."†[77]

Latin was utterly inescapable in Bergman's world—and this was, recall, in the 1770s, long after Latin had ostensibly died off in the sciences. Sweden was somewhat of a special case, for Latin was obligatory at Uppsala as late as the 1840s, both as a continuation of a distinguished tradition of Latin learning in Northern Europe, and a recognition of the fact that some vehicular language was necessary to communicate with European peers.[78] (Local flora in Central Europe were often bilingual—Latin and German—and even trilingual, including Czech, blending an international scholarly conversation about taxonomy with a local discussion about the locations and uses of certain plants.[79]) Latin was used as a necessary language of correspondence across borders, even by such vernacular luminaries as Voltaire, and individuals continued to stock their libraries with it. As a rough rule, libraries in the Enlightenment collected one-third in French, one-third in Latin, and the rest in the vernacular of that particular land—eighteenth-century Europe had *two* universal languages.[80] Only through an overemphasis on the moderniz-

*"Doleo quod oryctologiam lingua Suecica exarare velis, paucissimi Chemici Linguæ Suecicæ gnari sunt, quorum tamen maxime interest veritates novas cognoscere, et promulgare[. . .]."

†"Dignum hoc opus, quod in plures Lingvas traduceretur, ut fama tam illustris, doctique Viri ad plures gentes perveniret[. . .]."

ing narratives of Lavoisier and Priestley can one say, with one historian of Latin: "Chemistry, then, came onto the scene so late that Latin was never really relevant."[81] (This author, as it happens, is a Swede.)

Bergman's grandest project was as Latinate as his mentor Linnaeus's, and for good reason: starting in the mid-1770s but culminating with his *Sciagraphia Regni Mineralis* (1782), Bergman sought to develop and extend a classification of minerals in terms of classes, genera, species, and varieties that was explicitly modeled on plants. Classes, for example, contained "salts," "earths," and "metals," and genera of salts included acidic and alkali salts. And so on systematically down the chain, so that every chemical substance would have a unique name that would express its relationships with others.[82] Although this system quickly fell by the wayside—eclipsed by Guyton de Morveau's modern nomenclature based on Lavoisier's chemistry and later the systematic ordering of substances in the periodic table (as described in the following chapter)—it is impossible not to admire the scale and subtlety of Bergman's thinking. It seems equally impossible to imagine such a binomial nomenclature in anything but Latin, as Bergman himself put it in one of his final publications:

> In establishing entirely new names, I desire that their origins be Latin. This language is, or at least was, the vernacular of the learned: now it is dead and is not subject to constant changes. Therefore, if the reform is conducted in this language first, and afterward in living languages on the same model (as much as the spirit of each will allow), this same reform will be conducted more easily. For this very reason chemical language can attain general agreement in all places, which promises no small benefit not only in reading foreign works, but also in translating.*[83]

Always already Latin, it seems. But here is the intriguing part: Bergman had attempted this in Swedish first. From 1775 to 1784, while con-

*"In stabiliendis nimirum nimirum [sic] novis nominibus, ut a Latinis initium fieret, opto. Est haec lingua, vel saltim fuit, eruditorum vernacula: jam mortua quoque nullis quotidianis est obnoxia mutationibus. Si igitur in hac primum reformatio peragitur, in vivis postea ad eumdem modulum, quantum genius cujuslibet permiserit, eadem facilius perficietur. Hoc ipso lingua Chemica ubique locorum generalem adquirere potest convenientiam, quod non tantum in legendis exterorum operibus, sed etiam in transferendis, haud exiguam pollicetur utilitatem."

structing his system, he repeatedly tried Swedish names for these substances, and his manuscripts are littered with abandoned efforts. In the end, he found it substantially easier to generate the requisite adjectives from nouns using Latin's resources than the etymologically and grammatically distinct tools of Swedish, which forced him to generate lengthy descriptive phrases instead of pithy participles. Although his notes are often in Swedish, he rigorously translated everything into Latin—for both external reasons of foreign communication and *internal* ones related to the incapacities he perceived in his native tongue.[84]

The Ordinariness of Latin

In 1977, a survey of the growth of scientific literature declared in passing that the end of Latin was "a major misfortune."[85] Certainly, by 1850, Latin was largely exiled from the community of scientific languages aside from specialized functions like botanical nomenclature. But, as we have seen, the death of Latin was often significantly exaggerated. There exists an interesting tension between the real history of Latin and the history of Latin as it was imagined by natural philosophers beginning to enter Scientific Babel. There had indeed been a time when Latin functioned as the universal language of scientific communication in Europe, but that reign was both shorter and more contested than is commonly recalled by wistful nostalgists. On the other hand, Latin as a real language continued to be useful for scholars like Bergman long after it had ceased to be universal.

The underlying message of the polite interchanges of eighteenth-century chemists to the sage of Uppsala is that if Latin was not universal, if it was just another language that could just as easily be French or English, then it was—in an important sense—*not Latin*. It had lost the power that imagined Latin had held since the Renaissance, the same power imagined Sanskrit and imagined Classical Chinese held for their respective regions. And once Latin was "demoted" to being a scientific language like any other—a peer of Danish, Dutch, English, French, German, Italian, and Swedish—then what was the point of taking all the trouble to memorize declensions and deponent verbs?

This whirlwind history of Latin up to the beginning of the nineteenth century sets the stage for the history that follows in two senses. First, scientists experienced a very real sense of loss: from the growing inaccessibility of past writings that were cloistered in the citadel of the Latin language to the growing feeling by speakers of so-called minor languages that they were now second-class participants in the scholarly conversation.[86] At the same time, the list of languages in which people thought it fitting to do science was too long for most scholars to seriously consider learning all fluently. (Bergman's astonishing capacity for reading in multiple tongues was not widely shared in his own day, nor is it now.) By the middle decades of the nineteenth century, therefore, the folk memory of Latin among scientists and the general impracticality of treating all vernaculars equally resulted in a compression of scientific languages down to the triumvirate—a fitting Latin name!—of English, French, and German. Three is not as good as one, perhaps, but surely it is better than eight, or even four. Unless, of course, you were not lucky enough to be born or raised fluent in one of the three. What would become of you if you lived and thought in a language outside the triumvirate, and unlike Bergman you could not use Latin or another "neutral" tongue? In the next chapter, we begin our history in earnest with a controversy about knocking on the door of European science from the linguistic netherworld of Tsarist Russia.

CHAPTER 2

The Table and the Word

Лет пятнадцать тому назад, когда я много ездил и жил в Западной Европе, мне ни разу ни от кого не приходилось ничего подобного слышать; России боялись иные, многие ее почему-то не любили, никто ею не интересовался, о ней говорили столько же, как об Индии, Австралии. Теперь видна несомненная перемена—Россию знать желают, верят, что так или иначе она не одною своею физическою силою, а своими народными идеалами окажет рано или поздно свою долю влияния на судьбы цивилизации.*

D. I. MENDELEEV, 1877[1]

Dmitrii Mendeleev had every reason to be happy in the summer of '69—1869, that is. It all started with an idea he had back in February, sitting in his apartments in St. Petersburg, the capital of Imperial Russia, as he dove deeper into the writing of the second volume of his textbook, *The Principles of Chemistry* (*Osnovy khimii*). He had packed the manuscript of volume 1—what would eventually be several hundred printed pages—off to the publisher in December 1868, and now he was working on organizing volume 2. At first, it was hard going, since there were so many elements (fifty-five!) that needed to be treated in the same space with which he had covered only eight elements in loving detail in the first volume. Then he started comparing the atomic weights of groups of chemical elements. One thing led to another, and on 17 Feb-

*"About fifteen years ago, when I traveled and lived a great deal in western Europe, I never came to hear anything similar from anybody; some were afraid of Russia, many for some reason didn't care for it, no one was interested in it; people spoke about it as much as about India, Australia. Now an indubitable change is visible—people desire to know Russia, they believe that somehow or other, not only by its physical strength but also its people's ideals, it will exert sooner or later its measure of influence on the fate of civilization."

			Ti = 50	Zr = 90	? = 180.
			V = 51	Nb = 94	Ta = 182.
			Cr = 52	Mo = 96	W = 186.
			Mn = 55	Rh = 104,4	Pt = 197,4
			Fe = 56	Ru = 104,4	Ir = 198.
			Ni = Co = 59	Pl = 106,6	Os = 199.
H = 1			Cu = 63,4	Ag = 108	Hg = 200.
	Be = 9,4	Mg = 24	Zn = 65,2	Cd = 112	
	B = 11	Al = 27,4	? = 68	Ur = 116	Au = 197?
	C = 12	Si = 28	? = 70	Sn = 118	
	N = 14	P = 31	As = 75	Sb = 122	Bi = 210?
	O = 16	S = 32	Se = 79,4	Te = 128?	
	F = 19	Cl = 35,5	Br = 80	J = 127	
Li = 7	Na = 23	K = 39	Rb = 85,4	Cs = 133	Tl = 204.
		Ca = 40	Sr = 87,6	Ba = 137	Pb = 207.
		? = 45	Ce = 92		
		? Er = 56	La = 94		
		? Yt = 60	Di = 95		
		? In = 75,6	Th = 118?		

FIGURE 2.1. The first published version of Mendeleev's periodic system, dated 17 February 1869 (according to the old Russian calendar—1 March according to the Western European one). Mendeleev produced this tabular version while composing his textbook, *The Principles of Chemistry*. D. I. Mendeleev, *Periodicheskii Zakon: Klassiki Nauki*, ed. B. M. Kedrov (Moscow: Izd. AN SSSR, 1958), 9.

ruary (by the Russian old-style calendar; 1 March by the Western European one) he sent a cleaned-up sheet to the printers for them to offset so he could mail it to various chemists both in Russia and in Europe.[2] The resulting image, Figure 2.1, when rotated 90° clockwise and reflected in a mirror, clearly shows us what it is: Mendeleev's first published periodic system of chemical elements. This system, suitably expanded, revised, and reformatted, now hangs in every chemistry classroom on the planet and is widely known as the "periodic table" in English.

Mendeleev needed a title for his printed sheet, and he dubbed it, in Russian, "Attempt of a system of elements, based on their atomic weights and chemical affinity."* He requested that the printer produce 150 copies for distribution to colleagues in Russia—a whopping proportion of the number of active chemists in this relative newcomer to advanced chemical research. But he realized that although Western Europeans—

*"Опыт системы элементов, основанной на их атомном весе и химическом сходстве."

the arbiters of chemical credit—might be able to figure out what the image represented by staring at it for a while, the title would be meaningless, and so he also requested fifty copies with an alternative French title, a translation from the Russian: "Essai d'une système des éléments d'après leurs poids atomiques et fonctions chimiques."[3] French was a good call: although he was uncomfortable in all foreign languages, this was Mendeleev's best, and along with German and English it was one of the three languages that all chemists were expected to be able to read.

There was, however, a slight problem with one word in the French. He called what he was producing *a system*—*une système*—but in French he used the *feminine* indefinite article with the masculine noun. What happened here? Mendeleev initially wanted to call his image a *classification* (in Russian распределение/*raspredelenie*), and that noun is in fact feminine in French, so he used the appropriate article. When he replaced the noun, he neglected to repair the article. This is a completely understandable mistake. Russian notoriously lacks direct and indirect articles. It was challenging enough to decide whether one needed *the* or *a/an* before a noun; once Mendeleev had correctly figured it out, he never went back to correct the gender. In Mendeleev's first foreign publication on his periodic system, there was a mistake of one word in the translation. Later in 1869 one other translation of Mendeleev's periodic system appeared, also with a single mistake, and this time there was hell to pay. That error triggered one of the most vehement and inflammatory priority disputes of the nineteenth century in any science.

Priority disputes—arguments over who had come to a particular finding first—are endemic to science, and many of the landmark discoveries in its history are scorched by such conflicts: consider the calculus, conservation of energy, and evolution by natural selection, to name just three prominent examples.[4] The periodic system was the single most important discovery of inorganic chemistry in the nineteenth century—and quite possibly of chemistry in general, in any century—and credit for such an achievement would bring professional status, historical immortality, and national prestige. It was a prize to be fought over, and Dmitrii Mendeleev is now universally awarded the laurels.[5] But, as noted almost fifty years ago in what remains the most comprehensive history of the periodic system, by J. W. van Spronsen, "once the time was ripe, the periodic system of elements was discovered almost simultaneously in the most leading countries of Europe and North America." He apportioned the credit among no fewer than *six* individuals: Alexandre-Émile Béguyer de Chancourtois, William Odling, John Newlands, Gustavus Hinrichs, Lothar Meyer, and Dmitrii Mendeleev.[6] There is no question that the most vociferous contest at the time was between the last two.

This chapter chronicles the priority dispute between Lothar Meyer and Dmitrii Mendeleev, but it is not fundamentally *about* that dispute. Instead, I wish to use the story of this chemical conflict that became a Russian-German standoff to dramatize the clash between Russian and German as scientific languages. By roughly 1850, the cacophony of new languages brought about by the demotion of Latin explored in the previous chapter—the murmuring of Dutch and Swedish alongside the more prominent French and English—had softened and compressed into the three major scientific languages of the triumvirate: English, French, and German. Among themselves, these three languages comprised the vast majority of publishing in the natural sciences, and in particular in chemistry. As the next most significant languages, Italian and Latin (still), slid into obscurity, how could any language break into this tight club of three? Of course, one could simply publish in Czech or Greek, but that would not help with regard to the all-important issue of *credit*, the animating force behind every priority dispute. In order to count as a significant language of science, it was not enough simply to be written in, others had to be persuaded to *read* it. And if almost no practicing, established scientists outside of your own country knew your language—or even, in the case of Russian, your alphabet—how could you make them pay attention?

This chapter and the one that follows are about this problem of the introduction of a new language of science, told from the perspective of the "marginal" community (in this instance, that living in the largest country on the globe, i.e., Russia) trying to make their publications in their native tongue count from the perspective of Anglophones, Francophones, and Germanophones. This has only succeeded twice since the creation of the triumvirate: in the mid-nineteenth century with Russian, and in the mid-twentieth century with Japanese. There are two stages to the process of inclusion, divided here between this chapter and the next; they happened in the case of Russian in a rather counterintuitive inverse order. First, the Russians had to make the Europeans take notice, to convince them that there was something being produced in Russian that was worth reading—or, at the very least, that a publication in Russian served as adequate announcement of a discovery. And second, chemists had to actively construct Russian as a scientific language, make it able to "hold" science by endowing it with a nomenclature and other linguistic elements that made the process of mutual translation between it and the triumvirate more straightforward.

More straightforward, but never completely easy, as the fate of Mendeleev's periodic system illustrates. Tracing that story brings us directly to the vagaries of publication in Russian in the second half of the nineteenth century. The Mendeleev-Meyer dispute was not simply a controversy about publication (specifically, the *language* of publication); it quickly escalated into a nationalist border war between two scientific communities. Seeing how these groups interacted provides a glimpse at the intricately textured fabric of European chemistry and enables us to see how much can hinge on a single word, one which slipped past the translator's pen with barely a second glance. It would be over a decade before the dust settled.

A Small Mistake

The February 1869 leaflet was all well and good, but it was hardly sufficient. If Mendeleev really wanted to receive credit for his system of elements, his classification that he believed would enable him to correct previous misconceptions about atomic weights and also perhaps to predict the properties of yet-undiscovered elements, he knew he needed to do more than print one table underneath a single line of text. He still had to explicate the system, publish articles in scholarly journals that explained his process of reasoning and drew out the implications; anything less would be a chemical curio, a mere rearrangement of data. Mendeleev was a young chemist on the make in the Imperial capital, and he thought this system might make a splash if pitched just right, and for that he needed the imprimatur of two different audiences: Russian chemists, especially those in St. Petersburg who would be involved in decisions about professional advancement; and the international chemical community, who would not even notice anything published in Russian. That meant not one publication, but at least two.

Publishing in Russian had become almost trivially easy since the establishment of the Russian Chemical Society the previous autumn, complete with its own Russian-language *Journal of the Russian Chemical Society*, then in its first year. The printed minutes in that journal of the meeting of 6 March 1869 (18 March on the Western calendar), the first monthly meeting after his formulation of his periodic system, announced as its first point: "N. Menshutkin"—the secretary of the Society, editor of the journal, and Mendeleev's colleague at St. Petersburg University—"reports on behalf of D. Mendeleev an attempt at a system of elements, based on their atomic weight and chemical affinity. In the absence of D. Mendeleev"—who was visiting cheese cooperatives as a consultant—"discussion of this report is deferred to the following meeting."*[7] That announcement was followed by articles in both the April and August issues, expanding upon and deepening the implications of this new system of elements.

Taking care of the second community was almost as easy, with a ready outlet for translations of Russian into German in the form of the *Zeitschrift für Chemie*. One of the editors of this Göttingen-based journal was Friedrich Konrad Beilstein, whose name advertised his German ancestry but who was in fact born in St. Petersburg and was entirely bilingual in Russian and German. He had often promoted the idea of the *Zeitschrift* as a venue for publishing the work of Russian chemists before the Russian Chemical Society's own organ came into existence, and there was no reason Mendeleev should not avail himself of that offer now.[8] Mendeleev took his ten-page article from the April issue, shrunk it to a page-long abstract, and handed the Russian text over to Beilstein, who arranged for a translation and sent it off to Germany.

In the summer of 1869, Lothar Meyer had been making very good progress in his own chemical research, until he was taken aback by a letter from St. Petersburg. His close friend Friedrich Beilstein had sent him a translated abstract from Petersburg featuring a system of chemical elements, and asked Meyer to see to it that this was placed in the *Zeitschrift*.[9] Meyer did such tasks routinely, working as Beilstein's man-on-the-ground from his own position at Karlsruhe Polytechnic. Meyer transmitted the piece to the printers, but he could not have failed to despair. He had been working on such a system of elements—almost identical to the one produced by this "Mendelejeff." In the first edition of his widely read textbook, *Modern Theories of Chemistry* (*Die modernen Theorien der Chemie*), published in 1864, he had explored the correlations between families of elements, including a "table [which] gives

*"Н. Меншуткин сообщает от имени Д. Менделеева опыт системы элементов, основанной на их атомном весе и химическом сходстве. За отсутствием Д. Менделеева обсуждение этого сообщения отложено до следующего заседания."

	4 werthig	3 werthig	2 werthig	1 werthig	1 werthig	2 werthig
	—	—	—	—	Li = 7,03	(Be = 9,3?)
Differenz =	—	—	—	—	16,02	(14,7)
	C = 12,0	N = 14,04	O = 16,00	Fl = 19,0	Na = 23,05	Mg = 24,0
Differenz =	16,5	16,96	16,07	16,46	16,08	16,0
	Si = 28,5	P = 31,0	S = 32,07	Cl = 35,46	K = 39,13	Ca = 40,0
Differenz =	$\frac{89,1}{2}$ = 44,55	44,0	46,7	44,51	46,3	47,6
	—	As = 75,0	Se = 78,8	Br = 79,97	Rb = 85,4	Sr = 87,6
Differenz =	$\frac{89,1}{2}$ = 44,55	45,6	49,5	46,8	47,6	49,5
	Sn = 117,6	Sb = 120,6	Te = 128,3	J = 126,8	Cs = 133,0	Ba = 137,1
Differenz =	89,4 = 2.44,7	87,4 = 2.43,7	—	—	(71 = 2.35,5)	—
	Pb = 207,0	Bi = 208,0	—	—	(Tl = 204?)	—

FIGURE 2.2. Lothar Meyer's table of elements from the first edition of *Modern Theories of Chemistry* (1864). Notice the strong similarity to the Mendeleev table, which was first composed five years later. Lothar Meyer, *Die modernen Theorien der Chemie und ihre Bedeutung für die chemische Statik* (Breslau: Maruschke & Berendt, 1864), 137.

such relations for six groups of elements well characterized as belonging together."*[10] It was not complete, that's true, but one can see from Figure 2.2 that it indeed was quite similar to Mendeleev's, and dated five years earlier. In fact, in 1868, he had developed a complete table of elements—published only posthumously by his student and friend Karl Seubert—which he was slowly writing up.[11] And now he had been scooped.

Or had he? There was something odd about the abstract published in the *Zeitschrift für Chemie*, something missing . . . Oh, that was it! In the one-page abstract, spread across the bottom half of one page and the top half of its verso, Meyer reviewed the series of numbered points where Mendeleev drew out the implications of this table of elements. The very first of them read: "1. The elements ordered according to the magnitude of their atomic weights show a phased (*stufenweise*) change in proper-

*"Tabelle giebt solche Relationen für sechs als zusammengehörig wohl charakterisirte Gruppen von Elementen."

ties."*[12] Meyer had suspected for some time that the system of elements was in fact *periodic*, displaying a repetition of properties that recurred much like a sine wave; it seemed that Mendeleev had noticed only a step-wise or phased change in the properties, not the precise character of that relationship. Meyer took out his pen and continued revising his essay on his own system of elements, to be published in the most prestigious chemical journal of the day, the Munich-based *Annalen der Chemie und Pharmacie*, known universally as *Liebig's Annalen* after its founder and long-term editor, Justus von Liebig. Citing Mendeleev's *Zeitschrift* piece generously, Meyer noted that the Russian had observed that when "one orders the atomic weights of all elements without arbitrary selection by their magnitudes in a single row, this row splits into sections, and these fall into an unchanging succession one after another."† Mendeleev's contribution was important, but Meyer's emendation was more significant, because in his rendition "we take from the table that the properties of the elements are mostly *periodic* functions of the atomic weights,"‡ and that "[o]ne immediately sees from the course of the curve [Figure 2.3] that the volume of the elements, just as their chemical behavior, is a periodic function of the magnitude of their atomic weights."§[13]

Opening his own copy of the *Annalen* back in Petersburg, now it was Mendeleev's turn to sit stunned. What did Meyer mean that *he* was the one to introduce the word *periodic*? You could see right in the original April 1869 article that Mendeleev had considered periodicity *the* crucial feature of the table. In that very first publication, in the enumerated list of conclusions at the end of the article, the corresponding first item read: "1. The elements, arrayed by the magnitude of their atomic weights, present a distinct *periodicity* of properties."¶[14] It was even itali-

* "1. Die nach der Grösse des Atomgewichts geordneten Elemente zeigen eine stufenweise Abänderung in den Eigenschaften."

† "man die Atomgewichte aller Elemente ohne willkürliche Auswahl einfach nach der Grösse ihrer Zahlenwerthe in eine einzige Reihe ordnet, diese Reihe in Abschnitte zerlegt und diese in ungeänderter Folge an einander fügt."

‡ "entnehmen wir aus der Tafel, dass die Eigenschaften der Elemente grossentheils *periodische* Functionen des Atomgewichtes sind."

§ "Man sieht aus dem Verlaufe der Curve sofort, dass die Raumerfüllung der Elemente, eben so wie ihr chemisches Verhalten, eine periodische Function der Grösse ihres Atomgewichtes ist."

¶ "1. Элементы, расположенные по величине их атомного веса, представляют явственную *периодичность* свойств."

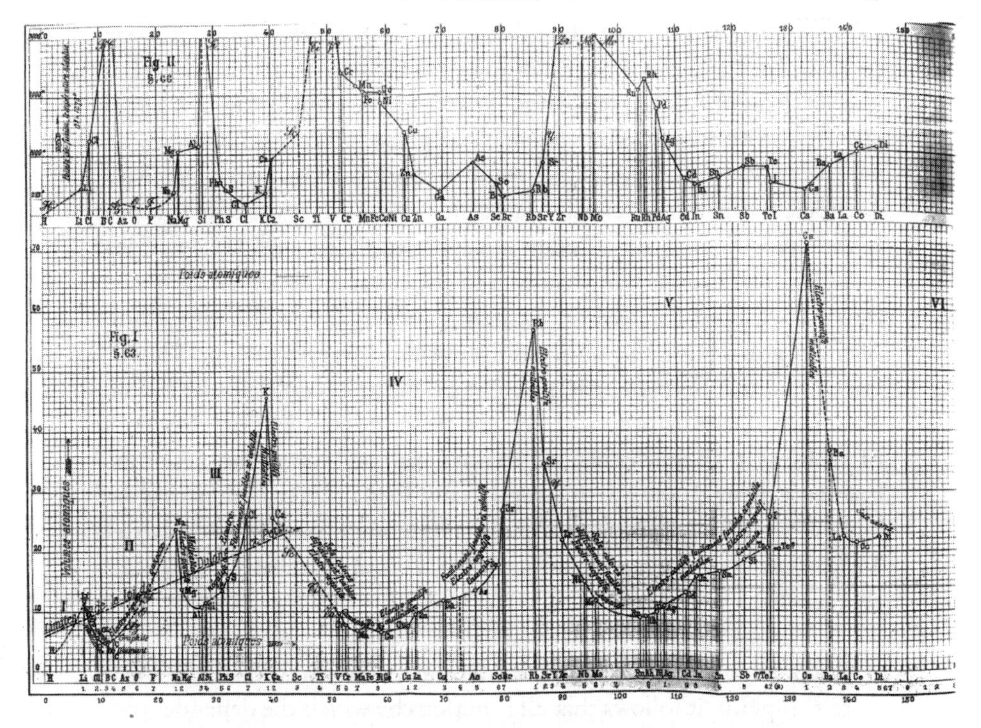

FIGURE 2.3. Lothar Meyer's atomic-volume curve, which presents the periodicity of chemical elements in a manner quite different from the now-standard table. Lothar Meyer, "Die Natur der chemischen Elemente als Function ihrer Atomgewichte," *Annalen der Chemie und Pharmacie*, Supp. VII (1870): 354–364, insert.

cized. How had the word *periodic* come to be rendered as *stufenweise* instead of *periodische*?

Mendeleev blamed Beilstein.[15] Beilstein was flooded by Russian-language abstracts, handed to him with the dual request that he both arrange for their translation *and* publish them as rapidly as possible—two charges that could not both be met, since careful translation took time. Beilstein had handed Mendeleev's abstract to A. A. Ferman, an assistant then working in his laboratory at the Technological Institute in St. Petersburg, and asked him to translate it. With speed as the chief goal, he raced through it, not considering the word "periodic" to be of particular importance and substituting "phased" instead, a choice he confessed to as an audience member at a lecture on the priority dispute in 1911, long after Mendeleev, Meyer, and Beilstein were all safely dead.[16]

The damage had been done, and Meyer was in print claiming to have made a central innovation on Mendeleev's system of elements, now universally called a "periodic system" in both German and Russian. In 1870, Mendeleev was even willing to cite Meyer and grant him limited credit *in Russian*.[17] Yet when it came to the high stakes of his massive reprise of the periodic system to be placed in *Liebig's Annalen* the year after Meyer's, Mendeleev would be more circumspect. As usual, Mendeleev wrote the lengthy piece in Russian, but this time he had his trusted friend Felix Wreden render it carefully into German, just as Wreden almost certainly translated the cover letter to the *Annalen*'s new editor, Emil Erlenmeyer. Insisting once again on the importance of the original Russian publications, even in this letter Mendeleev declared that the German article before Erlenmeyer could not be considered final: "Despite its size the present article does not report the course of my ideas in all the details, which are developed more completely and gradually in my Russian articles and in my 'Principles of Chemistry,' and which I would happily share with the German public."* [18] In the article Mendeleev hit the concept of periodicity—and, crucially, the *word*—repeatedly:

> From the foregoing, as well as from other surveys introduced by me to this point, it follows that all functions by which the dependence of properties upon the weight of the atom are expressed mark themselves as *periodic*.[. . .] Thus the periodic law can be expressed in the following manner: *the properties of the elements* (also as a result the simple and compound bodies formed out of them) *find themselves in a periodic dependence from their atomic weights*.† [19]

This, he hoped, ought to the settle the issue of credit and priority. Meyer did not completely agree, and while in the second issue of his *Modern Theories* textbook, published in 1872, he granted Mendeleev the lion's

*"Trotz ihres Umfanges giebt vorliegende Abhandlung meinen Ideengang doch nicht in allen den Details wieder, welche in meinen russischen Abhandlungen und in meinen 'Grundzügen der Chemie' vollkommener und allmäliger entwickelt werden und welche ich gern dem deutschen Publicum mitgeteilt hätte."

†"Aus dem Vorhergehenden, sowie aus anderen von mir bis jetzt ausgeführten Zusammenstellungen folgt, dass alle Functionen, durch welche die Abhängigkeit der Eigenschaften von dem Gewicht der Atome ausgedrückt wird, sich als *periodische* kennzeichnen.[. . .] Daher kann das periodische Gesetz folgendermassen ausgedrückt werden: *die Eigenschaften der Elemente* (folglich auch der aus ihnen gebildeten einfachen und zusammengesetzten Körper) *befinden sich in periodischer Abhängigkeit von deren Atomgewichten*."

share of the credit and fulsome praise, in the 1876 third edition he insisted that he himself had contributed a great deal to the development of the system, and that Mendeleev's "schema at that time [i.e., 1869] contained in itself still much arbitrariness and lack of regularity, which were later eliminated."*[20] They really ought to share the credit. Aside from minor sniping in articles across the 1870s, the issue lay quiet, but smoldering.

Until Adolphe Wurtz, professor of chemistry at the Sorbonne and the most distinguished chemist in France, decided to douse the whole affair in kerosene. In 1877, Wurtz wrote privately to Mendeleev to express "my opinion on your admirable works on atomic weights, which I consider the most important progress that the atomic theory has made for a long time."†[21] In his history of atomism, published two years later in French, Wurtz upped the ante by publicly emphasizing Mendeleev's invention of periodicity and his Russian identity: "Recently, the works of M. Mendéléff have opened a new day on the relations which exist between the atomic weights of simple bodies and their properties. The latter are a function of their atomic weights, and this function is *periodic*. That is the proposition put forward by the Russian chemist."‡[22] Wurtz was certainly entitled to his opinion, at least when publishing in French. In January 1880 the *Berichte* of the German Chemical Society published a letter from the French chemist that complained about the German translation of his *La théorie atomique*. Apparently, the German translator, a certain C. Siebert from Wiesbaden, had permitted an unauthorized preface and textual emendations—without seeking Wurtz's permission—that gave Lothar Meyer a greater share of credit in the periodic system. Wurtz sent a letter to the loudest megaphone in the German chemical community he could find in order to state his views that this revisionist position was "not well founded."§[23]

Lothar Meyer responded to the salvo twice. First, he wrote a letter

*"Sein damaliges Schema enthielt indessen noch manche Willkür und Unregelmässigkeit, die später ausgemerzt wurde."

†"mon sentiment sur vos admirables travaux sur les poids atomiques, que je considère comme le progrès le plus important que la théorie atomique ait fait depuis longtemps."

‡"Dans ces derniers temps, les travaux de M. Mendéléff ont jeté un jour nouveau sur les relations qui existent entre les poids atomiques des corps simples et leurs propriétés. Celles-ci sont fonction des poids atomiques, et cette fonction est *périodique*. Telle est la proposition énoncée par le chimiste russe."

§"nicht wohl begründet."

of his own to the council of the Society, declaring that he had been irritated by the similarity of Wurtz's book to his own *Modern Theories* and his publisher insisted on inserting a correction. (Wurtz considered this defense preposterous.)[24] At the end of his letter, however, Meyer also added a note about credit: "Occasionally I had also suspected that Mr. *Würtz* had not entirely correctly distinguished Mr. *Mendelejeff*'s and my contribution to the development of the most recent atomic theory from each other.[. . .] Since this is now touched upon, I want to make this historically entirely clear to him in a note to the *Berichte*."* [25] Meyer's second rebuttal, dated 29 January 1880 from his final post at the University of Tübingen in southern Germany, declared that any "unprejudiced judge"† could look at his first edition of his *Modern Theories* and see that the essence of the periodic system was already present. He then noted that the original abstract in the *Zeitschrift für Chemie* had left an important point ambiguous:

> In the accompanying text it was said that the elements ordered according to the magnitude of their atomic weights showed a phased (*stufenweise*) change in their properties, that the magnitudes of the atomic weights determine the properties, that certain atomic weights are in need of correction and that the discovery of new elements was predictable; in addition to still other less important comments. Mr. *Mendelejeff* published these points of view in any event before me and probably altogether for the first time.‡ [26]

He thus granted Mendeleev credit, but insisted that periodicity was his own innovation, lamenting only that the editors of the *Annalen* had not

*"Gelegentlich hatte ich auch erwähnt, Hr. *Würtz* habe Hrn. *Mendelejeff*'s und meinen Antheil an der Entwickelung der neueren Atomlehre nicht ganz richtig gegen einander abgegrenzt.[. . .] Da dieser einmal berührt ist, will ich ihn in einer Note in den Berichten historisch völlig klar stellen."

†"unbefangener Beurtheiler"

‡"Im begleitenden Texte war gesagt, dass die nach der Grösse des Atomgewichtes geordneten Elemente eine stufenweise Abänderung der Eigenschaften zeigen, dass die Grösse des Atomgewichtes die Eigenschaften bedinge, dass einige Atomgewichte der Berechtigung bedürftig und die Entdeckung neuer Elemente vorherzusehen sei; daneben noch einige weniger wichtige Bemerkungen. Diese Gesichtspunkte hat also Hr. *Mendelejeff* jedenfalls vor mir und wahrscheinlich überhaupt zuerst veröffentlicht."

granted him enough space in 1870 to elaborate upon the differences between their two theories.

Mendeleev was furious. In an annotation to his bibliography that he penned late in life, he noted of his reaction that "I cannot stand this polemic of priorities, but the Germans forced me to answer."* [27] Mendeleev devoted the bulk of his own retort to Meyer, published in the same volume of the *Berichte*, to translated quotations from various original Russian publications. Borrowing a rhetorical device deployed in his textbook, *The Principles of Chemistry*, Mendeleev confined most of his own commentary on the significance of these translations to the footnotes, observing that "The word *periodicity* is emphasized in the original,"† and that the repetition of this word throughout the article "clearly shows that I at the very beginning (March 1869) considered periodicity as the fundamental property of the system of elements I had offered. Here it is clearly seen that I did not borrow this word from Mr. *L. Meyer*."‡ He concluded by noting that the citation to the original was prominently displayed in the *Zeitschrift* publication, and "could have been known therefore to Mr. *L. Meyer*,"§ and thus that "Mr. *L. Meyer* did not have the periodic law in mind before I did, and introduced nothing new afterward."¶ [28]

This fight was thus still going on over ten years after it had begun. Meyer had hoped that his historical rejoinder to Wurtz—that he considered admirably dispassionate and objective—would have taken care of this mess. After all, he had already ceded most of the credit to Mendeleev, only despairing that the Russian had not cited his 1870 work more generously. But Mendeleev wanted *all* of the credit, and his claim to that hinged on evaluating the status of that April 1869 publication as a scientific publication. It was longer, more detailed, and crucially *earlier* than the *Zeitschrift* abstract, but it was also, Meyer observed, written in Russian. This, he believed, was an important difference:

*"Эту полемику приоритетов—я терпеть не могу, но меня немцы принуждали отвечать."

†"Das Wort *Periodicität* ist in dem Original unterstrichen."

‡"zeigt deutlich, dass ich ganz im Anfange (März 1869) die Periodicität für die Grundeigenschaft des von mir gegebenen Systems der Elemente hielt. Hieraus ist deutlich zu ersehen, dass ich dieses Wort nicht Hrn. *L. Meyer* entlehnt habe."

§"also Hrn. *L. Meyer* hätte bekannt sein können."

¶"dass Hr. *L. Meyer* vor mir das periodische Gesetz nicht im Sinne gehabt und nach mir nichts Neues hinzugefügt hat."

> I had found what I wrote in December 1869 about the periodicity of properties before the published abstract from Mr. *Mendelejeff*'s work in the *Zeitschrift für Chemie* in that same year came to my attention. Naturally however I only claimed for myself what that piece did not contain and what seemed to me to need improvement in it. Mr. *Mendelejeff* now claims that his articles which had then appeared in the *Russian* language contained everything that I had improved and introduced, and reproached me for not getting hold of his original articles. It seems to me an excessive demand that we German chemists read, besides those articles appearing in the Germanic and Romance languages, also those in the Slavic languages, and should monitor the accuracy of the German reports about their contents.*[29]

Now it was out in the open. Mendeleev had published, but he had published in Russian. In an important sense, this meant it did not count. Here was the bedrock issue behind this fight over priority: the status of Russian as a scientific language. Could the Russians consider articles written in their incomprehensible tongue and national journals as *equivalent* to those printed in established languages like German, French, and English—or even Italian?

Let Them Read German

The answer to that question requires a step backward, to the history of Russians' attempts to establish scientific publishing in their own language and on their own terms. It was a lengthy trek. Suppose you were

*"Was ich im December 1869 über die Periodicität der Eigenschaften schrieb, hatte ich gefunden, bevor mir der im demselben Jahre in der Zeitschrift für Chemie veröffentlichte Auszug aus Hrn. *Mendelejeff*'s Arbeit zu Gesichte kam. Natürlich aber habe ich nur das für mich in Anspruch genommen, was dieser nicht enthielt und was mir an ihm der Verbesserung bedürftig schien. Hr. *Mendelejeff* giebt nun an, dass seine damaligen in *russischer* Sprache erschienenen Abhandlungen alles das enthalten haben, was ich verbesserte und hinzufügte, und macht mir zum Vorwurfe, dass ich mir nicht seine Originalabhandlungen verschafft habe. Mir aber scheint es eine zu weit gehende Forderung, dass wir deutschen Chemiker, ausser den in germanischen und romanischen, auch noch die in slavischen Sprachen erscheinenden Abhandlungen lesen und die deutschen Berichte über ihren Inhalt auf ihre Genauigkeit prüfen sollen."

a chemist at the heart of the Russian Empire, St. Petersburg, in 1861, in the months after the Emancipation of the serfs in February, or perhaps even after the liberalization of the Tsarist censorship in 1865. Where would you publish original research? Russian-language technical journals were few and far between. In 1804 the St. Petersburg Academy of Sciences began publishing a *Technical Journal* (*Tekhnicheskii zhurnal*) for a few years, but its main outlet remained its *Bulletin*, which printed pieces in French and German for the first half of the nineteenth century (in the previous century, the obligatory language had largely been Latin), and in any event you would need the endorsement of an academician to publish there, which set a pretty high bar. The *Mining Journal* (*Gornyi zhurnal*) began publication in 1825, and for the rest of the century remained a significant outlet for works in applied chemistry and metallurgy. The problem was that almost no one read it or cited it, even among the elite scientists of the capital, let alone Western Europe. There were other experiments in both Petersburg and Moscow in the 1820s and 1830s, but they remained devoted more to popularization than to original research.[30] The solution seemed simple: write your article in French or German (or, very rarely, English), and send it abroad for publication.

In 1859, two chemists based in Petersburg—Aleksandr N. Engel'gardt, a talented organic chemist with rather substantial economic resources from his patrimonial estate, and Nikolai N. Sokolov, an ambitious theorist with an affinity for Auguste Comte's philosophy of Positivism—made an effort to remedy the situation.[31] First, they set up a private chemical laboratory on Galernaia Street, not far from the Winter Palace where the Hermitage Museum now sits, so researchers could conduct their research, for a fee. This relieved some pressure on the University and Academy laboratories, the latter of which was closed anyway to non-academicians. Then, the two of them established *N. Sokolov and A. Engel'gardt's Chemical Journal* (*Khimicheskii zhurnal N. Sokolova i A. Engel'gardta*), the first Russian-language journal explicitly and exclusively devoted to the science of chemistry. The price for a year, composed of twelve separate issues, was an affordable five rubles, and home delivery in St. Petersburg or Moscow was available for an additional ruble a year (other addresses commanded extra fees). The first issue was graced by a high-minded epigraph by distinguished historian Augustin Thierry, offered, *naturellement*, in French: "There is something in the world which is worth more than material pleasures, more than fortune, more than health itself—it is the development of science."* [32]

In their introduction to the first issue, Sokolov and Engel'gardt professed the highest of motives: the creation of a Russian chemical community. The journal was an essential part of that, not because there was no chemical information reaching the Russian public, but rather that there was too much, and not necessarily of the highest quality. The journal "will give our public the opportunity above all to toss out from the majority of the diverse essays on chemistry all the rubbish, all the unnecessary, part of it even harmful, which is unfortunately published in enormous quantity in all the literatures, and to select only that which has indubitable merit in some respect."† [33] Readers would be supplied with original works by Russian chemists, translations of important chemical works from other languages, selected abstracts and summaries, and news of interest to chemists.

It turned out to be pretty much a disaster. Initially, the two editors published their dissertations serially in the journal, and a few other Russian chemists, such as Mendeleev, submitted original work. But only a few. As the first year transitioned into the second, an increasing portion of the journal was devoted to lengthy articles summarizing the research of foreign chemists, often with several articles mashed up into one single review essay.[34] Even more problematic, the editors were obliged to fill their issues with translations of articles written by *Russian* chemists but published abroad in journals like *Liebig's Annalen*.[35] But, of course, Russian chemists could already read the originals in French and German—and, more importantly, so could the European chemists they perceived as their primary audience—so there was less and less demand for the *Chemical Journal*. Even those who did use it, like Mendeleev, on a postdoctoral jaunt to Heidelberg, complained that it was difficult to obtain copies while abroad; Sokolov shrugged it off, noting only that "the sending of Russian books abroad is attended here, as they say, by unusual difficulties."‡ [36] In the end, the fiery Sokolov decided enough

* "Il y a au monde quelque chose qui vaut mieux que les jouissances matérielles, mieux que la fortune, mieux que la santé elle-même, c'est la dévouement à la science."

† "даст возможность сверх того нашей публике отбросить из множества разнообразнейших сочинений по химии весь хлам, все не нужное, часть даже вредное, публикуемое к сожалению в огромном количестве во всех литературах и выбрать только то, что иметь несомнительное достойнство в каком нибудь отношении."

‡ "пересылка русских книг заграницу сопряжена у нас, как говорят, с необыкновенными трудностями."

was enough: he dissolved the private laboratory, donated its material resources to St. Petersburg University, and moved to a teaching job there (which he held until 1864, when he stormed out of the capital as well). The journal collapsed in 1860, after only two years of publication. As far as Sokolov was concerned, the Russian chemical community he was trying to summon into being had failed him. The *Chemical Journal* had foundered because Russian chemistry did not exist.

The evidence does not support Sokolov's pique. As the pages of his own journal attested, there were plenty of active researchers generating original findings in both experimental and theoretical chemistry, and they were eager to publish. It was simply that, when given a choice in the 1860s about where to do so, Russian chemists overwhelmingly chose to publish in German. And not just in any journal, but overwhelmingly in one relatively marginal chemical periodical that we have already encountered: the *Zeitschrift für Chemie.*

The *Zeitschrift* was not originally supposed to be a chemical journal, and it was certainly never intended to cater to Russians. When it was established in Heidelberg in 1858 by the quartet of August Kekulé, Gustav Lewinstein, Friedrich Eisenlohr, and Moritz Cantor as the *Kritische Zeitschrift für Chemie, Physik und Mathematik*, it was as a review journal, publishing critical commentary on recent publications in a wide variety of fields. Kekulé—soon to become one of the founders of the structure theory of organic molecules and eventually a titan of German chemistry—defended the venture to the grand man of chemical publishing, Justus von Liebig, by claiming that "through detailed abstracts a service will be rendered to the public and that only thus can a dam be placed against the incessantly increasing slime-literature."* [37] Instead of erecting that barrier, it soon joined the slimy ranks; three of the four original editors abandoned the journal by the following year, and Gustav Lewinstein was joined on the masthead by pharmacist-turned-chemist Emil Erlenmeyer, who had just begun a lectureship in organic chemistry at the local university. Over the next five years, the journal, now a specialist journal renamed the *Zeitschrift für Chemie und Pharmacie*, would become so heavily identified with Erlenmeyer (Lewinstein soon decamped as well) that many library catalogs would simply refer to it as "Erlenmeyer's Zeitschrift."[38]

*"durch eingehende Rezensionen dem Publicum ein Dienst geleistet und daß nur so der fortwährend zunehmenden Schmier-literatur einigermaßen ein Damm gesetzt werden kann."

The identification with Erlenmeyer was a mixed blessing. He was a talented theoretical chemist but a rather obnoxious editor. He published original pieces in the *Zeitschrift*, but he also reprinted abstracts of articles from other journals, and these he would pepper with sarcastic editorial comments, appendices, and interlinear exclamation points of disdain, earning him considerable enmity from the German chemical community.[39] On the other hand, he socialized extensively with the sizeable group of Russian chemists who spent postdoctoral research visits in Heidelberg—most famously Mendeleev, but also the chemist-cum-composer Aleksandr Borodin and dozens of others—and published German-language articles by them in great number.[40] (For his services to the roughly sixty Russians who passed through his small laboratory on Karpfengasse in Heidelberg, the Tsarist government awarded him the Order of St. Anna in 1865.[41])

The affiliation with the young Russians—and their evident affection for the man they dubbed "Eremich"—almost certainly prolonged the life of the journal, as Germans abandoned Erlenmeyer to his sneering. Although it was cumbersome to obtain the *Zeitschrift* within Russia (you had to make special arrangements with booksellers, in addition to the problems with the Russian mails), roughly 150 Russians subscribed to the journal by 1865, dwarfing German and West European orders. As Beilstein would remark the following year, "Erlenmeyer's Zeitschrift was more popular in Russia than in Germany."*[42] The finances of the journal were suffering, and Erlenmeyer was desperate to offload it. As he wrote to Aleksandr Butlerov, then professor of chemistry at Kazan but shortly to move to St. Petersburg: "Indeed, dear friend, I would like to induce you to consider whether you would not want to take this up yourself and make it into a Russian journal, but one that is printed in the German language. Perhaps you could thus unite a Russian chemical society that made the *Zeitschrift* into its organ."†[43] Butlerov passed, but Erlenmeyer eventually found his successors in three young chemists at Göttingen: Hans Hübner, Rudolf Fittig, and Friedrich Konrad Beilstein.

*"Die Erlenmeyer'sche Zeitschrift war in Rußland verbreiteter als in Deutschland."

†"Doch, lieber Freund, möchte ich Ihnen zu bedenken geben ob Sie dieselbe nicht in die Hände nehmen und zu einer russischen Zeitschrift machen wollen, die aber in deutscher Sprache gedruckt ist. Vielleicht können Sie damit eine russische Chemikergesellschaft verbinden, die die Zeitschrift zu ihrem Organ macht."

It was the choice of Beilstein, who would move back to his native city of St. Petersburg in 1866 to assume a post at the Technological Institute, that would cement the Russians further to the *Zeitschrift*, with implications for the fate of the periodic system.[44] He was an inspired choice to navigate the changing face of European chemistry: a native speaker of German and Russian, he was fluent as well in English and French, and managed reasonably deftly in Swedish and Italian to boot.[45] He was also incredibly industrious and a gifted organic chemist, both of which stood him in good stead as he and his colleagues attempted to revive the periodical Erlenmeyer had handed on to them. "My God!" he lamented to Kekulé on 3 November 1865. "If I had been able to guess that one would earn for so much hard and bitter work so much unhappiness, dissatisfaction, trouble, and ingratitude, I would have sent *Erlenmeyer* home when he offered me the continuation of his rag last year."* [46] The Göttingen triumvirate began to rebrand the periodical as one that published more quickly than the leading journal, *Liebig's Annalen*, and so became the venue of choice for certain chemists seeking rapid publication to vouchsafe their priority in chemical discoveries. That was one reason why Mendeleev had chosen it.

But not the main reason. Beilstein carried on his duties as an editor for the *Zeitschrift* when he left Göttingen, but he acquired an additional responsibility: "I take everything Russian, since I remain the correspondent for Russia."† [47] Much like Erlenmeyer had, Beilstein realized that the support and contributions of Russian chemists, who had no national chemical journal of their own in the mid-1860s, was crucial for the financial solvency of the journal. As he wrote in a revealing letter to Butlerov in January 1865, shortly after assuming the role of editor:

> I will in conclusion emphasize again that the 'Zeitschrift' possesses in my person a warm representative of Russia's interests. I wish that the Russian chemists not just laboriously work themselves to death with a Russian edition of their works (for you, who write German so expertly, this is truly not necessary!). But many might thus put off the publication of works, and thus I beg that they send me only the

*"Bei Gott! Hätte ich ahnen können, daß man für diese harte und saure Arbeit soviel Unglück, Unfrieden, Ungemach und Undank ernten würde, ich hätte den *Erlenmeyer* heimgeschickt, als er mir voriges Jahre die Fortführung seines Würstblattes antrug."

†"Alles russische nehme ich an, da ich Correspondent für Russland bleibe."

> *Russian* articles. I wish to bear the burden of a correct translation. [. . .] Chemists speak only *one* language and thus one should also know in German what newly appears in Russia.*[48]

He practiced what he preached—in fact, he had been doing so for years. When Aleksandr Engel'gardt neglected to publish his researches anywhere but in his doomed *Chemical Journal*, Beilstein summarized them in German and placed a report into the *Zeitschrift*.[49] He did the same for Mendeleev's 1864 Russian doctoral dissertation on alcohol-water solutions, adding that he hoped the author would publish a more extended version of his findings in another language, thereby "making his classic work also known to the remaining public."† [50] It is clear from his correspondence that he worried extensively over the quality of the translations he commissioned or performed himself, and the *Zeitschrift* under his editorship continued to be the German-language periodical of choice for Russian chemists, the only national community to so favor it.

Unfortunately, it was not enough. In 1871, the *Zeitschrift* closed up shop, ending this experiment in transnational chemistry. As always, there was plenty of blame to go around, but the editors consistently fingered one culprit. In 1867, even before the unification of the country that would come to be called "Germany," the German Chemical Society was founded, and soon began publishing its journal, the *Berichte*. The *Berichte* also had to compete with the *Annalen*, and saw a niche in rapid publication of shorter articles, the very same strategy undertaken by the *Zeitschrift*. It was bigger, however, and more prestigious, and subscribers to the latter began leaching away. "There remains no doubt: the *Zeitschrift für Chemie* can no longer be conducted *the way* it has been until now," Beilstein wrote to Erlenmeyer in 1871. "Through the successful activity of the Berliner *Berichte* one of the chief tasks of

* "Ich will zum Schluß noch hervorheben, daß die 'Zeitschrift' in meiner Person einen warmen Vertreter der Interessen Rußlands besitzt. Ich wünsche, daß sich die russischen Chemiker nicht erst mühsam abplagen mit einer russischen Redaktion ihrer Arbeiten (bei Ihnen, der so gewandt deutsch schreibt, ist es freilich nicht nöthig!). Aber Mancher könnte dadurch die Publikation von Arbeiten aufschieben, u. darum bitte ich mir denn nur *russisch* die Abhandlungen zuzusenden. Für eine correkte Übersetzung will ich schon Sorge tragen.[. . .] Die Chemiker reden so *eine* Sprache u. darum soll man auch in Deutschland wissen, was in Rußland neu erscheint."

† "seine klassische Arbeit auch dem übrigen Publicum bekannt machen."

the *Zeitschrift*—to publish rapidly—is effectively solved."* [51] The blame, that is, lay with the Germans, who were centralizing cultural authority along with political authority in Berlin.

There was much truth to this account, but Beilstein and his fellow editors neglected another competitor to the *Zeitschrift*, one which peeled off its most loyal adherents. The Russian Chemical Society was created the year after the German, and the following year the *Journal of the Russian Chemical Society* suddenly appeared, described in the Society's charter thus: "this publication will include the works of Russian chemists, printed in the Russian language."† [52] Unlike the doomed venture of the *Chemical Journal* a decade earlier, the *Journal* has continued, under a number of name changes, down to the present, becoming one of the most successful periodicals in the history of chemistry. It was not at first obvious that things would turn out this way.

In its first year, the *Journal* printed a total of eighty copies, including the sixty issues reserved for Society members, most of whom were concentrated in St. Petersburg.[53] This meant, in short, that no one outside of the same small circle of Russian chemists was reading it. Mendeleev lamented the state of affairs in 1871, no doubt influenced by his recent tangles with Lothar Meyer, and he suggested that "[i]n view of the fact that many of the works printed in the Society's journal remain partly unknown abroad, partly known [only] through short extracts,"‡ the Society should be careful to send copies to be reported in the *Jahresbericht*, the German annual report of chemical publications.[54] Mendeleev's worry was that Germans remained unaware of Russian publications; his contemporary Vladimir Markovnikov of Moscow University was more concerned about the *Russians*. As he wrote to his mentor Butlerov in 1874: "Tell me, please, why have all the Petersburg chemists begun again to publish their works in foreign journals, and even earlier than in Russian? Why on Earth do our Society and Journal exist? I find

*"Bei Gott! Hätte ich ahnen können, daß man für diese harte und saure Arbeit soviel Unglück, Unfrieden, Ungemach und Undank ernten würde, ich hätte den *Erlenmeyer* heimgeschickt, als er mir voriges Jahre die Fortführung seines Würstblattes antrug."

†"что его издание будет включать труды русских химиков, печатаемые на русском языке."

‡"В виду того, что многие из работ, напечатанных в журнале Общества, остались заграинцею частью неизвестными, частью известными по кратким извлечениям."

this completely tactless and, if this continues, I'll quit the Society."* [55] (To remedy this problem, the Society even created prizes such as the Zinin/Voskresenskii prize and the Sokolov prize, both established in 1880, which were to be awarded only to works printed in Russian.[56])

Chemists working in Russia were keenly aware of local sensitivity on this question. For example, Beilstein wrote to Erlenmeyer, now an editor at the *Annalen*, that he hoped the latter could delay a forthcoming article on naphthalene: "Namely, I would not want this article to appear earlier by you than in our Russian journal. My patriotic friends would raise a stink that I did not provide the national organ with original articles."† [57] Or, as Butlerov's student Aleksandr Popov wrote to his advisor: "Would you approve of my intention to place in our chemical journal my works which I am producing here in Bonn and which at the same time will be printed in German journals? I intend to send for our journal more detailed descriptions than for the Germans."‡ [58] If anything the debates over the periodic system only highlighted these concerns, besides the general problems with the Russian mail and the tardy publication of several early issues of the *Journal*—occasionally Russians had to learn what was in their own journal by reading the abstracts in the *Berichte*.[59]

Yet Russians now seemed willing to back their own journal in their own language, and thus the *Zeitschrift* lost its prime clients. Markovnikov suggested a division of labor in 1870—"It is proposed to publish the works of Russian chemists by the degree of the accumulation of materials; shorter reports, made at the meetings of the Society, should be printed in the *Zeitschrift*"§—but it was too little, too late.[60] Beilstein,

*"Скажите, пожалуйста, почему это все петербуржские химики начали опять публиковать свои работы в иностранных журналах, и даже раньше, чем на русском языке? К чему же существует наше Общество и Журнал? Я нахожу это совершенно бестактным, и, если так продолжится, то выйду из Общества."

†"Ich möchte nähmlich nicht, daß diese Abhandlung früher bei Ihnen als in unserem russischen Journal erscheint. Meine patriotischen Freunde würden mir Krakehl machen, das vaterländische Organ nicht mit Original-Abhandlungen zu versehen."

‡"Одобрите ли Вы мое намерение помещать в наш хим. журнал мои работы, которые я произвожу здесь в Бонне и которые в то же время будут напечатаны в немецких журналах? Я намерен для нашего журнала посылать более подробные описания, чем для немецких."

§"Положено издавать работы русских химиков по мере накопления материалов; краткие же сообщения, делаемые в заседаниях Общества, печатать в Zeitschrift'e."

for one, was frantic that this insistence on publishing in Russian would doom the Russians to neglect, and turned to Erlenmeyer in 1872 with an impassioned plea:

> In any case I would like to make you aware how much it would lie in the interests of the readers of the *Annalen* if you wanted to give a little attention to the *Journal of the Russian Chemical Society*. Up to now I have enabled the traffic through abstracts. Since the New Year, however, following the news from *Hübner*, I have laid aside my pen [at the *Zeitschrift*]. The *Annalen* must bring the works out completely. Now however the Russians have all become great patriots: they no longer want to write up their articles in foreign languages. Only a few, e.g. Menshutkin, are so considerate as to worry about a translation themselves. Thus it is predictable that much useful work will be lost. *You* will earn a great merit if you tame this evil.* [61]

Erlenmeyer was willing to help, but only if the responsibility for translations was assumed by the Russians. "I am of the view to ask the authors themselves to send us their articles in German or even French. It is greatly preferable to me, if the people concerned send their things themselves; they thus at the same time assume the responsibility for what stands written."† [62] After all, Erlenmeyer was observing the priority dispute unfolding between Meyer and Mendeleev in the pages of his own *Annalen*. He would hate to be blamed for something like that.

*"Jedenfalls möchte ich Sie darauf aufmerksam machen, wie sehr es im Interesse der Leser der Annalen läge, wenn Sie dem *Journal der russischen chemischen Gesellschaft* einige Aufmerksamkeit schenken wollten. Bis jetzt habe ich den Verkehr durch Auszüge vermittelt. Seit Neujahr habe ich aber, infolge der Nachrichten von *Hübner*, meine Feder niedergelegt. Die Annalen müßten die Arbeiten vollständig bringen. Nun sind die Russen aber große Patrioten geworden: sie wollen ihre Abhandlungen nicht mehr in fremden Sprachen abfassen. Nur wenige wie z.B. Menschutkin sind so liebenswürdig selbst für eine Übersetzung zu sorgen. Daher ist vorauszusehen, daß manche nützliche Arbeit verloren gehen wird. *Sie* werden sich um Viele ein großes Verdienst erwerben, wenn Sie diesem Übel steuern."

†"ich die Absicht habe die Herren Autoren selbst um Einsendung ihrer Abhandlungen in deutscher oder französischer Sprache zu bitten. Es ist mir viel lieber, wenn die betrff. ihre Sachen selbst einsenden, sie übernehmen damit zugleich die Verantwortung für das was geschrieben steht."

Solomon's Baby

There was probably no way to avoid a priority dispute about the periodic system of chemical elements. There were so many people approaching some version of an arrangement of the elements along the two axes of weight and chemical properties, that any two of them might have found themselves struggling to assume the credit. But history did not unfold in an imagined parallel universe, and instead of a different priority dispute—or, however unlikely, no dispute at all—European chemists witnessed a sustained decade of angry sallies and counter-thrusts over the proper attribution of the discovery to either Dmitrii Mendeleev or Lothar Meyer.

Just as there could have been many different contenders for priority—others, such as John Newlands, who repeatedly attempted to claim credit for himself, might not have been summarily ignored by all—the Meyer-Mendeleev conflict could have unfolded in a number of different ways. It could, for example, have been triggered by the discovery of the three not-yet-discovered elements whose properties Mendeleev predicted: gallium (Mendeleev's eka-aluminum) in 1875, scandium (eka-boron) in 1879, and germanium (eka-silicon) in 1886. Or chemists might have focused on Meyer's curve of atomic volumes, and encouraged a range of graphical presentations of the relationships between the elements. But those alternative histories also did not come to pass. Instead, we see Mendeleev and Meyer sparring with each other about credit largely as self-defined "Russian" and "German" chemists. The history suggests very strongly that this particular nationalist inflection, not altogether rare in this period, was accentuated by the faulty translation in the first German-language article Mendeleev published on his system in 1869. It was, in short, a reflection of debates over scientific languages, concentrated in a single word: periodic.

The emphasis on language was, to some extent, derived from the nationalist ideologies then sweeping across European culture, from which science was hardly exempt. This was the age of the unification of Germany, the creation of the French Third Republic, the *Risorgimento* in Italy, the Great Reforms in Russia, and many other smaller-scale clashes stemming from the entrenchment of the nation-state as the primary mode of the European political order (at least in Western Europe). But one does not need to look to such dramatic developments to locate the roots of the Mendeleev-Meyer conflict. There was, rather, a simple conjuncture of events in the late 1860s as Russian chemistry was beginning to transition from being a subsidiary of German chemistry into an established feature of the Tsarist polity. Emblematic of this transition was the few years of overlap when Russians published *simultaneously* in the *Journal* of the Russian Chemical Society and the *Zeitschrift für Chemie*. Russian science was not yet prominent enough to command attention when published in the Russian language, and so Mendeleev felt compelled to print his findings in German as well; the disconnect between the Russian and German versions motivated the subsequent hostility with Meyer.

That hostility was never really resolved on a personal level. Mendeleev kept an exhaustive archive of correspondence, and yet one finds there only two substantial items filed under Lothar Meyer's name. The second of these was a note from Meyer's widow informing Mendeleev of her husband's death in 1895—an indication that the Petersburger was on the list of people to be personally informed of the sad event.[63] In the earlier communication, a letter from 1893 that is the only personal correspondence between the two in the archive, Meyer informs Mendeleev that the distinguished Leipzig chemist Wilhelm Ostwald had commissioned two issues of his *Klassiker*—pamphlets of primary sources on monumental chemical discoveries—on the development of the periodic system. Meyer edited the first, on the "precursors" who had noticed smaller patterns among the elements before the 1860s.[64] Since he was uncomfortable calling himself a "classic," Meyer delegated the second volume on himself and Mendeleev to his student Karl Seubert. Meyer now asked Mendeleev to send copies of articles, especially "your article in the 1st volume of the Russ. Society, from which I had recently received an actual translation through Beilstein's mediation."*[65] Translated into German, of course. One can detect a subtle friction in Meyer's careful phrasing. We have no record of Mendeleev's response, but Seubert's volume containing the pieces was published in 1895.[66]

By that point, the controversy between the two had reached a semistable equilibrium. After the heated exchange in the *Berichte* of the German Chemical Society in 1880, Meyer and Mendeleev never again crossed swords directly. The tension was, however, palpable, and an outside group decided to step in and resolve it by fiat. In 1882, the Royal Society of London, Britain's premier scholarly association, jointly awarded the two men the coveted Davy Medal "[f]or their discovery of the peri-

*"Ihre Abhandlung in 1. Bd. der russ. Gesellschaft, von der ich durch Beilsteins Vermittlung kürzlich eine wirkliche Uebersetzung erhalten habe."

odic relations of the atomic weights."[67] This award was later dubbed by Seubert in his 1895 volume "a most just and beautiful decision,"* and it seemed to have calmed matters considerably.[68] A nonpartisan national organization opting for a middle path seemed to codify a consensus developing even among nationally committed observers. For example, when Butlerov gave lectures (in Russian) on the history of recent chemistry in 1879–1880, he also divided the credit between the two, and Nikolai Menshutkin continued this pattern in 1895 when he announced Meyer's death at a meeting of the Russian Chemical Society—with Mendeleev himself presiding.[69] Interactions between the men now dubbed "co-discoverers" remained officially correct on the few occasions when they interacted, as in one contemporary description of them together on the dais at the 1887 Manchester meeting of the British Association. Here too, language played its role, when "there was a call for a speech from Mendeléef, he declined to make an attempt to address the section in English." He knew that this was beyond his linguistic capacities, so the Russian just stood up and bowed. But then Meyer, seated next to Mendeleev, rose, and—to avoid misunderstanding—declared: "I am not Mendeléeff." But a moment later, "speaking in faultless English, asked permission to address the section in German, and then proceeded, on behalf of Mendeléeff and other foreign chemists present, to express the pleasure they had derived from listening to the Presidential address."[70] At that time, as in 1880, Meyer got the last word. But after Meyer's death in 1895, Mendeleev was left to shape the history, at which point he relented on his exclusion of Meyer from any credit and included him within his narratives of the system—but only as a "strengthener" of the system, not as a full-fledged co-discoverer.[71] And it is Mendeleev's post-Meyer allocation of credit that is dominant today.

Perhaps the real victory was not who discovered the periodic system, but which languages were seen to "count" among the scientists of Europe. In no small part due to Mendeleev's emphasis on the importance of reading his original writings in Russian to adjudicate priority—and no doubt the impressive quality of those works themselves—Western European scientists began to take notice of the Russian-language works published in the *Journal*. Foreign correspondents would report, in translation, on the major activities discussed at meetings of the Russian Chemical Society. The Belgians, for example, began publishing the *Journal*'s table of contents in their own journal, in French, in 1875.[72] At the

* "gerechteste und schönste Entscheidung."

twenty-fifth anniversary celebration of the Russian Chemical Society, President of the Chemical Society of London Henry E. Armstrong sent a congratulatory letter: "Notwithstanding the great difficulties which your language imposes, your english [sic] colleagues learn from time to time of your labours, the name of your Society and a record of its work regularly appearing in our volume of abstracts of chemical papers."[73] At the same meeting where this letter was read out, again under Mendeleev's presiding eye, Menshutkin lauded his writings on the periodic system: "These works, printed in Russian, now become an achievement of universal science, thanks to abstracts about them in foreign scholarly societies."*[74] To be sure, Western Europeans were not signing up to learn Russian in droves, but a few did indeed try to master the Slavic tongue, and many of the others at least now came to understand that they could not simply dismiss writings in the language, as Lothar Meyer had, as not registering in the scientific literature.

This was surely no small part of Mendeleev's reasoning, in 1899, when he wrote—in French—that "as a Russian, I am proud of having participated in the establishment of the periodic law."†[75] For something had indeed happened in the previous forty years that marked the distance the Russian chemical community, and the scientific community in general, had traveled since the abortive efforts at founding a chemical periodical in 1859. In 1890, Russian historian Vasilii I. Modestov wrote about what he could now call *Russian science*: "We know that during just these past twenty five years *Russian science was created*, a science which begins to garner to itself both in this and that area a recognition which before did not exist."‡[76] For this recognition to happen, Russian science had actually to be written *in Russian*, and that entailed a deliberate effort to modify the ancestral tongue so that it had the capacity and flexibility to express scientific thoughts. Much as Latin had to adapt itself to Greek under Cicero's nimble pen, so the Russians had to reform their language in the light of the scientific languages wafting toward them from the West: the triumvirate of English, French, and German.

*"Эти труды, печатаемые по-русски, становятся теперь достоянием всемирной науки, благодаря рефератам о них в иностранных ученых обществах."

†"Voilà pourquoi, en ma qualité de Russe, je suis fier d'avoir participé à l'établissement de la loi périodique."

‡"Мы знаем, что в течение только-что истекшего двадцатипятилетия *создалась русская наука*, которая начинает получать в той, то в другой области себе признание, чего прежде не было."

CHAPTER 3

Hydrogen Oxygenovich

> Другое дело если б, например, он встретился с Либихом, не зная, что это вот Либих, хоть в вагоне железной дороги. И если б только завязался разговор о химии и нашему господину удалось бы к разговору примазаться, то, сомнения нет, он мог бы выдержать самый полный ученый спор, зная из химии всего только одно слово «химия». Он удивил бы, конечно, Либиха, но—кто знает—в глазах слушателей остался бы, может быть, победителем. Ибо в русском человеке дерзости его ученого языка—почти нет пределов.*
>
> F. M. DOSTOEVSKY, 1873[1]

Everyone says that Russian is a difficult language. Even centuries before Russia began to be a significant player in European politics—largely a consequence of Peter the Great's 1721 victory over Sweden, replacing one great northern Empire with a much vaster one—diplomatic and mercantile writings by Westerners bemoaned its complexity, abstruseness, and general impossibility. As seen in the previous chapter, Russian was considered far beyond the pale for European scientists in the late nineteenth century, individuals who routinely mastered English, French, and German (and others besides, if one of these three was not their native tongue)—not to mention the Latin they still carried with them as a badge of educational purgatory—so much so that the priority dispute over the periodic system can be largely understood as a struggle

* "It would be another matter if, for example, he met [chemist Justus von] Liebig, not knowing that this was indeed Liebig, say in a railroad carriage. And if a conversation about chemistry were to begin and our gentleman succeeded in joining in, then, there is no doubt, he could sustain the fullest scholarly debate, knowing about chemistry only the single word 'chemistry.' He would astound Liebig, of course, but—who knows—in the eyes of listeners he might emerge the victor. Because there are almost no limits to the audacity of a Russian person in his scholarly language."

to make Western scholars pay attention, at least in domains like chemistry, to writings issuing from St. Petersburg, Moscow, and Kazan.

What was so hard about Russian? Most obvious is the alphabet, so unfamiliar to those raised with scripts derived from the Romans. This barrier, however, is easily surmounted, especially for one with a passing familiarity with the Greek alphabet—a trait common to classically educated elites from the nineteenth century, and even more so to chemists with some training in mathematics. Many of the letters are either identical to Latin ones (а = a, м = m), or simply modifications of the Hellenic system (п = p, д = d). True, there are fascinatingly odd letters to capture specifically Slavic sounds (ж = zh, щ = shch), but surely some of these are at least as straightforward as the consonant clusters by which Latin-scripted Polish renders these same sounds.

The writing system does not touch the linguistic heart of Russian itself, which indeed poses some serious challenges to the grammarphobic student. I do not propose a detailed exposition of the structure of Russian, but some familiarity with the general characteristics of the language will help elucidate two major aspects of our story of scientific languages: first, why Western scientists (and others) found the language so alien, despite its membership in the Indo-European language family; and second, the particular obstacles that Russian chemists faced in constructing a scientific nomenclature in their language that would match up to the extant vocabularies in English, French, and German. Since all of the problems that come up in this chapter have to do with terminology and nomenclature, I'll just focus on nouns.

Each Russian noun has one of three grammatical genders—masculine, feminine, and neuter. This is something of a shock to Anglophones, but it is matched by the same three genders in German. And, unlike German, one can (usually) identify the gender by inspection: if it ends in a consonant, it is masculine; if it ends in -а or -я (another way of writing "a"), it is almost always feminine (but there are also feminine nouns that end in a soft sign, ь, which cause all sorts of grammatical mischief); if it ends in -о or -е, it is neuter. The gender matters because adjectives have to agree, as do verbs in the past tense, and also because gender governs the declension into cases. Russian, like Latin, is an inflected case language, meaning that the endings of nouns change on the basis of their grammatical function in a sentence. If Pushkin walked into a room, we would simply write "Pushkin." If I punched Pushkin, transforming him into a direct object, we would write "Pushkin*a*," in the accusative case. If I told Pushkin dirty jokes, he becomes "Pushkin*u*," noticeably dative. You get the picture. There are six cases (nominative, accusative, genitive, prepositional, instrumental, and dative), one more than Latin, two more than German, and five more than English (approximately; the 's structure to indicate possession is the vestige of an old English genitive). English gets by without cases by rendering sentences in fixed word order: "Tolstoy kicked Pushkin" has a manifestly different meaning from "Pushkin kicked Tolstoy." In Russian, one marks this difference by changing the endings of the nouns, which means the word order can be flexible. If you are trying to read a chemical article to learn whether to add the acid to the salt or vice versa, this is an important distinction.

Those are the only features of the language you need know to understand what follows. This chapter is about how one builds a language—not an entire language (that's the next chapter), but the subset to be used for science. I use the word "build" deliberately, because scientific languages have to be quite consciously constructed. No language—not Latin, not German, not English—"naturally" holds scientific concepts. There are many features of a language that have to be adapted to contain science, but in this chapter I focus on *lexical* changes. Science, and in particular chemistry, is a human activity that both requires a large number of names for objects in the world you want to describe, and requires those names to be precisely defined so that one can generalize from them. Likewise, scientists deploy a whole slew of abstract concepts (think of "potential" or "compound") that need to be carefully distinguished from their everyday meanings. The core reason why scientific languages require so much construction is that modern science focuses upon novelty: new objects in the world, new ideas, new theories. In chemistry, some of these particular problems receive unusual saliency. If you find an entirely new chemical element, it needs a name to differentiate it from all previous elements, and ideally its name would indicate that it *was* an element, part of a system. (In English, we have for over a century used the suffix *-ium* for this purpose.) Those names have to become common currency in the relevant community, or else they are useless for communication. How does this happen?

Over the course of the nineteenth century, Russian chemists gradually proposed, debated, and then adopted various chemical nomenclatures, assimilating the language in the process to French or German models for how one should "speak" or "write" chemically. The process was immeasurably complicated by the tremendous discoveries and conceptual transformations that tore through the science in this same period, rendering both chemical French and chemical German unstable. The Russians had two tools in building their scientific language: Russian itself, which has always displayed a Protean capacity to absorb words and even syntax from other languages; and knowledge of foreign languages, from which Russian chemists could appropriate and adapt ideas.

With those two resources, we come to the heart of this chapter: translation. I tell two stories in what follows. First, how Western concepts were translated into Russian to form the rudiments of a chemical nomenclature. The crucial point here was timing: Russians began developing a systematic *inorganic* nomenclature at precisely the moment that Antoine Lavoisier's chemistry had reformulated the language of chemistry in France, and they approached *organic* nomenclature in lockstep with a reform of that subject in Western Europe. Russian debates were thus part of a broader reconstruction of European scientific languages. From translation in, we then move to translation out, exploring the flow of translated textbooks *from* Russian *into* German, the opposite of the usual traffic. The success of this venture not only demonstrates the end of the struggle to establish Russian as a "legitimate" language of science, but also shows a nomenclature stabilized to a point where intertranslatability was relatively straightforward. This was a long way from 988 AD, and the conversion of Vladimir of Kiev to Orthodox Christianity.

The Making of Modern Russian

Today, Russian exists. It is a language associated with the world's largest country, and textbooks and college courses proclaim that they will teach it to you. (Often, quite well.) But it would be a mistake to think of languages as existing in the same way as, say, that church down the street, or even like a lamb that will one day grow into an adult sheep. Languages are not single entities that either stand unaltered through the ages, or organisms that grow from childhood to maturity (and, sadly, sometimes die). Languages are in constant interaction, flowing into each other, diverging into dialects and shifting vowel patterns; and the erection of firm boundaries around the edge of a certain portion of speech behavior and declaring "This is Russian" or "This is Ukrainian" is the outcome of a series of intellectual and political decisions that do not always correspond to clear-cut distinctions in actual usage. Since the eleventh century AD—a millennium ago—we have records of a language that has, over time, become what we know as Russian, the dominant Slavic language today in terms of number of speakers.[2] The process of becoming, however, was not quite linear.

Russian is a Slavic language, a member alongside Belorussian and Ukrainian of the East Slavic branch of that Indo-European language group. Slavic also has Western (Polish, Czech) and Southern (Bulgarian, Serbo-Croatian) branches, covering the broad eastern expanse of the European continent. These languages are obviously related to each other but are not necessarily mutually intelligible (much as German, Dutch, and Swedish are clearly related members of the Germanic language family, but fluency in one hardly conveys command of the others). The origins of the Slavic family are murky. We have reliable information from the sixth century AD about the presence of speakers of what we would now call Slavonic languages in the Balkans, but the languages themselves were not very strongly separated even in the ninth and tenth centuries, when our information becomes more reliable. The fact that they were so closely related at that time meant that various groups could use a common written language: Old Church Slavonic.[3]

In 863, Prince Rostislav of Moravia—in what is today the Czech Republic—sent a request for Christian missionaries to Byzantine Emperor Michael III in Constantinople, to assist his people in resisting foreign religious intrusion. (The foreigners were what we would today call Catholics.) Two monks, Constantine and Methodius, were dispatched, and one of their charges was to develop a script for the various Slavic tongues they encountered, derived from the Greek alphabet. On his deathbed, Constantine took the name Cyrill, and the later evolution of his Glagolithic script still bears his name.[4] The writing system was designed to render a written language based on a Macedonian dialect of Bulgarian, and it was adopted first by Western Slavs and then moved east. The language so written—Old Church Slavonic—became, in the words of one historian of Russian (as translated by Mary Forsyth), "a kind of common literary language in the medieval Slav world."[5] It functioned, in many ways, like Latin occasionally did: as a purely written language for the liturgy and theology. Unlike Latin, it was apparently never used for speech, but it enabled epistolary communication among linguistically diverging groups who shared Slavonic even as they used what were becoming Old Russian or Polish in their everyday interactions.[6] But, since it was *not* Latin and *not* Greek, it also had the effect of insulating this eastern region from the explosion of classical learning in contemporary Europe. Old Church Slavonic was both a unifying force and an isolating one.

In the region between Kiev, Novgorod, and Moscow, various dialects were spoken that are now labeled "Old Russian." Beginning in the eleventh century, we find ecclesiastical writings in Russian, and in the following centuries these were joined by legal and business documents—the so-called chancellery language—and then literature proper.[7] Many historians of Russian tend to speak of the language as literally trapped by Old Church Slavonic, arguing that "the subsequent history of both Russian language and Russian literature has been in a sense a long process of emancipation from the initial and paralysing influence of Byzantine culture working through the medium of Bulgarian."[8] According to this rather essentialist vision of language, Russian's development into a proper language was stalled by repeated incursions of Byzantine influence, most notably the "Second South Slavic" influence of the fourteenth and fifteenth centuries, a consequence of learned immigrants from the Balkans flooding Moscow.[9] According to this traditional account, Russian had to work to "liberate" itself from these backward-looking influences. Ironically, it did so by assimilating a different set of foreign models.

In the early seventeenth century, Moscow became the center of a sizable group of foreign merchants, who imported foreign books on topics ranging from medicine to mining to law. A printing boom followed, helping to standardize Russian. (There had been limited printing earlier in Muscovy; the first Russian printed book to carry a date appeared in 1564.) This foreign learning often arrived via Poland or highly Polonized Ukrainian and Belorussian regions, and many of the new Russian words of Latin or German origin in fact entered through Polish mediation.[10]

Russian has always been saturated with loan words. Common Slavic, which is the basis for East Slavic, already had Iranian and Germanic loans, the former donating terms for religion and the latter for materials and administrative organization. Scandinavian words poured in from the north beginning in the ninth century, often related to fishing and nautical matters. Abstract terms came either directly from Greek or through Old Church Slavonic mediation (and to a much lesser degree from Latin). Not all the imports were Indo-European: the Mongol invasions of the thirteenth century brought another torrent of linguistic borrowings, often related to finance, administration, trade, and communications—including the Russian words for "money" and "pocket."[11] Thus the appropriation of Polish (and through it Latin and German) was not a new phenomenon. The flow back and forth shaped Russian, but not in any specific direction.

This aimlessness changed decisively at the turn of the eighteenth century, as Tsar Peter the Great (reigned 1682–1725) undertook a determined program of modernizing certain aspects of the administration and military of the Russian lands, in the process relocating the capital from Moscow to the brand-new city of St. Petersburg. Peter initiated what one scholar has called the "polytechnicalization of language," which brings us back to the main line of our story about scientific languages.[12] Not only did he reform the alphabet in 1708, removing some of the more Slavonic features (Vladimir Lenin would introduce a final alphabetic simplification in 1918), he commissioned a massive series of translations of foreign texts to train the Russian nobility impressed into his service. Peter issued instructions to avoid Slavonic words and use everyday speech for translations, forcefully chiding translators who strayed from this directive. The tensions were at times unbearable. A certain Volkov, finding himself unable to render some passages of de la Quintinye's *Instructions sur les jardins fruitiers et potagers* into Russian, committed suicide.[13]

Peter's second impetus to the creation of a scientific Russian language was his 1724 establishment of the Imperial Academy of Sciences in St. Petersburg. Peter had several goals for his Academy; the promotion of the Russian language was not one of them. Russian was not treated as a medium of scholarly discourse. The official language of publication was Latin—imported into Russia *specifically* to be used for science, as there was no domestic Catholic religious tradition preceding it—at the very moment when the dominance of this language was slipping among European scholars. Mathematician Christian Goldbach was appointed Secretary of the Academic Conference explicitly because of his command of the language, although German and French often slipped into the minutes and, given the Central European origin of most of the academicians, German was the language of conversation. Latin remained obligatory for the presentation of treatises, a source of constant irritation. Not all appointees to the presidency of the Academy, a patronage position, understood Latin. In 1734, the notoriously pro-German Empress Anna selected Baron Korff for the post, but he knew almost no Latin so minutes were kept in German. Count Razumovskii's accession in 1742 brought Latin back, but Count Orlov's appointment in 1766 entailed the return of German, citing Korff as precedent. (In 1773, everyone gave up, and the minutes were taken in French.) As if this were not confusing enough, Russian was obligatory for all administration, necessitating translation to and from German when communicating with academicians, most of whom knew little or no Russian.[14]

Latin remained the biggest stumbling block. The Academy was accompanied by a *gymnasium*, intended to train domestic pupils to become the next generation of technical specialists. It was deemed essential that they learn Latin, but Russians lacked pedagogical texts in their native tongue and no one believed the market was large enough to create a Russian textbook, so Russophone students needed to first master *German* and then learn Latin through that language. This gave native speakers of German (mostly from the Baltic region) a major advantage. While they graduated from Latin in roughly three years, it took Russian students up to *fifteen* to complete the same course.[15] Latin continued to be seen as a force retarding Russian advancement, and Nikolai Popovskii, rector of the *gymnasium* at the newly established University of Moscow, conducted his lectures entirely in Russian specifically to contest the monopoly of this foreign language.[16] It took a while to catch up. It was not until *October 1859*, in fact, that Heinrich Lenz, dean of the physico-mathematical faculty of St. Petersburg University—and clearly of German ancestry himself—felt emboldened to petition his supervising ministry: "[T]he Faculty finds that from now on, there is decisively no need in particular instruction in the Latin language in the category of the natural sciences."* [17]

As the sciences were beginning to move toward Russian as a language of communication within Russia, the nature of Russian itself was changing radically. Beginning in the middle of the eighteenth century with the writings of the first ethnic Russian member of the Academy of Sciences—poet, chemist, and polymath Mikhail Lomonosov (1711–1765)—one of the most momentous transformations in the history of Russian took place, gradually reforming not just the vocabulary but even the syntax and word order of Russian to resemble Western European languages, especially French.[18] By the early nineteenth century, a modern Russian entirely intelligible to a speaker of today's language

*"то Факультет находит, что впредь, в особенном преподавании латинского языка в разряде естественных наук не имеется решительно никакой надобности."

had emerged, and alongside it a scientific style shaped by libraries full of foreign texts.[19] But Russian scientists still needed a language to communicate with Western academicians, and Latin was no longer workable. The solution had to be something contemporary, and the obvious choice seemed to be German.

The Universal Language of the Slavs

A commonplace about Russian culture in the nineteenth century has it that the elite all spoke French. Like many commonplaces, this is not false, but it obscures the important role of German for learned conversation, especially in scientific circles. One obvious reason for this was the presence of large numbers of Russian subjects who were native German speakers among the bureaucratic and academic elite.[20] Last chapter we already met Friedrich Konrad Beilstein, born in St. Petersburg and fluent in both Russian and German. Among friends or at home, he preferred German.[21] Oral German was not confined to private settings. In 1854, Carl Julius Fritzsche, a German-born chemist who since the 1830s had been a mainstay of academic culture in St. Petersburg and used Russian with ease, volunteered to give a series of charity lectures on chemistry to raise funds for the Crimean War; his Russian-language petition to the state specified that the lectures would be in German, presumably to bring in a greater audience.[22]

The eclipse of Russian within the Empire was particularly acute in the Baltic regions. Karl Klaus, professor of chemistry at the German-language University of Dorpat (today, Tartu in Estonia), felt constrained to correspond with Aleksandr Butlerov in German. (Regrettably, the only versions of these letters I could locate were in Russian translation, so that is what is reproduced in the footnotes.) "As you see, I wanted to write you in Russian and would have done so, of course, in case of necessity," he wrote in 1853, "because here, in Dorpat, I risk completely forgetting that little of the Russian language that I learned by the sweat of my brow; however, in view of the fact that you understand German as well as you do your native language, I will write to you in German on this occasion in order not to waste excess time."*[23] (In later

*"Как видите, я хотсл написать Вам по–русски и сделал бы это, конечно, в случае необходимости, потому что здесь, в Дерпте, я рискую полностью забыть то немногое из русского языка, что я выучил в поте лица своего; однако ввиду того,

letters he did write in Russian, both to please Butlerov and because "the Russian language is more pleasant for friendly relations."*[24]) Butlerov had written to see if he could defend his dissertation at Dorpat—such venue-shopping was common practice among Russian scientists—but Klaus informed him that this would be impossible, for "none of the members of the department would be able either to read your dissertation, or conduct a disputation with you, since none of them commands the Russian language.[. . .] You would have to translate your dissertation or have it translated into German or French and conduct the disputation either in German or in French."†[25] This was at a leading institution in the *Russian* Empire.

Nontheless, the main advocates of the use of German turned out to be not Baltic Germans but Russians themselves, especially when speaking with non-Russian Slavs. For example, as Dmitrii Mendeleev's close friend and Petersburg University geologist Aleksandr Inostrantsev recalled about his time abroad in Prague:

> In the evenings my wife and I usually went to dine and observe certain "Slavonic Evenings," as they were called at that time. Sometimes an especial interest was presented here when, after dinner, around a mug of beer, a general conversation began, and sometimes even speeches, although these were difficult for us to understand due to our poor knowledge of the Czech language. To acquaint us with these conversations and speeches the masters of ceremonies of this club introduced us to two young people, Vanžura and Patera, who spoke Russian not badly, and when they were absent it was necessary sometimes to speak with your neighbor in, as one ironically says, the common Slavic language, i.e., in German.‡[26]

что Вы понимаете немецкий язык так же хорошо, как и свой родной язык, я на этот раз, чтобы не тратить лишнего времени, буду писать к Вам по-немецки."

*"русский язык приятнее для дружеских сношений."

†"никто из членов факультета не смог бы ни прочесть Вашу диссертацию, ни вести с Вами диспут, поскольку никто из них не владеет русским языком.[. . .] Вам придется перевести или дать перевести Вашу диссертацию на немецкий или французский язык и вести диспут либо по-немецки, либо по-французски."

‡"По вечерам мы с женою обыкновенно ходили ужинать и повидать некоторых в так называемую в то время «Славянскую беседу». Особенный интерес иногда представляло здесь то, что после ужина, за кружкой пива, начинали общий разговор, а иногда и речи, хотя нами и трудно понимаемые в силу плохого знания

Likewise, in 1867 Aleksandr Borodin, who taught chemistry at the Medico-Surgical Academy not far from Inostrantsev, wrote to his musical patron Milii Balakirev atop a manuscript of some Czech musical themes: "I gave the title in German, since in general the German language is for Slavs the international language. I am convinced of this every day, sitting in Petersburg. Not long ago I had a conversation with a Czech in German[...]."* [27] The tone is ironic, but it speaks to a deeply lived reality. The situation extended beyond Slavs. In 1902, Nikolai Menshutkin, secretary of the Russian Chemical Society, boasted that at a meeting of "northern" (read: Scandinavian) scientists and physicians in Helsinki—at that time part of the Russian Empire—roughly 20% of the thousand attendees were Russians, heavily concentrated in the chemistry section. The prominence of Russians made the Scandinavians relent in their habit of using Swedish or Danish, forcing all papers in chemistry to be delivered either in German or with German summaries.[28] Russian chemists, when abroad, were Germanophone.

Menshutkin's glee gives the impression that German was almost second nature, and they could alternate with ease between their "international language" and their native one. Countless asides in correspondence and memoirs attest, however, as physiologist Ivan Sechenov noted in his autobiography, that "[i]gnorance of languages among the majority of our students represents a great misfortune."† [29] Attentive professors sometimes resorted to extreme measures to get their students up to linguistic speed. For example, as chemist Ivan Kablukov recalled about an interchange with his teacher, Moscow professor Vladimir Markovnikov:

> I went to V. V. Markovnikov and asked:
>
> —How am I supposed to study organic chemistry?
>
> He said:

чешского языка. Для знакомства с этими разговорами и речами распорядители этого клуба познакомили нас с двумя молодыми людьми, Ванжурой и Патерой, недурно говорившими по-русски, а когда их не было, приходилось иногда говорить с соседом, как иронично говорят, на общеславянском языке, т.е. по-немецки."

*"Заглавие я дал немецкое, так как вообще немецкий язык для славян есть международный язык. Я в этом убеждаюсь каждый день, сидя в Петербурге. Недавно я беседовал с одним чехом на немецком языке[...]."

†"Незнание языков у большинства наших студентов представляет большое зло."

—Prepare ethyl acetoacetate.

—And how should I prepare it?

—Take the German journal *Liebigs Annalen der Chemie* and read about it there.

I didn't know German. One could either take the French or the German track in the gymnasium. I took the French. What to do? I took the article, took a dictionary, and began to read. In the end, it is impossible to recommend this method, which is of course very difficult, but sometimes it works.

I read everything, prepared ethyl acetoacetate, and this was the spur to further research.[*][30]

Markovnikov himself had difficulties with German. Upon arriving in Berlin, he looked around and discovered, as if it were a surprise, "to my displeasure here they speak German, and you know how weak I am in this language."[†][31] The solution? "With chemists I conversed here in French because," he wrote to his advisor Butlerov, "having begun with [Adolf von] Baeyer in German, soon was obliged to shut my trap."[‡][32] Letters transmitting self-translated articles almost always contained a statement to this effect: "At the conclusion of my writing I find myself obliged to ask your forgiveness in a possibly too clumsy handling of the German language."[§][33] (The original German is equivalently awkward.)

[*] "Пришел я к В. В. Марковникову, и спрашиваю:

—Как мне заниматься органической химией?

Он говорит:

—Приготовьте ацетоуксусный эфир.

—А как его приготовит?

—Возьмите немецкий журнал «Liebig's Annalen der Chemie» и прочтите там об этом.

Я немецкого языка не знал. В гимназии можно было проходить или французский, или немецкий. Я проходил французский. Как же быть? Я взял статью, взял словарь и начал читать. В конце концов, такой способ, конечно, очень труден, рекомендовать его нельзя, но, иной раз, приходится.

Я все прочел, приготовил ацетоуксусный эфир, и это явилось толчком к дальнейшему исследованию."

[†] "но на мое несчастье здесь говорят по-немецки, а Вы знаете, как я слаб в этом языке."

[‡] "С химиками я объясняюсь здесь по-французски, ибо, начав с Байером по-немецки, вскоре принужден был задать столбняка."

[§] "Zum Schlusse meines Schreibens finde ich mich genöthigt Sie um eine Entschuldigung in einer vielleicht zu sehr umständlichen Behandlung der deutschen Sprache zu bitten."

Dmitrii Mendeleev—erstwhile victim of translation mistakes—found himself similarly incapacitated both in speech and in writing, notwithstanding two years spent living in Western Europe (1859–1861) and enormous quantities of foreign books kept in his library.[34] While in Heidelberg, he relied on a Russian named Baksht to translate his articles into German; in conversation, he muddled through or resorted to halting French.[35] This barrier was most inconvenient in personal correspondence, where the intimacy of the medium placed the burden of composition entirely on his own shoulders. When writing to Wilhelm Ostwald, then at Leipzig but by birth a Baltic German subject of the Tsar, he could resort to his native language: "I write in Russian because I want to answer quickly and translation is a lengthy affair."* [36] But most German professors did not know the language, and so when he had no amanuensis he had to inflict broken German, full of Russianisms and misspellings, on his interlocutor. Writing to August Kekulé from his dacha in Boblovo, he lamented: "In the countryside, where I now live, no one knows German and you know how weak I myself am in this language. I want however to try to relate to you everything that is important concerning my views about the constitution of Benzol—according to my opinion."† He concluded with an obligatory apology: "Forgive me my German. From Petersburg I would be able to write better—here is by me no person, also no dictionary."‡ [37] He was no more comfortable with Erlenmeyer, whom he had known for years:

> Ooof! I have finished. I understand nothing and believe that you cannot understand my letter any more.
>
> If my letter actually remains not understood, then I write French. I can certainly find a translator, but in so delicate [a] business I want to involve nobody else.§ [38]

*"Пишу по-русски, потому что желаю ответить немедля, а перевод—дело длинное."

†"Im Dorf, wo ich wohne, kein Mensch kennt deutsch und Sie wissen wie schwach bin ich selbst in diese Sprache. Ich will doch probiren Ihnen zu erzeilen bezüglich meine Ansichten über d. Constitution v[on] Benzol alles was ist die wichtigste—nach meine Meinung."

‡"Entschuldigen Sie mir meine deutsche Sprache. Von Petersburg ich mögte besser schreiben—hier ist bei mir keine Mensch, doch keine Dictioner."

§"Uf! Ich habe geendet. Ich verstehe nichts und glaube, dass Sie könen von meine Brief nicht mehr verstehen.

Wenn wirklich mein Brief bleibt unverstanden, dann schreibe ich französisch. Ich

His German was pretty good compared with his English. He was able to read the language and often referred to British articles, but of the three dominant languages of science, this was by far his weakest. When he gave a prestigious Faraday Lecture in London in 1889—on links between the periodic law and Isaac Newton's laws—he was invited to write his address in Russian. It was then translated into English by Vassili Ivanovich Anderson, chair of the Mechanical Section of the British Association for the Advancement of Science, who had been born and educated in St. Petersburg. That text was read at the meeting by Sir James Dewar; Mendeleev sat on the dais.[39] The page proofs for the published version of the lecture arrived to him at Boblovo, but all of his peers had dispersed to their summer addresses. He delegated the corrections to his colleague Menshutkin, seemingly comfortable in all languages of the triumvirate: "I received from Prof. [Henry] Armstrong the attached proofs of an article about the periodic law[. . .]. You know that I cannot command the details of the English language, and therefore I resolved to ask you to read through and correct it wherever you consider it appropriate."* [40]

Mendeleev's struggles with English paled in comparison with the troubles Englishmen had reading his Russian. Some British chemists had indeed studied the language. Alexander Crum Brown, one of the architects of structure theory in organic chemistry, wrote Mendeleev in much excitement about the latter's imminent arrival in Edinburgh in 1884. "As far as language is concerned, I speak German—not well, but all the same entirely comprehensibly—and French only with the greatest difficulties. I even had certain successes in reading Russian. Butlerov sent me already a long time ago his 'Introduction to a Complete Study of Organic Chemistry,' and I read rather a lot from it before the German edition appeared," he wrote. "I also have your 'Principles of Chemistry,' and I also read something from it. But I am afraid that we will depend on German and French in our conversations, because I am sure that

kann gewiss ein Übersätzer finden, aber in so delicate Geschichte will ich niemand andere hernehmen."

*"я получил от проф. Армстронга прилагаемую корректуру статьи о периодическом законе[. . .]. Вы знаете, что я не могу владеть подробностями английского языка, и потому решаюсь просить Вас прочесть и исправить, что где сочтете надобным."

you are much better acquainted with the English language than I with Russian."* [41]

The case of fellow Scottish chemist William Ramsay was, however, more typical. Ramsay was solidly working through the literature on atmospheric gases—he would receive the 1904 Nobel prize in chemistry for his discovery of the inert gases—and he came across some references to Mendeleev's volume of studies on the topic, *On the Expansion of Gases*, published in 1875. They met at the Faraday Lecture and the topic had come up (presumably in German), so Ramsay wrote to follow up, explicitly stating that he wrote in German because it was likely the best common tongue.[42] In the next missive, Ramsay thanked Mendeleev for the book, with sadness noting the tome's impenetrable language. "It will be for me hard work to read the book, but I want to try," he wrote in German. "With the help of a dictionary and a grammar, I hope in any event to be able to spell out the sense."† [43] It seems he did not get terribly far, for in 1892 he wrote Mendeleev on the same theme, this time in French: "Does your memoire exist only in the Russian language? Or could one find a translation, or even an abstract, in some Western journal?"‡ [44] No such luck. Ramsay observed that "I see in the text some numbers which guide me, and I will do my best to understand your beautiful work."§ [45] The unity of notation, mathematical formulas, and scientific nomenclature—albeit in Cyrillic—surely made the task easier. Such interchangeability of nomenclature, however, was not easily achieved.

* "Что касается языка, то я говорю по-немецки—не хорошо, но все же совсем понятно, по-французски же с большими затруднениями. Я даже сделал некоторые успехи в чтении по-русски. Бутлеров прислал мне уже давно свое «Введение к полному изучению органической химии», и я прочел оттуда довольно много, прежде чем появилось немецкое издание. У меня имеется также Ваши «Основы химии», и я также прочел из них кое-что. Но я боюсь, что мы будем зависеть от немецкого и французского языка при наших разговорах, потому что я уверен, что Вы занкомы с английским языком гораздо лучше, чем я русским."

† "Es wird für mich eine schwere Arbeit sein das Buch zu lesen, doch will ich es versuchen. Mit Hilfe eines Dictionärs und eines Grammatiks, hoffe ich jedenfalls die Sinne heraus buchstabieren zu können."

‡ "Est ce que votre memoire n'existe que dans la langue ruse? ou peut on trouver une traduction, ou bien un resumé dans quelque journal occidental?"

§ "je vois dans la texte des chiffres qui me dirigent, et je ferai mon mieux de comprendre votre belle ouvrage."

Chemical Name-Calling

In 1870, surely one of the most unusual proposals in the history of chemical nomenclature was put forth in the minutes of the Russian Chemical Society. N. A. Liasovskii (a chemist of no lasting reputation or legacy) suggested that Russians should change their naming conventions of chemical compounds to feature "combinations of the sort of the Russian *patronymics* and *family names*; for example, for potassium chloride [KCl_2] to adopt the name *potassium chlorovich* or *potassium chlorov*, for potassium hypochlorite [KOCl] *potassium chlorovich acidov* or *potassium chloro-acidov*, for potassium chlorate [$KClO_3$] *potassium chlorovich three acidov* or *potassium chloro-three-acidov*." The idea was to use the resources built into the Russian language to open up conceptual possibilities closed to the Germans and French. "Upon the introduction of numbers for several valencies into such names," he continued, such procedures "present the advantage of very simple transmission of formulas into names, constructable in the same order in which the elements enter into formulas, therefore the combinations, similar to those generally deployed in the Russian language, could be easily assimilated."*[46] Of course there were also drawbacks. Liasovskii noted that this convention would produce almost identical names for salts and acids—rather important concepts to distinguish. He did not dwell on the rather more obvious disadvantage: this system was unintelligible to European chemists; even if they could read Russian, it would not translate. The proposal dropped like a stone. In trying to be more "natural" to Russian itself, Liasovskii's system was the wrong kind of artificial.

All chemical nomenclatures are artificial. Noted Danish linguist Otto Jespersen observed in 1929 that "[i]f you look through a list of chemical elements you will find a curious jumble of words of different kinds." To select just English, you note quite ancient traditional words, such as *gold* or *iron*; such metals were named so long ago that just about every language has its own idiosyncratic terms for them (*Gold*, *Eisen*;

*"сочетания в роде русских *отчеств* и *фамилий*, например, для хлористого калия принять название *калий хлорович* или *калий хлоров*, для хлорноватисто-кислого калия—*калий хлорович кислов* или *калий хлоро-кислов*, для хлорноватокислого калия—*калий хлорович трех кислов* или *калий хлоро-трех-кислов*. При введении в такие названия чисел для нескольких паев, они представляют выгоду весьма легкой передачи формул в названия, составляемые в том же порядке, в каком элементы входят в формулы, причем сочетания, подобные общеупотребительным в русском языке, могут легко усвоиваться."

or, *fer*; золото/*zoloto*, железо/*zhelezo*). Then, there were words derived from Greek or Latin roots, such as the *oxygen* that Lavoisier coined. (The Germans rejected this term, preferring to calque it as *Sauerstoff*, acid substance.) Finally, Jespersen noted, a consensus solidified around the suffix *-ium*, although even here the stem could be derived from a place (ytterbium), a country (germanium), a planet (selenium), and so on.[47] And that was just for the elements. When considering the many categories of compounds that populate organic chemistry (esters, ketones), the situation was still more fraught.

Chemical nomenclature *must* be artificial. New substances are constantly being discovered or created, and it is impossible to provide a finite list of names, or rules for naming, to cover all eventualities. On the other hand, random naming won't do. How would a student learn the order that underlay chemical transformations? And then how would you translate your findings for international chemists, the only group that could vouchsafe the validity of your knowledge claims? Liasovskii's system was as perplexing to his peers as to us today, because it moved away from intertranslatability with European nomenclatures, and today chemists have delegated the rights of naming—within certain strict parameters—to the discoverer as determined by the International Union of Pure and Applied Chemistry (IUPAC), which we shall encounter in chapter 6. So, while each language generally retains traditional words for traditional elements (such as the metals *gold* and *mercury*), international scientists have resolved the fundamental problem of naming through convention, coordinated by institutions. In the latter half of the nineteenth century, however, these institutions did not yet exist. In the realm of chemical naming, the Russians were not at all "backward" with respect to Western Europe—they embarked on a modern nomenclature in lockstep with their international peers, and they faced the same frustrations.

Aside from the traditional metals (like copper) and other substances common among apothecaries and metallurgists (like sulfur), chemical nomenclature in Russian remained decidedly spartan until the end of the eighteenth century. This is hardly surprising: if even the Academy of Sciences refused to use Russian in their treatises, where would the demand for Russian names come from? By the 1770s, however, university courses began to use Russian, and theoretical chemical works started appearing in Russian translation. In order to teach those courses and translate those books, Russian scientists began to debate and develop a nomenclature suited for Russian. The importance of suitability is sometimes lost on English speakers. In English, *copper* can be either a noun or an adjective (as it is in *copper* sulfate). But in French or Russian, one cannot simply use a noun as an adjective without any morphological tinkering—one either has to introduce a preposition (sulfate *of* copper) or add a suffix that turns the root into an adjective. (German helpfully allows *Kupfersulfat*.) But which suffix to use? Arguments about the choices battered Russian chemistry for over a century: pick the right one, and you had a neat system that eased education and theorizing; opt for the wrong one, and you generated an unholy mess.

The 1770s was an excellent moment for Russians to begin searching for an inorganic nomenclature, because this was precisely the moment that Lavoisier and his colleagues began to overhaul the entire system of French names—and, by knock-on effect, German and English names. Russians had access to these books either in the original or in translation comparatively rapidly; between 1772 and 1801, twelve translated books on the new chemistry appeared. Fedor Politkovskii, fresh from a two-year trip to Paris, lectured on the subject in 1783 in Moscow. If scholars were prepared to read French—and who wasn't?—the Academy library also had a copy of Lavoisier's *Opuscles physiques et chimiques*, sent by the author himself in 1774. German commentaries followed, and by 1801 almost the entire (very small) Russian chemical community was committed to Lavoisier's new chemistry.[48] They only needed to figure out how to talk about it.

As noted in 1870 by Fedor Savchenkov before the Russian Chemical Society in one of the occasional debates about chemical nomenclature, the dawn of the nineteenth century saw "a rather close translation of French names, introduced on the basis of principles adopted in the nomenclature."*[49] But the perception that Russians were simply mimicking the French was more a surface appearance. In an 1810 article on Lavoisier's principles, academician Iakov Zakharov had cautioned against the French system: "The French naming system is now adopted in all of Europe. Languages that for the most part or in entirety descend from Latin had no difficulties at all in introducing the very same words into their languages, it was only necessary to substitute the final syllable for one appropriate to the properties of that language." That included the Spaniards, Portuguese, Italians, and even the English. But "[t]he Russian language with all of its branches has a completely dif-

*"довольно близкий перевод французских названий, введенных на основании начал принятых в номенклатуре."

ferent quality, just as do German and others."* The Germans adapted the French system to their language, and "we should also follow this example."† [50] He went on to develop an integrated system that used native Slavonic prefixes (*pere-*, *do-*) to mark levels of oxidation. In the end, what was happening was a greater convergence of nomenclatures *syntactically*, even as they differed lexically.

By 1836, Academician Hermann Hess articulated a synthetic version of these earlier proposals, and his framework has survived mostly unchanged down to the present.[51] Although the essentials of inorganic nomenclature were established early enough, it was not until 1912 that the Russian Chemical Society officially sanctioned this system. The delay was in part because of the vexed nomenclature of *organic* compounds. After the introduction of the structure theory of organic molecules in the 1860s, the field boomed, and chemical periodicals announced the discovery of hundreds, even thousands, of new compounds, byproducts of the new pharmaceutical and artificial dyestuffs industries. The problem of arbitrary, individualized naming beset Western Europeans as well. Eminent British chemist Edward Frankland complained that "[e]very young chemist here seems to think that he does something both highly important and original, if he can invent some slight modification in the nomenclature of chemical compounds. Hence in the place of the tolerably uniform old system of names, all sorts of systems and various have sprung into existence, and all uniformity has been lost. Much the same state of things appears now to prevail in Germany where the language also lends itself much less readily to the new system."[52] The situation continued to deteriorate into the late 1880s and early 1890s, when Alsatian chemist Charles Friedel convened a group of chemists in Geneva to develop a new international nomenclature to tame the new compounds.[53]

As far as the Russians were concerned, "international" meant a club of British, French, and Germans. The only Russian involved in the discussions was Friedrich Beilstein, and many Petersburgers were not will-

*"Французское имязначение принято ныне во всей Европе. Языки, кои по большей части или со всем от Латинского происходят, не имели никакого затруднения ввести те же самые слова в их язык, нужно было только переменить окончательный слог свойству того языка приличной.[. . .] Со всем другое качество имеет язык Росиийской со всеми его отраслями, равно как Немецкой и другие."

†"Сему примеру должны последовать и мы."

ing to consider him truly "Russian."[54] On 8 October 1892 (Old Style [O.S.] date), the Russian Chemical Society held an inconclusive discussion about the new proposals. Beilstein encouraged chemists to adapt to the new regulations, which would standardize European chemical publishing; by no means was it intended to change daily practice: "No one imagines eliminating the old names, nor introducing the new nomenclature into conversational language."* A purely written convention? Mendeleev would have none of it, declaring "that only a language that has worked itself out historically is a living language. Such is the natural, international language of chemistry—the language of formulas. Translation of the language of formulas to oral and written speech is a difficult affair, and it is doubtful that, given the large number of words necessary for this, new names might summon up exemplary notions and therefore would be appropriate for oral and written speech."† (Of course, formula-writing was also a convention, and a relatively recent one at that.) In the end, Nikolai Menshutkin postponed any decisions.[55] Like inorganic nomenclature, the issue remained unresolved for decades. Intertranslatability among European chemistries reached its height not in grand theoretical debates over naming conventions, but in mundane textbooks.

Translating Textbooks

Russian chemists could find a reasonable amount of textbook literature in their native language as they entered the 1860s. It was not, however, originally Russian. The first chemical textbook composed in Russian was the *Handbook to the Teaching of Chemistry* (*Rukovodstvo po prepodavaniiu khimii*), published in two parts in 1808, and mainly directed to doctors, teachers, and mining officials. From then until the 1830s there was essentially no good chemistry textbook in Russian until Hermann Hess obliged with his *Foundations of Pure Chemistry* (*Osnovanie chistoi khimii*) in 1831.[56] The demand for these books was rela-

*"Уничтожать старые названия не имеется в виду, равно как и вводить новую номенклатуру в разговорный язык."

†"что язык лишь исторически выработавшийся есть живой язык. Таков естественный, международный язык химии—язык формул. Перевод же языка формул на устную и письменную речь—дело трудное и сомнительно, чтоб при большом числе необходимых для этого слов новые названия могли вызывать образные представления и потому годились бы для устной и письменной речи."

tively small, and they were quickly outdated by the pace of theoretical and empirical advances. Most of the chemical literature, as in the age of Peter the Great, was translated from the West. Especially popular was the *Schule der Chemie* by Julius Adolph Stöckhardt (in Russian, Shtekgardt), which came out in three editions in 1859, 1862, and 1867.[57] The flow of chemical textbooks was most decisively *into* Russian—and certainly never *back* into German.

Until 1868, that is, when Aleksandr Butlerov's *Introduction to a Complete Study of Organic Chemistry*, originally published in 1864, appeared in German translation and made quite an impact on theorizing in the metropole of organic chemistry. This was an unusual course of events, and its origins lay in Kazan, where Butlerov diligently taught organic chemistry to all levels of university students. Butlerov had experience with translated textbooks, and in his own classes he used a translation of Carl Gotthelf Lehmann's *Handbuch der physiologischen Chemie*, supplemented by lectures based on Justus von Liebig's organic chemistry text, which he used in German (for it had not been translated into Russian).[58] In the late 1850s, Butlerov began diverging from the extant textbooks, developing his own notes into a fuller presentation of organic chemistry with a new set of foundational principles.

We now call this framework "structure theory" and it is based on the concept of tetravalent carbon and mutually bonded carbon chains. Credit for it is usually assigned to August Kekulé, although there are many other claimants—including Butlerov.[59] The latter began to fully formulate his theory only while on a trip abroad in 1861, and he began to test his theories in his lectures in 1861–1862, from which regrettably no lecture notes survive. In any event, he could not have given the full course of lectures, since Russian universities were closed for the second half of that academic year owing to student unrest, and Butlerov himself fell ill. Lecture notes do survive from the 1862–1863 course, and from these it is clear that Butlerov thought of his project as the composition of an introductory textbook, but one that would reformulate the fundamental principles of organic chemistry—a reasonable strategy in an age when original findings and theoretical innovations were quite often first introduced in textbooks.[60]

This *Introduction* was published—as with much European publishing of that day—in several separate fascicles, dated January 1864, May 1865, and October 1866, at which point it was also released as a single bound volume. The first fascicle closely resembled the lecture course from 1862–1863 and concerned the general theoretical picture; the later fascicles discussed empirical data, made predictions for new experiments, and discussed applications. The book was an immediate sensation among Russian chemists, and they had substantial difficulties securing copies not only in secondary cities like Kiev and Kharkov, but in St. Petersburg itself.

As Russophone chemists came to appreciate what Butlerov had accomplished, they continually implored him to think about a broader audience. For example, Karl Schmidt, a chemist at Dorpat University, wrote to Butlerov (in German, although regrettably I could find only a Russian edition of it) urging translation. "With the publication of this book for the *West* in German or French you will earn the gratitude of many young chemists," he enthused. "You write so well that for you it does not comprise any work at all and considering that this subject summons widespread interest, it will be easy to find a publisher. I am convinced that many of our Western colleagues, like me, *would meet* your work *with joy and gratitude*, if they could read it freely[...]."*[61] Schmidt at least could manage with the Russian; Western Europeans could not. Adolphe Wurtz in Paris wrote to Butlerov in 1864 that he was looking forward to receiving the book: "In this case I would find your work on my return and if only it were not written in Russian, I would read it with the interest deserved by that which leaves your pen."†[62] Likewise, August Hofmann, the dean of Berlin chemistry, lamented to Markovnikov upon receiving from him a copy of the Russian edition that he was unable to read it.[63] Even Petr Alekseev, a chemist in Kiev, decried the ghetto of the Russian language: "And I am very, very sorry that your essay is not printed in a single foreign language."‡[64]

Finding a publisher for the German translation was not as easy as Schmidt had surmised. As Butlerov complained to Emil Erlenmeyer: "What concerns the publication of my 'Introduction' in German, I am very little to blame that it is not yet done. I could not find a pub-

*"Изданием этой книги для *Запада* на немецком или французском языке Вы заслужите благодарность многих молодых химиков[...]. Вы пишете так хорошо, что для Вас это не составит никакого труда и, учитывая, что предмет этот вызывает всесторонний интерес, легко найти издателя. Я убежден, что многие наши западные коллеги, подобно мне, *встретили бы* Вашу работу *с радостью и признательностью*, если бы они могли ее свободно прочесть[...]."

†"Dans ce cas je tourverai votre ouvrage à mon retour et pourvu qu'il ne soit pas écrit en russe, je le lirai avec l'intérêt que mérite ce qui sort de votre plume."

‡"И очень, очень жаль, что Ваше сочинение не издано ни на одном иностранном языке."

lisher[. . .]."* [65] He even contemplated publishing it in France, or in German in St. Petersburg, while Butlerov's student Markovnikov shuttled around Germany remonstrating with publishers to no avail. Reflection on the mechanics of the publishing industry, as provided by Nikolai Golovkinskii, another student emissary, explained why:

> I am doing what I can in searches for a German publisher, but I can only manage a little. It is difficult for a German bookseller to decide to publish the book of a Russian author if the local professors do not approve such an unusual matter. And how is one to attain the professors' approval if the book is unknown to them?† [66]

Finally, Beilstein arranged a contract with Quandt & Handel, the publishers of his own *Zeitschrift für Chemie*, and Butlerov contracted a local teacher in Kazan named Risch to undertake the translation.[67] Butlerov allowed the publishers to change the title of the book a bit to *Lehrbuch der organischen Chemie: Zur Einführung in das specielle Studium derselben*, and the complete version appeared in 1868. Butlerov insisted that "German edition translated from the Russian"‡ appear prominently on the title page. He did not want his readers to forget what language the original was written in.

Risch, on the other hand, received no credit. Perhaps it had something to do with the quality of the work? After all, as Markovnikov wrote to Butlerov as the book was about to appear: "Concerning the language of the translation, the publishers said that it is a bit heavy. On the other hand, the same could be said also about the original." The problem was not language, but rather adapting the book to meet the expectations of a new audience. Markovnikov continued: "Of course, the latter cannot have any influence on the merit of the German edition, but in general it seems to me that you ought to revise your book a bit in order that it would meet an entirely good reception among the public here.[. . .] Don't forget that your readers will be Germans, accustomed

*"Was der Veröffentlichung meines 'Einleitung' in der deutschen Sprache betrifft, so bin ich sehr wenig schuld, daß es bisher noch nicht gemacht ist. Ich konnte keinen Verläger finden[. . .]."

†"В поисках издателя немецкого текста я сделаю, что могу, но могу я немного. Трудно, чтобы немецкий книгопродавец решился издать книгу русского автора, если его не одобрять на такое непривычное дело местные профессора. А как добыть это одобрение профессоров, если книга им неизвестна?"

‡"aus dem russischen übersetzte deutsche Ausgabe"

in general to the learning of facts by heart and not to general conceptions which stretch far beyond the limits of what is known."* [68] It was essential, for this grandeur of vision, that Butlerov alone be the focus of the translation, and that meant (among other things) suppressing the translator's due.

Credit, in fact, was what this translation had always been about: demonstrating *in German* that he had published the main elements of structure theory at least simultaneously with Kekulé, and in some cases earlier. It was the same path Mendeleev took in 1869 with his periodic system, with results that we have already witnessed—and, importantly, it was a path Mendeleev had *not* taken with his own textbook of organic chemistry, published in 1862 to great acclaim in St. Petersburg. Beilstein, of all people, did more than anyone else to publicize this innovative textbook, the first to fully integrate "type theory" into its pedagogical presentation. (Type theory was soon to be vanquished by structure theory, but Mendeleev couldn't have known that.) Beilstein insisted that Erlenmeyer publish a review in the *Zeitschrift für Chemie*, joking that "this article will especially interest the large Russian colony which you have set up in your house. Your laboratory is becoming as it were a center for the present Russian emigration."† [69] He summarized the book and laid out the scale of Mendeleev's insights, concluding: "In Russia the chemical literature was constrained until now almost exclusively to translations of the better-known German and French works. The book before us deserves thus special consideration as an original work, if it had not already awoken our interest through its characteristic and solid treatment."‡ [70] But, of course, no German ran out and picked up a Rus-

*"Относительно языка перевода издатели выразились, что он несколько тяжел. Впрочем, то же можно сказать и об оригинале. Конечно, последнее не может иметь влияния на достойнство немецкого издания, но вообще, мне кажется, Вам следует кое-что переработать в Вашей книге для того, чтобы она встретила вполне хороший прием в здешней публике.[. . .] Не забудьте, что Вашими читателями будут немцы, привыкшие вообще к зазубриванию фактов, а не к общим соображениям, простирающимся далеко за пределы известного."

†"Vielleicht interessirt diese Abhandlung speziell die große russische Kolonie, die sich in Ihrem Hause niedergelassen hat. Ihr Laboratorium wird gewissermaßen ein Centrum der augenblicklichen russischen Völkerwanderung."

‡"In Russland beschränkte sich bisher die chemische Literatur fast ausschliesslich auf Uebersetzungen der bekannteren deutschen und französischen Werke. Das vorliegende Buch verdient daher schon als Originalarbeit eine besondere Berücksichtigung, wenn es nicht schon durch seine eigenthümliche und gediegene Bearbeitung unser Interesse erregte."

sian grammar in order to read this book. Later, with the growing visibility of Russian chemistry due in no small part to the controversy over the discovery of the periodic system, certain Western chemists did in fact study the language enough to parse a technical article with the aid of a dictionary. But that had not happened yet.

Butlerov learned the lesson. As he reflected in his own textbook, Mendeleev's work, "[t]he only and excellent original Russian textbook of organic chemistry," was "not widely distributed in Western Europe, doubtless only because it still has not found a translator."* [71] Deploying the full range of resources at his disposal—the expressive scope of modern Russian, his own excellent language skills, and an emergent international nomenclature—Butlerov managed not only to rescue his book from obscurity and make a bid for priority, but also to grant a level of dignity for Russian publications. "From the moment of the translation of these textbooks into foreign languages begins the reverse flow of chemical pedagogical literature from the East to the West," observed Paul Walden, a bilingual Riga-born chemist; "Russia, which for a long time provided itself with translated literature from the West, now itself begins not only to produce its own chemical literature, but emerges also in the character of a competitor on the Western literary market, abounding with rich and excellent chemical literature."† [72] Markovnikov beamed that Butlerov's "articles and especially the translation of his book into German greatly enabled [his theory's] assimilation and distribution among Western scientists."‡ [73] Adding Russian as another language of science—although clearly subsidiary to the triumvirate—seemed to benefit everyone. But how much further could one expand the quantity of scientific languages?

*"Единственный и отличный, русский оригинальный учебник органической химии Менделеева,—учебник, не распространенный в Западной Европе, без сомнения, только потому, что для него не нашлось еще переводчика."

†"С момента перевода этих учебников на иностранные языки начинается обратное течение химической педагогической литературы с востока на запад; Россия, стало продолжительное время снабжавшаяся переводной литературою с запада, ныне сама начинает не только производить собственную химическую литературу, но выступает даже в качестве соперницы на западном литературном рынке, изобилующем богатой и превосходной химической литературой."

‡"статьи и особенно перевод его сочинения на немецкий язык немало способствовали ее усвоению и распространению между западными учеными."

CHAPTER 4

Speaking Utopian

En Eŭropo oni ordinare pensas, ke en la tuta Ĥinujo oni parolas nur unu lingvon—la ĥinan. Estas vero, ke la loĝantoj de Pekino, kiel ankaŭ la loĝantoj de Kantono, Ŝanĥajo, Futŝano aŭ Amojo parolas ĥine, sed de la dua flanko estas ankaŭ vero, ke la plej granda parto de la loĝantoj de unu el la diritaj urboj povus kompreni la loĝanton de alia urbo ne pli bone, ol ekzemple la Berlinano la Londonanon aŭ la Parizano la Holandanon.*

JAN JANOWSKI[1]

"And the whole earth was of one language, and of one speech," begins the most famous story in the Western tradition on the problem of communication (Genesis 11:1–9), as translated by the good people working for King James I of England. We know how it goes: they began to build a really tall building, and the Lord was displeased. ("Behold, the people is one, and they have all one language; and this they begin to do: and now nothing will be restrained from them, which they have imagined to do. Go to, let us go down, and there confound their language, that they may not understand one another's speech.") It probably did not happen exactly that way, but it is hard to argue with the description of the ensuing confusion.

A decade after the carnage of World War I had been unleashed on the peoples of Europe, as a shaky international order emerged that only later would be endowed with the sad epithet "interwar," scientists and scholars took stock of the transition to the brave new world that seemed in so many ways different from what had preceded the guns of August.

* "In Europe one ordinarily thinks that in all of China one speaks only one language: Chinese. The truth is that the residents of Peking, as also the residents of Canton, Shanghai, Fujian or Amoy [Xiamen] speak Chinese, but on the other hand the truth is also that the greatest part of the residents of the said cities could understand the residents of another city no better than for example a Berliner a Londoner or a Parisian a Dutchman."

The informal linguistic truce that had existed between the triumvirate of English, French, and German—slightly disrupted by the emergence of Russian as a language of chemistry in the late nineteenth century—was likely to be sundered in the coming decades. "To-day, with the recrudescence of many minor nationalities, and the revived national feeling of some larger units, caused by the Great War," noted Roland Kent in 1924, "we may be facing an era in which important publications will appear in Finnish, Lithuanian, Hungarian, Serbian, Irish, Turkish, Hebrew, Arabic, Hindustani, Japanese, Chinese."[2] Luther Dyer, a year before Kent, was still more pessimistic:

> A decade or so ago, a reading knowledge of English, French and German enabled these isolated scholars to keep fairly abreast the latest developments. To-day he needs Italian, Spanish, Dutch, the Scandinavian and Slavonic languages; one may even add the Japanese. The chemists of Italy and Sweden are doing important work just as the chemists of England and Germany.[3]

Babel was already here.

Was this true? Let us set aside for a moment the part about the contemporary Babel, and examine Dyer's "decade earlier"—that is, before the outbreak of the War—when apparently the triumvirate sufficed. Here is an analysis (by Austrian physicist Leopold Pfaundler) of the situation in 1910:

> It is required or supposed that every scholar or man of science should know at least German, French, and English. For the majority of German scholars and men of science this may hold good, but in the case of the French it is less true, and in the case of the English least of all. The knowledge of these three languages is, however, no longer sufficient, and that for the following reasons.
>
> In the first place, several other languages must be taken into account, for many Italians write only Italian, many Dutchmen only Dutch, whilst numerous Russians, Poles, Czechs, Hungarians, Scandinavians, and Spaniards employ only their national languages.[4]

According to Pfaundler, then, even the triumvirate failed to hold outside of Germany, since French and British scientists seemed locked into fewer than three scholarly languages. Babel only made the bad present worse.

What if we go back even earlier? According to Louis Couturat and Léopold Leau, who penned a magisterial volume on "auxiliary languages"—that is, languages used as tools between peoples of different tongues—never was the need for such a language more salient. "Its necessity emerges even more evidently," they noted in 1903, "from the development of means of communication: what is the good of transporting oneself in a few hours to a foreign country, if one cannot understand the inhabitants nor make them understand you? And what is the good of telegraphing from one continent to another, or telephoning from one country to another, if the two correspondents do not have a common language in which they can write or converse?"*[5] Even twenty years before Dyer's diagnosis of a contemporary Babel, it seems the dream of almost universal scholarly communication through the medium of three dominant languages—a shadow of Latin, but a robust shadow nonetheless—was not the case. And we could go further back. The scholars of Europe (and also North America) seemed perpetually unable to make themselves understood. This problem became salient with the emergence of Russian as a scientific language, and it had only gotten worse.

The savants of Europe at the turn of the century considered almost no problem more severe than this conundrum of too many languages flooding the fragile community of scientists. There were two causes for concern. The first was the inability of scientists to actually create science in such a world. Once again, Louis Couturat provides a succinct statement of the worry: "Briefly: to keep themselves acquainted with the special scientific work and studies which interested them, all savants would have to be polyglots; but to become polyglots they would have to abandon every other study, and therefore they would be almost destitute of knowledge of their special subjects."[6] Productivity would dwindle to nothing, and the march of progress would be checked.

The second problem was worse: scientists might wall themselves up in the monoglot echo chambers of their native languages. As the Danish linguist Otto Jespersen reiterated in 1928, the "nationality movement"

*"Sa nécessité résulte encore plus évidemment du développement des moyens de communication: à quoi bon pouvoir se transporter en quelques heures dans un pays étranger, si l'on ne peut ni comprendre les habitants ni se faire comprendre d'eux? À quoi bon pouvoir télégraphier d'un continent à l'autre, et téléphoner d'un pays à l'autre, si les deux correspondants n'ont pas de langue commune dans laquelle ils puissent écrire ou converser?"

had vitiated the sufficiency of the triumvirate to negotiate the lands of scholarship. "Even small nations want to assert themselves and fly their own colours on every occasion, by way of showing their independence of their mightier neighbours." Of course, they could now do so with the ease of printing and the spread of literacy, but at what cost? "But what is a benefit to these countries themselves, may in some cases be detrimental to the world at large, and even to authors, in so far as thoughts that deserved diffusion all over the globe are now made accessible merely to a small fraction of those that should be interested in them."[7]

Neither worry was new. At an 1888 meeting of the American Philosophical Society, based in remote Philadelphia, the death of Latin and then the triumvirate was lamented. Looking at their own library holdings, the Americans observed that "[e]very little principality claims that it should print what it has to tell the world of science in its own dialect, and claims that the world of science should learn this dialect. Thus we have on the list of our scientific exchanges publications in Roumanian and Bohemian, in Icelandic and Basque, in Swedish and Hungarian, in Armenian and modern Greek, in Japanese and in Portuguese, without counting the more familiar tongues."[8] Responsible scientists could not keep up, and irresponsible scientists selfishly generated knowledge that—by virtue of its incarceration in Hungarian (to pick an offender at random)—had ceased to be knowledge because nobody *knew* it.

How had communication been possible in the early nineteenth century? Through English, French, and German. These tongues were indeed associated with powerful nation-states, but they were also something more. Nonnative speakers had learned these languages en masse in order to communicate with others. They were, each of them, *auxiliaries*, and thus facilitated communication across the crazy-quilt of European speech. Looking back even further, the memory of Latin as a language of scholarship in the medieval and early-modern periods pointed to the same solution. Latin had also been an auxiliary, and perhaps a better one than the triumvirate. For starters, it had been singular: one needed to learn only one language, not three, in order to absorb the findings of contemporary science. Latin was also, in the sixteenth century, nobody's native language, and so no inherent advantage was granted to any particular people over the others.

The solution to the fin-de-siècle Babel was thus obvious: scientists needed a universal auxiliary to communicate their findings. This chapter and the next follow the dramatic quest for such an auxiliary during the thirty years that preceded the outbreak of World War I. In principle, *any* language could serve as an auxiliary—German, Navajo, Slovenian—and thus the selection was a weighty decision, for it would lock in scholarly communication for the foreseeable future. Two developments converged at the turn of the century: the quest of scientists to find a universal, ideally *neutral*, auxiliary; and the emergence of a plethora of constructed languages. Far from aberrations lurking in the quirky margins of European thought, during the first decade of the twentieth century "artificial" or constructed languages such as Esperanto were appropriated as a perfectly sensible cure for the disease of linguistic proliferation that scientists of all stripes had diagnosed.

The Logic of the Auxiliary

This story requires that you take constructed languages seriously, at least for a moment. For most people, this is a hard sell, marked by the general term commonly used to describe these languages: "artificial languages." "Artificial" literally means "made, constructed," but it carries a distinct pejorative connotation of "fake, inauthentic"—think of artificial Christmas trees. Actually, trees are a useful analogy, because they demonstrate the logic behind the implication. There are *natural* trees out there, ones that you might chop down in December and decorate; the existence of the natural is what makes the "artificial" seem "fake."

But language is not quite like that. The modern languages are not "natural" in the way a plant is (assuming, of course, that we know what we are talking about when we talk about natural plants, which after millennia of artificial selection, antibiotic treatments, pesticides, and now genetic modification, is somewhat of an open question). What do we mean by "natural" languages? As Jespersen pointed out many years ago, "very much in the so-called natural languages is 'artificial,' and very much in the so-called artificial languages is natural[...]."[9]

First, the artificial in the natural. Chemical nomenclature is a classic example. We would hesitate to banish German or Russian from the camp of natural languages, but the chemical nomenclature in both of those languages was carefully constructed from a mélange of foreign words, ancient roots, and lexemes native to the language. Artificiality—in the sense of deliberately made by humans—is evident in a variety of more general cases: Modern Hebrew, the transformation of Hindustani into Hindi and Urdu, the revival of Irish, the purging of English words from French, and so on. English does not have an officially sanctioned body to manage the language as French does with the Académie

Française, but the countless pressures that move even English in certain directions and not in others are similarly not always products of "natural forces." We would do better to follow Jespersen's terminology and think of such languages as "ethnic" rather than "natural," for they too are groomed and modified through conscious effort to adapt them to modern conditions.

Likewise, the natural in the artificial. Of course, artificial languages have to be made from something, and they are usually made from ethnic languages. Phonemes are taken from here, syntax from there, patterns that one wants to eliminate (such as the persistent irregularity of "to be" in many languages) are noted and then extirpated, and so on. But there is an even clearer way to appreciate the blurriness of the boundary: there are roughly one thousand native speakers of Esperanto. Yes, Esperanto. Typically, they have been raised in households where Esperanto was the only language common to the parents. Nothing is wrong with these people; their linguistic behavior is not stunted or defective in any way. They use the language for the entire range of human experience just as do users of ethnic languages, and Esperanto has acquired new features because of these "denaskuloj" (lit., "from-birthers"). In other words, Esperanto is now a "natural language." Wrap your mind around that for a minute. Rather than "artificial," "planned languages" or "constructed languages" are more accurate descriptors, and I will opt for the latter.[10]

But back to the dilemma of our scientists at the dawn of the new, exciting twentieth century. Babel surrounded them at precisely the moment when there was so much science booming across Europe that they wanted to access. What were they to do? Proposals of various stripes abounded, and the general consensus settled fairly rapidly—and with remarkably little controversy—on the obviousness of the need for a constructed language to serve as the universal auxiliary. Since this conclusion was so self-evident to them, and is so counterintuitive to us, it helps to trace their logic.

The most straightforward proposal would be to use an ethnic language as an auxiliary. After all, those languages already existed, already had a body of speakers who use them every day, and—if you selected a language that was used broadly for scientific work, like English, French, or German—contained a scientific nomenclature ready to hand. This is, of course, what has happened today, with the ubiquitous use of English in the sciences, but in 1900 such an outcome was unfathomable. The French would never tolerate German; the Germans would never tolerate English; the English would tolerate nothing at all; and none of the rising nationalist movements would submit to any of these three. Compromises seemed equally doomed. R. P. Peeters suggested in the first years of the century a division of disciplines by language; say, the mathematicians would use English, but the philosophers would agree to use German, the chemists French, the naturalists Russian, and so on. But what about interdisciplinary discussions?[11] No, the auxiliary could not be an active ethnic language. As Couturat, a strong proponent of the constructed-language solution, put it in 1910, "the solution by the national languages is the real chimera and utopia; and the solution by artificial languages seems the only practical option."*[12]

Of course, there were more national or ethnic languages to choose from than those then used as native languages; in fact, using Latin—a "dead" language—offered promise to some. Next to artificial languages, some modification of Latin seemed the most likely auxiliary around 1900 (and continued to be advocated into the 1930s, and somewhat beyond).[13] But here, too, there were problems. Latin is, first of all, complicated: deponent verbs, sequence of tenses, the elaborate subjunctive, five declensions, three genders, and other terrors of schoolchildren and altar boys the world over. And it was also, well, *dead*. Even though much of the world's scientific terminology is derived from Latin and Greek roots, the key word is "derived." Latin was adapted to modern conditions, not modern conditions to Latin. Even though all three words are Latin in origin, a whole new vocabulary for "internal combustion engine" would have to be generated de novo or ex nihilo. Finally, the proposal of reviving Latin always labored under aspersions of conservatism and dogmatism, fed by the perpetual association of the language with the Catholic Church.

It was a commonplace in the first decade of the twentieth century that if Latin were to become the universal auxiliary, it would have to be simplified in some way, stripped of the irregularities and quirks beloved of philologists. The most successful (relatively speaking) of these "neo-Latins" was proferred by distinguished Italian mathematician Giuseppe Peano in 1903 under the name "Latino sine flexione": "Latin without inflections," also known as Interlingua (but not to be confused with Alexander Gode's more popular Interlingua of 1951, which we will encounter in chapter 8). Albert Guérard, one of the chief historians of constructed

*"la solution par les langages nationales est réellement la chimère et l'utopie; et la solution par la langue artificielle apparaît comme la seule pratique."

languages, believed that Peano had indeed "placed the whole question, for the first time, on a strictly scientific basis."[14] Peano insisted that one should use Latin nouns (mostly) in the nominative singular and verbs (mostly) in the infinitive, and let English's strict word order take care of meaning. (Although, to be honest, he wasn't particularly rigorous about this, as examples will show, with ablatives and conjugations littering the purity of the constructed language.) You could expect scholars in 1903 to have some familiarity with Latin, and so it should produce no difficulty. As he began his manifesto on the subject (published in the mathematical journal he edited):

> Lingua latina fuit internationalis in omni scientia, ab imperio Romano, usque ad finem saeculi XVIII. Hodie multi reputant illam nimis difficilem esse, iam in scientia, magis in commercio.
>
> Sed non tota lingua latina est necessaria; parva pars sufficit ad exprimendam quamlibet ideam.[15]

I will leave that untranslated. Either it makes sense to you, or it does not—and that gives you a feel for what most of Peano's contemporaries also thought about Latino sine flexione (with the exception of some Italians, who rallied behind the system).[16]

If scientists needed a universal auxiliary because of the emerging Babel, and they could not use a living language, and also not a dead one, the only option would seem to be to make a new language tailored to the purpose. These new languages differed from the seventeenth-century efforts of John Wilkins and his colleagues; the important distinction, coined in 1856 in a paper at the Société de Linguistique, is between *a priori* languages and *a posteriori* ones. *A priori* constructed languages created everything from scratch: all the words were new (and thus divorced from the illogicality of historical connotations), much of the grammar and syntax was new, and they had to be learned from whole cloth. These were the first constructed languages, and owing to their extreme difficulty they have never been very popular. Usually only the constructor knew how to use it, and then not always well. An exception in terms of popularity to some degree was François Sudre's "Solresol" (1827). Using a total of seven phonemes (do, re, mi, fa, sol, la, ti), Sudre constructed a language which could be sung according to the musical scale. He packed theaters with displays where adepts would translate an audience message into Solresol and play it on an instrument, and he would decode it properly. Seven phonemes, however, were not all that many, and so words quickly became very long. A neat parlor trick, but not a viable language. As Couturat and Leau puzzled in 1903, "[o]ne has trouble explaining the relative success of this language, the most impoverished, the most artificial, and the most impractical of all the *a priori* languages."*[17] Solresol was the first and last *a priori* language to win broad public attention.

If a constructed universal auxiliary was to be had, it was going to have to be *a posteriori*: built upon ethnic languages but stripping them of the exceptions and complexities that bogged down students of traditional tongues—in the manner of Peano's *a posteriori* simplification of Latin, but only more so. The major argument for this was . . . *science itself*. The nomenclature of science, however much it varied among various languages, was in essence already international, and although scientists feared the Babel to come, it was the case in 1910 that they "can read foreign scientific literature much more easily than newspapers or novels written in the same languages."[18] Using the internationality of science as scaffolding, one could build a tool to facilitate communication without transgressing national sensitivities. For, as Jespersen noted, the ethnic languages would still have their role to play:

> [The auxiliary] must necessarily remain an intellectual language, a language for the brain, not for the heart; it can never expect to give expression to those deep emotions which find their natural outlet through a national language. There will always be something dry and prosaic about it, and it is a mistake to try to translate very deep poetry in it, for it will be capable of rendering only those elements of poetry which might as well have been expressed through a paraphrase in native prose.[19]

To be sure, there were naysayers, those who pointed out that the advocates of constructed languages approached the problem of scientific Babel by proposing to scientists "*that they add on to other foreign languages yet another foreign language to be learned*. And one calls that simplification, the saving of energy!"†[20] But these people were just that—naysayers—and the fin-de-siècle debate hinged less on whether

* "On a peine à s'expliquer le succès relatif de cette langue, la plus pauvre, la plus artificielle et la plus impraticable de toutes les langues *a priori*."

† "*daß sie zu den andern Fremdsprachen noch eine neue Fremdsprache hinzulernen*. Und das nennt man Vereinfachung, Ersparung von Energie!"

constructed languages were a good idea, but rather on which one should be adopted. Europeans at 1900 already knew that it was possible to have tens of thousands, even hundreds of thousands, of individuals of different nationalities communicating through a *neutral, constructed* language. They knew about Volapük.

Worldspeak

Volapük (in Volapük: *vol* = world, *a* = genitive ending, *pük* = speak, language; hence "language of the world" or "worldspeak") was one of the most astonishing linguistic developments of the late nineteenth century, and its story would make a gripping subject for a novel (and in fact already has).[21] The idea for creating this universal language, according to one report, came to the Roman Catholic priest Johann Martin Schleyer while he was sleeping on the night of 31 March 1879 at his parish near Lake Constance in southern Germany. In late 1880 he published his grand framework, which he introduced with characteristic enthusiasm:

> Through the magnificent **worldwide postal system** a tremendous step forwards to this beautiful goal is made. Also with reference to money, measures, weight, divisions of time, *laws* and **language** . . . the *brotherhood* of man should **unite** more and more! To this **language union** on a magnificent scale the present short work will give the first impetus.* [22]

There are two points worth noting in this passage, one typical of the golden age of language construction and the other rather unusual. The common trope was to draw inspiration from contemporary innovations in communications and transportation technologies, and the standardizations that followed in their wake. Peculiar about Volapük was Schleyer's avowed intention to create not an auxiliary, but a new universal language to supplant all the world's tongues. In the motto of what would

* "Durch di großartige Weltpost ist ein gewaltiger Schritt zu disem schönen Zile vorwärz gemacht worden. Auch inbezug auf Geld, Maß, Gewicht, Zeiteinteilung, *Geséze* und Sprache . . . sollte sich das *Brudergeschlecht* der Menschen merundmer einigen! Zu diser Spracheinigung im großartigsten Maßstabe will vorligendes Werkchen den ersten Anstoß geben." The unusual orthography and emphasis is Schleyer's.

later become a movement: *Menade bal, püki bal*, to one mankind, one language!

The idea spread like wildfire. Textbooks cropped up in a host of languages, and commercial enterprises promoted Volapük to ease international correspondence.[23] At first spreading in German-speaking lands, by the middle of the 1880s Volapük had taken France by storm. Les Grands Magasins du Printemps, the great Parisian department store, offered courses on it, training no fewer than 121 new speakers of the universal language of the future.[24] By the peak of the movement in 1888, advocates argued that 210,000 people had studied the language, and a significant proportion of them continued to use it. Even granting exaggeration by Volapük boosters, the penetration of the language was impressive.[25] Yet more astonishing than numbers was its geographical distribution. Samples of Volapük newspapers either partially or entirely written in the new language stored in the archives of the American Philosophical Society in Philadelphia range from China, Denmark, Turin, Oregon, Zurich, and Prague. The charter of the North-American Volapük Club (*Volapükaklub Nolümelopik*) registered well over fifty members, mostly based in New England.[26] The great French writer Ernest Renan observed these developments with wonder, bleakly quipping (in Albert Guérard's translation) that "a few generations hence, naught will remain of our writings, but a few selections with interlinear translation into Volapük."[27]

What was this wonderful language, this mechanism to undo the ravages of Babel? I have extracted a sample sentence, pretty much at random, from an 1888 textbook:

> Neläbo jimatel yagela pedlefof, nendas yuf äkanom pablinön ofe.[28]

Make sense? Volapük is based on a large collection of fundamental roots, derived *a posteriori* (mostly from English and German), suitably modified to meet Schleyer's criteria: they tended to be monosyllabic, they had to begin and end with a consonant (but not, for grammatical reasons, with *s*), every instance of the letter *r* was replaced by *l* out of consideration to East Asians (since Japanese, for example, does not make a distinction between the sounds), and there were other morphological transformations demanded by the need to avoid homonymy and other flaws of ethnic languages. Upon these roots, one attached prefixes and suffixes to decline them as nouns or conjugate them as verbs (Volapük has four cases and a full complement of tenses). By simple agglutination, therefore, one can express complex thoughts from rudimentary units.

Let's take our sentence word by word. "Neläbo" is an adverb (that's what the *o* signifies, and it is a negation of a fundamental concept *läb*, which means "luck." Hence, "unluckily." "Jimatel" is easier still. The prefix "ji" is pronounced "shi" (the *j* would be rendered in English as *sh*), and feminizes the root; *mat* is the root for *mate*, derived cleanly from the English, and *el* is a suffix that indicates an agent or person. So *jimatel* is a female person who is a mate, also known as "wife." "Yagela" has two suffixes we have met before: the genitive suffix indicating possession and the agentive suffix; *yag* is derived from the German *jagen*, "to hunt." So we are talking about "the hunter's wife." Now to our verb, "pedlefof." To conjugate a verb—in this case *dlefön*, "to hit"—for the third-person singular feminine, we add the suffix "-of" to the root *dlef*; by contrast, first-person singular would be "-ob" and first-person plural "-obs." The prefix *e* marks the perfect tense, and the additional prefix *p* makes it passive. (This suffixing and prefixing can go on for quite a while; famously, one contemporary calculated that each Volapük verb could come in 505,400 different forms.) We are done with the first clause.

The second clause begins with "nendas," which means "unless"—sometimes, you just have to memorize subordinate conjunctions! "Yuf" is the root meaning "help," and since it has no suffixes or prefixes, we can rest easy: it's in the nominative case, and is the subject of our clause. Then our verb "äkanom" is a conjugation of *kanön*, "to be able to," with the suffix indicating third-person singular masculine, in exact analogy to the first clause. Since "yuf" is masculine, we know that is the subject of the verb, and the prefix *ä* is the past imperfect tense. "Blinön" is the infinitive for "to bring," where you can see how the *r* of "bring" was changed to an *l* and Schleyer simplified the consonant cluster at the end. The prefix *p* is familiar—it makes things passive—and the *a* preceding is a dummy prefix to avert the unpronounceable consonant cluster *pbl*. By now, "ofe" might explain itself. We have already seen *of* as the particle for third-person singular feminine; here it has the suffix *e* which makes it dative, since "she" is going to be our indirect object. And now we have it:

> Neläbo jimatel yagela pedlefof, nendas yuf äkanom pablinön ofe.
> Unluckily the hunter's wife was struck, unless help was able to be brought to her.

Simple enough. I mean it: Volapük *was* easy. It certainly was a lot easier than many, possibly most, ethnic languages—easier than German, say, for an Italian.[29]

People flocked to it in droves. In 1884, just four years after the language was published, Schleyer convened the first Volapük congress at Friedrichshafen by Lake Constance. Three years later, the second Congress took place in Munich, and roughly two hundred attended, where they established the Universal Association of Volapükists (*Volapükaklub Valemik*), which in turn created the International Volapük Academy (*Kadem Bevünetik Volapüka*). Discussions at the first two meetings took place in German, the common language of most of the participants, but at the third congress in Paris in 1889, Volapük was the only language used; even the bellhops and waiters spoke it. The shift to Paris marked the internationalization of the movement, an achievement largely due to Auguste Kerckhoffs, professor of modern languages at the École des Hautes Études Commerciales, who founded the *Association française pour la propagation du Volapük* in 1886 and was appointed president of the Academy the following year. (Schleyer was named Grandmaster for life.) By 1889, there were 283 Volapük clubs world wide, 1,600 Volapük diplomas granted, 316 publications about it (182 of them in 1888 alone) in 25 languages—85 were written in German, but 60 appeared entirely in Volapük—and between 25 and 35 periodicals from around the globe.[30]

Then, quite suddenly, the bottom fell out. To be frank, there had been warning signs since the creation of the Academy at the Munich meeting. Kerckhoffs, who worked tirelessly to spread Volapük, proposed a series of reforms to make the elaborate language easier to learn. Schleyer dug in his heels: the language was perfect, and the success of the movement demonstrated it. He argued that he should have a veto on any reform proposed by the Academy, while Kerckhoffs was willing to grant him only a large say—three votes for him alone, among the seventeen members of the Grand Council. The debate came to a head at Paris with the discussion of the proposals, and by 1891 Kerckhoffs had resigned. By then, Volapükists had already begun to disappear with astonishing speed.[31] As Albert Guérard put it in 1922: "The strangest thing about Volapük was the suddenness of its collapse. In 1888–89, it seemed as though it would conquer the world: in 1890 it was dying."[32] (A caveat: With constructed languages, one has to be careful before declaring something extinct. There was a stray pamphlet defending Volapük in 1904; a one-man revival in 1931 by a Dutch physician named

Arie de Jong, who argued as late as 1956 that "V[olapük] will never be old-fashioned, just like every other I[nternational] L[anguage], precisely because V. offers all the desired guarantees of exactitude and neutrality"*[33]; and an attempt at revival in 1979, to mark the centenary. There is even a Volapük division of Wikipedia. Nonetheless it is probably safe to declare Volapük moribund.[34])

There are two key features of this story. The first is the fragility of the Volapük movement. With all the signs of health and vibrancy, how could it all end so quickly? Why did all the Volapükists leave, and where did they go? This question was also central for contemporaries. I will reserve the autopsy for the following chapter, where we will see how interpretations of Volapük's demise haunted the controversy over a constructed auxiliary for science. There is, however, another lesson here: for about a decade, despite the oddities of its morphology and syntax, Volapük *worked*. It was proof positive that people would use a constructed language to express themselves. Sure, it died, but perhaps the important lesson was that it had lived.

Hope Returns

Volapük triggered a great deal of enthusiasm in the academic world, which examined it as a possible solution to the Babel looming over the horizon. On 21 October 1887, the American Philosophical Society appointed a three-person committee chaired by D. G. Brinton to examine it. In their January 1888 report they announced that they found "something to praise and much to condemn in [Schleyer's] attempt." The Philadelphia scholars saw Volapük as contrary to just about every feature of Indo-European (known more commonly at the time as "Aryan") languages, considered by them the only civilized tongues: "Volapük is synthetic and complex; all modern dialects become more and more analytic and grammatically simple; the formal elements of Volapük are those long since discarded as outgrown by Aryan speech; its phonetics are strange in parts to every Aryan; portions of its vocabulary are made up for the occasion; and its expressions involve unavoidable obscurities." It just would not do. Nonetheless, a recent constructed language project might offer the requisite linguistic simplicity. "The plan of Dr. Samenhof," Brinton's committee concluded, "is especially to be

*". . . le V. n'a jamais été dépassé, voire égalé par aucune autre L.I. précisément parce que le V. présente toutes les garanties voulues d'exactitude et de neutralité."

recommended in this respect, and may be offered as an excellent example of sound judgment."[35] They recommended keeping an eye on it.

This plan, published in 1887 by Dr. Ludwik Lejzer Zamenhof (to use his preferred spelling rather than Brinton's German-inflected one), is by far the dominant constructed language in the world today, dwarfing its nearest competitors by orders of magnitude. You know it as Esperanto. Today most people who are not Esperantists consider it, frankly, borderline ridiculous. Then again, people who are not Esperantists typically do not know that much about it. It is the official language of the Bahai'i faith, is associated with pacifist and internationalist movements, was actively persecuted by both Adolf Hitler and Joseph Stalin, and was even used by the United States in the 1950s as the language of "Aggressor, the Maneuver Enemy" in large-scale simulations of combat against a foreign power so as not to unduly alarm any actual countries.[36] There is also a full-length 1966 movie in Esperanto called *Incubus*, featuring a young William Shatner. This hodgepodge of associations makes the language seem random and goofy, and I raise them here to get them out of our system. For in the years around 1900 Esperanto was not at all silly, and the most serious minds of Europe learned it, analyzed it, adopted it, or rejected it—they considered it a viable proposal to overcome the chaos of tongues.

Like Volapük, Esperanto was the creation of a single person, but the similarity ends there. For example, while Schleyer hoped to replace all ethnic languages with his superior creation, Zamenhof's goal was to provide a *second* language for everyone, a means of communication outside of the ethnic tongues—in other words, an auxiliary. Zamenhof, a Jewish oculist based in Warsaw (then one of the largest cities in the Russian Empire), published his plan for an "internacia lingvo" in a short Russian-language pamphlet that cleared the Imperial censor on 21 May 1887. He wrote it under a pseudonym, Doktoro Esperanto—"Dr. Hoping"—and the language eventually assumed the eponym. The book came to be known in the movement as the *Unua Libro* (First Book), and contains a brief introduction explaining the complete grammar of the language encoded in sixteen rules, a vocabulary of 900 roots, and Esperanto versions of the Lord's Prayer, a passage from the Bible, a sample letter, and some poems written by Zamenhof himself. He had been working on a project for a universal auxiliary, a neutral tongue that could serve for communication between the diverse nations of Europe, since his youth, and both his biography and his path to the language have been well documented in many admiring studies, so I will not rehearse them here.[37] More important for our purposes is understanding a bit about how the language works, how it spread, and how its users approached the issues surrounding scientific communications.

The rules of Esperanto are remarkably straightforward. Consider the first sentence from the epigraph of this chapter:

> En Eŭropo oni ordinare pensas, ke en la tuta Ĥinujo oni parolas nur unu lingvon—la ĥinan.
> [In Europe one ordinarily thinks, that in all of China one speaks only one language: Chinese.]

One of the first things that strikes you is that it looks like some mix of familiar languages. That's because Esperanto roots *are* drawn from six widely known source languages: English, French, German, Italian, Russian, Spanish. (The Russian is pretty understated.) Each word indicates its grammatical part of speech in its ending. If it ends in *o*, it is a noun; in *a*, an adjective; in *e*, an adverb. So we can see that "Eŭropo" is the noun *Europe*, but we could easily make the adjective "Eŭropa," meaning *European*, from the same root. "Ordinare" means *ordinarily*, in the same way "ordinara" would mean *ordinary*. To make a noun plural, you add a *j*, pronounced like an English "y," and adjectives must agree with nouns—"all of China" is "la tuta Ĥinujo," just like "all of the Chinas" (whatever that might mean) would be "la tutaj Ĥinujoj." If something is the direct object of a verb, as with "lingvon," we append an *n* to mark the accusative case, which we would also append to the adjective if there were one. That is just about everything you need to know about nouns. Verbs are even easier: if it ends in *as*, it is present tense for all persons and numbers; *is* for past; *os* for future. There are no exceptions, no irregularities. Armed with a dictionary and what I just described, you could make it through a great deal of Esperanto.

But the language has some idiosyncratic features, visible in the word "Ĥinujo." First, what's going on with that circumflex? There are five "hatted" letters in Esperanto, *ĉ*, *ĝ*, *ĥ*, *ĵ*, *ŝ*, which stand for hushing sounds as in *ch*at, *g*em, Ba*ch*, *j*oke, and *sh*ut. No other language has this orthography—it may look like the Czech *haček*, but it is upside-down. Besides the consonants, one vowel shares a peculiar, almost unique, diacritic: the ŭ, which represents a glide like the English "w." Also interesting is the suffix *uj*, which is attached to the root *ĥin* before the final *o* (which, recall, marks this as a noun). *Ĥino* means "Chinese person" in the same way *ĥina* means "Chinese" as in "Chinese food." The suffix means "container of," so in Esperanto *Ĥinujo* means "container of Chinese people," or the country China. There are a host of these suffixes—*id* means "descendant," *in* means "feminine," *eg* means "intensified," and so on—and they can also function as roots for words: *ujo* means "container," *uja* means "of or pertaining to a container." With these tools you can build an enormous array of words from a relatively small number of roots. Zamenhof deliberately designed it to be easy to learn.

At first, most of the people who took him up on it were Russians, which makes sense, considering that the *Unua Libro* was written in Russian and thus linguistically inaccessible to a large portion of Europeans. Leo Tolstoy, the famous novelist, penned a short endorsement of the language as extremely simple (he claimed it took him two hours to learn), although at a time when Tolstoy was persona non grata to the Tsarist regime this may have generated more trouble than it was worth (which wasn't much in any event).[38] In 1888, Leopold Einstein, who had founded the *Weltsprach-Verein* in Nuremberg in 1885 and was devoted to Volapük, received a copy of a translation of Zamenhof's booklet in 1888. Smitten, he shifted to Esperanto, taking the Nuremberg Volapükists with him. Einstein died the following year, but his disciple Chrystian Schmidt continued in his stead and began issuing *La Esperantisto*, the first periodical in Zamenhof's language.[39] In 1890 the journal was transferred to Zamenhof's supervision, and became an important force for cohesion in the movement, until 1895, when it was shuttered by the Tsarist censorship for publishing a translation of Tolstoy's banned *Faith and Reason*. The journal moved to Uppsala. (Esperanto was also big in Sweden.) The subsequent growth of the movement was stunning. By 1907 the *Unua Libro* had been translated from Russian into Polish, French, German, English, Hebrew, Yiddish, Swedish, Lithuanian, Danish, Bulgarian, Italian, Spanish, Czech, Latvian, Portuguese, Dutch, Hungarian, Estonian, Catalan, Flemish, Japanese, Greek, Ukrainian, and Arabic. There were at least 756 Esperanto organizations worldwide, 123 of them outside Europe, and 64 journals.[40]

The most important development in Esperanto's early history was its transition from a language dominated by Russians—who endowed the movement with some of its enduring idealism and shaped Esperanto literary style—to a movement centered in Paris in the 1890s. The shift in center of gravity is associated with Louis de Beaufront. For reasons discussed in the following chapter, de Beaufront later became a deeply unpopular figure in Esperanto circles, and the Esperantophile historiography tends to diminish his importance and cast aspersions on his character as "the most enigmatic, quite possibly the most pathetic, figure in the history of Esperanto."[41] To be fair, this is pretty easy to do, since to call de Beaufront's veracity questionable would be an understatement.

According to de Beaufront, he had been working on his own artificial language, dubbed "Adjuvanto," but when he came across Esperanto he realized its intrinsic superiority and threw all of his energies behind it, becoming, in the words of a 1907 Esperanto textbook, "the greatest and most fervent of all the apostles of Esperanto."[42] After 1905, in response to some personal attacks, he claimed he was a Marquis who had been forced by poverty to become a private tutor to the family of Count Chandon de Briailles.[43] We do not know whether Adjuvanto ever existed,[44] but we do know that in 1892 de Beaufront published a large Esperanto textbook in French and an important promotional leaflet in 1895. In 1898 he founded the *Société pour la propagation d'Espéranto* and the bilingual journal *L'Espérantiste*, marking the tipping point of the French ascendancy within the movement.[45] Soon, the powerful and stodgily conservative Hachette became the main publisher of Esperanto texts, locking in the language's unusual orthography.

De Beaufront's influence centrally shaped two significant decisions of the fledgling movement, both of which came to a head at the first world congress of Esperanto at Boulogne-sur-Mer in 1905: moral and political neutrality; and linguistic conservatism and stability. The eponymous Boulogne Declaration issued at that meeting, penned by Zamenhof, affirmed that Esperantists did not seek to supplant natural languages, and distanced the movement from religious or political ideas, an avowal de Beaufront insisted upon in the face of Zamenhof's own interest in promoting an ideal of universal brotherhood (*Homoaranismo*, strongly influenced by the Jewish ethical precepts sometimes known as "Hillelism").[46] Those important issues proved tangential to the question of scientific communication. Not so the fourth point of the Declaration, which read in part (as translated by sociologist of the Esperanto movement Peter Forster):

> Esperanto has no personal legislator and depends on no particular man. All opinions and works of the creator of Esperanto have, like the opinions and works of every Esperantist, an absolutely *private* character, compulsory for nobody. The only foundation, compulsory for all Esperantists, once and for all, is the booklet *Fundamento de Esperanto*, in which nobody has the right to make a change.[47]

Constructed languages face an intrinsic problem: how to keep the language fixed enough to build a community of speakers when each user tends to push the idiom in individualized directions. Ethnic languages have a body of literature and custom that stabilizes the tongue, an option closed to Esperantists. The *Fundamento* imposed a standard to prevent disintegration. It consisted of the grammatical part of the *Unua Libro* (i.e., the sixteen rules); the universal dictionary (*Universala Vortaro*) of basic translations into English, French, German, Polish, and Russian; and the 1894 collection of exercises produced by Zamenhof (the *Ekzercaro*). The Declaration elevated the *Fundamento* to a universal baseline or standard for Esperantists; one could add to the language, but one could not modify or take away from this core. It became, in the parlance of the movement, *netuŝebla*—untouchable. To enforce this, a 68-member Lingva Komitato (language committee) was established.

The Declaration locked down the language. In the *Unua Libro* of 1887, Zamenhof declared that he was "far from considering the language I have proposed as somehow perfected, so that nothing can be higher or better than it; but I tried, as much as I could, to satisfy all those demands which one could pose to an international language."* [48] He asked the public for comments, and only after considering them would he secure the language in "a final, permanent form. If someone were to consider these corrections unsatisfactory, he should not forget that the language even afterwards would not be sealed from all sorts of possible improvements, with the single difference that then the right of changing it will belong already not to me, but to an authoritative, generally recognized academy of this language."† [49] Even after those slight modifications were produced (in the *Dua Libro*, or "Second Book"), the issue of reform was still alive. Subscribers of *La Esperantisto* were asked in 1894 to vote on a slate of reforms in both August and—absent a clear result—again in November, with a majority of 144 opting to retain the language as it was, while the 109 reformist votes split three ways.[50] By

*"Я далек от того, чтобы считать предложенный мною язык чем-то совершенным, выше и лучше чего уже быть ничего не может; но я старался, насколько мог, удовлетворить всем требованиям, которые можно ставить интернациональному языку."

†"окончательная, постоянная форма. Если бы кому либо эти поправки казались недостаточными, тот не должен забывать, что язык и впоследствии не будет замкнут для всевозможных улучшений, с тою только разницей, что тогда право изменять будет принадлежать уже не мне, а авторитетной, общепризнанной академии этого языка."

1905, groundswells in favor of reform had begun to reemerge, not least because of problems treating scientific nomenclature within Esperanto.

"Without doubt," Richard Lorenz opined in 1910, "one of the most important conditions to be satisfied by an artificial international language is, that it should be capable of being employed in science."[51] Volapükists, for example, had produced translations of scientific texts and even original expositions of analytic geometry.[52] Yet the Esperanto community generated very little scientific literature until the foundation of *Internacia Scienca Revuo* (International Science Review) in Paris in 1904. This relative silence is somewhat striking, since the language had been created by a physician, and science was singled out as an important application in the 1887 inaugural document:

> Whoever has tried to live in a city populated by people of different nations, struggling amongst themselves, would have doubtless felt that enormous benefit an international language would present to humanity, one which, *not encroaching upon the internal life of peoples*, could, at least in countries with a multilingual population, be a state and a societal language. I think there is no need for me to expound, finally, on what enormous significance an international language would have for science, trade—in a word, at every step.*[53]

Nonetheless, until the advent of the *Revuo*, scientific terminology was noticeably lacking. Within its pages, several discussions about developing a chemical nomenclature in the constructed language demonstrate quite clearly the problems of building a scientific auxiliary from scratch, especially the tension between internationality and uniformity.

The *Revuo* had an auspicious beginning. Edited by Paul Fruictier, a physician who had begun learning Esperanto in 1900, it was published by Hachette and sponsored by a whole raft of distinguished names: not only Zamenhof, but also the French Physical Society, the International Electrical Society, and (mostly French) scientists including Marcel-

*"Кто раз попробовал жить в городе, населенном людьми различных, борящихся между собою, наций, тот почувствовал без сомнения, какую громадную услугу оказал бы человечеству интернациональный язык, который, *не вторгаясь в домашнюю жизнь народов*, мог бы, по крайней мере в странах с разноязычным населением, быть языком государственным и общественным. Какое, наконец, огромное значение имел бы международный язык для науки, торговли—словом на каждом шагу,—об этом, я думаю, мне нечего распространяться."

lin Berthelot, Henri Poincaré, Henri Becquerel, and William Ramsay. (The last two were Nobel laureates for 1903 and 1904 in physics and chemistry, respectively.) Fruictier announced the monthly's aspirations on the first page: "*Internacia scienca revuo* has the goal indeed to create and fix the special terms which are necessary to professional colleagues of various countries in order to communicate among themselves. *Internacia scienca revuo* will do that by a natural method, importing technical and popular articles, and analyses of the most interesting works."* [54] For example, readers were immediately treated to a translation of the preface to Poincaré's *Science and Hypothesis*, and later that year Fruictier serialized a translation of Russian chemist Dmitrii Mendeleev's hypothesis of a new chemical element for the world ether, which would supposedly occupy a slot in his famous periodic system. (Ivan Chetverikov, the translator, promised Mendeleev that "with the help of Esperanto, it will be as clear as a bright day for everyone.")† [55] As it happens, neither the hoped-for clarity nor the elemental ether quite materialized. In typical fashion, the translation included editorial footnotes explaining new Esperanto coinages for technical terms, and a collection of all new terms was published in the December 1904 issue.

After a year, Fruictier happily noted "the interest of many scientists in Esperanto."‡ [56] One might conclude this was optimistic exaggeration, but communications in other journals such as the American flagship *Science* indicated that scientists were indeed taking notice of the constructed language.[57] Fruictier continued to edit the journal until health issues and other obligations required him to step down in December 1906, to be replaced by the Swiss mathematician René de Saussure, brother of the renowned linguist Ferdinand. De Saussure injected new dynamism into the production, and lobbied for Esperanto activism with the wave of standardizations that was sweeping contemporary Europe.

De Saussure's helming of the *Revuo* was linked to the creation of the Esperanta Scienca Asocio (Esperanto Scientific Association) in January 1907, which he also directed. Much like the journal, the Asocio is evi-

*"*Internacia scienca revuo* celas ja krei kaj fiksi la terminojn specialajn, kiuj estas necesaj al diverslandaj samprofesianoj por komunikadi inter si. *Internacia scienca revuo* tion faros per natura metodo, alportante artikolojn teknikajn aŭ vulgarigajn, kaj analizojn de plej interesantaj laboroj."

†"при помощи эсперанто, будет ясно для всех как белый день.

‡"l'intereso de multaj sciencistoj pri Esperanto."

dence of the *normality* of discussions of Esperanto in the scholarly community during the first decade of the twentieth century. Far from being a joke, Esperanto was considered a perfectly plausible, even desirable, solution to the vexing issue of the auxiliary, given the broad agreement that neither living nor dead ethnic languages would serve. In perfect homology to other standardizing organizations or linguistic academies, the Asocio had a board of academics who collected and systematized technical vocabularies. As a 1907 Esperanto textbook enthused:

> This is perhaps the most practical step yet taken towards the standardization of technical terms, which is so badly needed in all branches of science. A universal language offers the best solution of the vexed question, because it starts with a clean sheet. Once a term has been admitted, by the competent committee for a particular branch of science, into the technical Esperanto vocabulary of that science, it becomes universal, because it has no pre-existent rivals[. . .].[58]

The Asocio also lobbied scientists to accept the use of Esperanto as an official language at international conferences (which sometimes worked), and to induce journals to accept Esperanto submissions (less so).

In the first issue of the *Revuo*, January 1904, only one article appeared among the various translations floating a proposal for a consistent Esperanto nomenclature for a particular science: chemistry. "Chemists certainly do not need a complete dictionary: it would be impossible because of the very large number of chemicals. But it is sufficient that they will have a key for word formation and certain examples using a template," R. van Melckebeke and Th. Renard noted. "We kept on working mainly on the principle of internationality, supporting ourselves on the German, English, and French languages, which are those used most generally among scientists."* [59] In short, they were building an auxiliary upon the assumption that chemistry already functioned as a kind of auxiliary. So far so good. Most of their attention was devoted

*"Kemiistoj ne bezonas certe plenan vortaron: ĝi estus neebla pro la tre granda nombro da kemiaĵoj. Sed sufiĉas, ke oni havos ŝlosilon por la vortfarado kaj kelkajn ekzemplojn uzotajn ŝablone.[. . .] Ni laboradis precipe laŭ la principo de internacieco, apogante nin sur la lingvoj germana, angla, kaj franca, kiuj estas la plej uzataj ĝenerale ĉe la scienculoj."

to inorganic chemistry, which was simpler than organic chemistry, but their progress was frustrated by the status of the *Fundamento*. One can see the problem perhaps most clearly with the chemical substance mercury: "Hg. *Hidrargo* is in the dictionary [the *Universala Vortaro*—MG]. *Merkuro* is used chemically in the English, German, and French languages."* [60] Could one change the *Fundamento* in order to adapt Esperanto as a scientific auxiliary?

This sally into the contentious world of chemical nomenclature drew more letters to the editor than any other article in the *Revuo*'s entire first year. General Hippolyte Sébert, one of the lions of the French movement, wrote in to the March issue with other modifications. Perhaps the elements known in English as "phosphorus" and "sulfur" should become "fosfo" and "sulfo," freeing up "fosforo" for the phenomenon of "phosphorescence"? Following this reformist spirit, he sided with van Melckebeke and Renard on mercury: "In the same way I would prefer *merkuro* to 'hidrargo,' because the first is known by the English, Italians, and French; however I am of a less firm opinion because of the possible confusion with the god Merkuro [Mercury]."† [61] Others rejected all the suggestions and stumped for a brand new elemental suffix: "iumo."[62]

The Russian community generally opposed reform, as was the case with D. Piskunov: "I prefer **hidrargo** *a*) in order not to confuse this idea with [the] idea of the planet; *b*) in order not to confuse it with the Roman god of commerce: **Hermeso**; *c*) because it will fit more to the abbreviated sign **Hg.**"‡ [63] Antoni Grabowski, a Polish Esperantist, had the most authoritative voice, insisting firmly on Zamenhof's terms, even before the Boulogne Declaration made such adherence functionally obligatory:

> *Scienca Revuo* does not have the goal of imposing on chemists besides the existing international collection of norms a new confusing one.[. . .] Dr. Zamenhof gave precise directions for supplementing our dictionary. What is in the dictionary shall remain without

*"Hg. *Hidrargo* estas en la vortaro. *Merkuro* estas uzata kemie en lingvoj angla, germana kaj franca."

†"Same mi preferus *merkuro* ol 'hidrargo,' ĉar la unua estas konata de angloj, italoj, ka[j] francoj; tamen mi estas malpli firmopinia, pro la ebla konfuzo kun la dio Merkuro."

‡"Mi preferas hidrargo *a*) por ne konfuzi tiun ideon kun ideo de l'planedo; *b*) por ne konfuzi ĝin kun nomo de roma dio de komerco: Hermeso; *c*) ĉar ĝi pli konvenos al sia signo mallongigita Hg."

> changes. One is to form derivatives from the existing roots or from newly introduced international roots. We shall accept international roots without confusing changes so that everyone will be able to learn them easily.* [64]

Van Melckebeke and Renard responded: "About that we remark, that if Dr. Zamenhof gave examples in the *Universala Vortaro*, by that he was not at all speaking about the rules which he gave for the adoption of technical words."† [65] The debate did not end decisively, but since the *Fundamento* came to stand for the chief standardizing force in the community, Grabowski's position seemed in the ascendant. Fruictier did not come down on either side, even publishing a completely different proposal in April 1904.[66] And so the situation remained until 1907: Esperantists proved remarkably nimble in generating ideas for adapting to science, but the anarchic social structure of their community—mirroring, in a sense, that of science itself—produced no standard.

Delegating the Auxiliary

Thus it was clear to many advocates for a constructed scientific auxiliary that any solution at the linguistic level had to come alongside a corresponding organizational framework. An opportunity for developing precisely such an authoritative body emerged in Paris in 1900 with the first meeting of the International Association of Academies. Established the preceding year and based in Vienna, the Association represented the impulse of scholars to coordinate between the multitude of scholarly societies that had proliferated in the late nineteenth century. Within the framework of the Association, a Delegation for the Adoption of an International Auxiliary Language was formed on 17 January 1901, with support from 310 member organizations (some of which were no grander than a local chamber of commerce) and 1,250 indi-

*"*Scienca Revuo* ne havas la celon altrudi al la kemiistoj anstataŭ la ekzistanta internacia nomaro ian novan konfuzantan.[. . .] Por la plenigo de nia vortaro D° Zamenhof donis precizajn direktilojn. Kio estas en la vortaro, restu sen ŝanĝoj. Devenaĵoj oni formu el la ekzistantaj radikoj aŭ de novaj enkondukotaj radikoj internaciaj. Radikojn internaciajn ni akceptu sen konfuzigantaj ŝanĝoj por ke ĉiu povu ilin facile ekkoni."

†"Pro tio ni rimarkas, ke se D° Zamenhof donis ekzemplojn en *Universala Vortaro*, li pri tio tute ne parolis en la reguloj, kiujn li donis por la alpreno de teknikaj vortoj."

vidual members of academies of university faculties. The charter of the Delegation began:

(1) It is desirable that an international auxiliary language should be introduced which, though not intended to replace the natural languages in the internal life of nations, should be adapted to written and oral intercourse between persons of different mother-tongues.
(2) Such an international language must, in order to fulfill its object, satisfy the following conditions:—
  (a) It must be capable of serving the needs of science as well as those of daily life, commerce, and general intercourse.
  (b) It must be capable of being easily learnt by all persons of average elementary education, especially those belonging to the civilised nations of Europe.
  (c) It must not be any one of the living national languages.[67]

Louis Couturat, a logician who specialized on the work of seventeenth-century polymath Gottfried Wilhelm von Leibniz, was appointed secretary of the Delegation, and in that capacity coauthored with mathematician Léopold Leau the most comprehensive contemporary account of constructed languages. Couturat began writing broadly on the need for a constructed auxiliary in order to generate more support (and therefore authority) for the Delegation, along the way recruiting the internationally famous Leipzig chemist Wilhelm Ostwald as a member of the Delegation on 26 October 1901.[68] He became its chair on 20 November 1906.

As the Delegation gathered information and listened to presentations by the advocates and inventors of various constructed languages, a pattern emerged. Even as early as 1903, a probable outcome loomed. It seemed, wrote Couturat and Leau, that "the more modern projects (and according to us the best) converge more and more upon a determined type."*[69] And this should not be too surprising, considering how many of these efforts were either sparked by the perceived crisis of scientific communication or at least given additional impetus from it. Scientists had become used to communicating internationally in *several* auxiliaries—English, French, and German—and the nomenclature

*"les projets les plus modernes (et selon nous les meilleurs) convergent de plus en plus vers un type déterminé."

and practices they had developed formed the backdrop for the Delegation's deliberations. As those conversations headed to a climax in 1907, one enthusiast for Zamenhof's creation could barely contain his excitement: "It is anticipated that the language chosen will be Esperanto."[70] An international auxiliary for science was at hand.

CHAPTER 5

The Wizards of Ido

> Ido advere ne esas tam richa kam la Angla, ne tam eleganta kam la Franca, ne tam forta kam la Germana, ne tam bela kam la Italiana, ne tam nuancoza kam la Rusa, ne tam hemala kam la Dana. Ma merkez bone, ke omna ta bona qualesi, quin on prizas e laudas en naturala lingui, trovesas nur kande indijeni parolas e skribas oli, ma ne en la boki ed en la plumi di stranjeri. Ed Ido povas tre facile esar plu richa kam la Angla parolata da Franco, e multe plu eleganta kam la Franca di ula Dano; ol esas plu forta kam la Germana di ula Italiano, plu bela kam la Italiana di la Angli, plu nuancoza kun la Rusa di la Germani, e plu hemala kam la Dana di la Rusi.*
>
> OTTO JESPERSEN[1]

It is easy to dismiss the enthusiasm over constructed languages for science as the clamor of a handful of zealous amateurs who had little real impact. Despite persistent calls for a constructed auxiliary—some of which we will observe in the ensuing chapters—we do not live in a world where scientists routinely converse in Esperanto. Nonetheless, it would be hasty to set aside these debates as an amusing detour in the history of the scientific fringe. Worries about the incipient cacophony of tongues were repeatedly voiced by a wide range of scientists across an array of both popular and scientific periodicals, even if many of these

*"Ido, in truth, is not as rich as English, not as elegant as French, not as strong as German, not as beautiful as Italian, not as nuanced as Russian, not as comfortable as Danish. But note well, that all these good qualities, which one appreciates and praises in natural languages, are found only when natives speak or write them, but not in the mouths or in the pens of strangers. And Ido can very easily be richer than the English spoken by a Frenchman, and much more elegant than the French of any Dane; it can be stronger than the German of any Italian, more beautiful than the Italian of the Englishmen, more nuanced than the Russian of the Germans, and more comfortable than the Danish of the Russians."

Cassandras were not devotees of Volapük or its ilk. To be sure, many of the advocates of these projects were marginal figures, but others were decidedly not. Consider, for example, mathematicians such as Giuseppe Peano and René de Saussure, or, most famously, chemist Wilhelm Ostwald. Ostwald was an eccentric thinker, granted, but he was also a scientist with a worldwide reputation, and his decades of advocacy for constructed languages bring us out of the realm of Ludwik Zamenhof and Johann Schleyer and into the heart of modern chemistry.

Ostwald was born in 1853 in Riga, which is the capital of what is today Latvia but at the time of his birth was a thriving Baltic port of the Russian Empire. Thus, much like Zamenhof in Russian Poland, Ostwald grew up in the midst of several languages, specifically Russian, Latvian, and German—Ostwald's native language and the dominant tongue of the elite servitors in the Tsarist bureaucracy. (Ostwald's father was not among their number; he was a master cooper.) Ostwald enrolled at Dorpat University (in today's Tartu, Estonia) in 1872 and quickly established himself as a talented young scientist. In January 1882 he began to teach at the Polytechnic School in Riga, and in August 1887 moved to Leipzig University, where he made his career. Starting from detailed experimental work in electrochemistry, he achieved his reputation through the theoretical framework of physical chemistry—the application of classical thermodynamics to chemical problems—that he developed with his Leipzig students Svante Arrhenius (a Swede, awarded the Nobel Prize in Chemistry in 1903) and Jacobus van't Hoff (a Dutchman, awarded in 1901). (Ostwald's turn came in 1909.) In August 1906, Ostwald retired to his estate, named "Energie," in Großbothen, roughly halfway between Leipzig and Dresden, to work on his many projects: developing his philosophy of energeticist monism (the notion that everything was essentially energy), initiatives to organize international science, and constructed languages.

The last comes as a bit of a surprise to those who know Ostwald for his eponymous law of dilution, yet it was an interest of long standing. In 1880, at the very dawn of Volapük, he was introduced to that language by one of his physics professors at Dorpat, Arthur von Öttingen, who used it in correspondence with foreigners.[2] Although Ostwald did not become a partisan then, his growing commitment to the principle of conservation of energy as a guiding doctrine for all areas of life drew him back to the issue. In the summer of 1901, while giving some lectures at Leipzig on the late eighteenth-century German philosophical movement *Naturphilosophie*, he devoted part of a lecture to the problem of the language barrier from the point of view of energetics.[3] By October of that year, Louis Couturat had drafted him to the Delegation for the Adoption of an International Auxiliary Language, and Ostwald's path toward becoming what one historian has called "the high priest of Ido" began.[4]

To get to Ido, Ostwald passed through Esperanto. In October 1904 he had published an article in *Internacia Scienca Revuo* arguing for an international language specifically because of the demands of science, and in that same month he left Germany to spend several months as an exchange professor in the United States.[5] He stumped for Zamenhof's language across the nation—for example, attending Harvard University's student Esperanto club shortly after his arrival—and he was one of the spurs for the rapid takeoff of the language in that country.[6] As the member of the Delegation with the greatest international reputation in any field, and as one of the world's most prominent scientists and popular philosophers, Ostwald doggedly continued his advocacy when he returned to Europe. As he wrote to his former student Arrhenius on 28 December 1906: "I now spend the greatest part of my energy on the question of the international auxiliary language."*[7] (He tried on several occasions to persuade Arrhenius to intercede with the crown prince of Sweden to support the Delegation's initiatives; after a few months, Arrhenius politely begged off: "For this reason I must also regrettably say that I have no time to spare for Esperanto, as much as I sympathize with the idea."†[8]) This frenetic activity was related to his decision to abandon his post at Leipzig, as he wrote to Charles Eliot Norton of Harvard around the same time: "I have given up my professorship and all my official duties and am living as a free lance, spending the better part of my time and energy for the propagation of the idea of the international auxiliary languages."[9]

Ostwald's support of a constructed auxiliary was intimately connected to his energetic philosophy. In one of his many pamphlets and speeches about the language barrier from the first decade of the century, he analogized the problem to building a house. Houses are constructed for certain ends, but what should one do when the situation changes and you need to adapt the residence for a specialized purpose?:

* "Den größten Teil meiner Energie wende ich jetzt an die Frage der internationalen Hilfssprache."

† "Aus diesem Grund muss ich auch leider sagen, dass ich keine Zeit für das Esperanto übrig habe, so viel ich mit der Idee sympathisiere."

> Admittedly we would not utterly tear down and destroy the old house, for too much of the life of our ancestors is stored inside. But could we not build alongside it a special house for special ends?[. . .] We could indeed, to speak again without parables, erect next to the native language a general, simple, commercial and scientific language, that could achieve the communication of peoples with each other even incomparably more effectively than the telegraph and railroad.* [10]

In much the same way, to use another favorite metaphor of Ostwald's, think about how much money is lost in conversion fees at each border crossing—a unified currency would simplify exchange and prevent waste. The standard railroad gauge also demonstrated the benefits of uniformity; most of Europe happily used the same one, and the waste at the Russian border caused by transferring to a different gauge was an apt analogy for the need for a universal auxiliary.[11] These examples are telling: they all draw from his energeticist philosophy, and they all concern actual *things*, not languages in the abstract. Ostwald was familiar with the several languages necessary for his chemical research, but he did not care for linguistics. He declared, for example, that grammar study "does not cultivate, but actually impairs, the power of logical and original thinking."[12] He wanted an auxiliary because it would save energy; he wanted a constructed one because he thought the problem with language learning was not the rules, but the exceptions.[13]

This chapter follows the career of Wilhelm Ostwald among the constructed language enthusiasts during the first decade of his involvement, which was also the first of the twentieth century. There are three reasons why this approach is particularly illuminating of the issue of scientific languages. First, in 1907 the Delegation (which Ostwald chaired) issued an endorsement of a particular constructed language named Ido, fracturing the Esperanto community in one of the most seminal events of that movement's history, and leaving aftershocks down to this day.[14]

* "Freilich werden wir das alte Haus nicht ganz und gar abreißen und vernichten, dazu steckt eben zu viel von dem Leben unserer Vorfahren darin. Aber können wir uns nicht daneben ein besonderes Haus für besondere Zwecke bauen?[. . .] Wir können sehr wohl, um wieder ohne Gleichnis zu sprechen, neben der Muttersprache eine allgemeine, einfache Geschäfts- und Wissenschaftssprache erbauen, die für den Verkehr der Völker untereinander noch unvergleichlich viel nützlicher wirken kann, als Telegraph und Eisenbahn."

Second, Ostwald allows us to step back from the abstraction of grammars to explore how personal the issues of language use and coordination were. And, finally, Ido offers one of the clearest examples where a constructed language not only was proposed as a way to help science, but also was argued for as being particularly *scientific*; Ostwald exemplified this feature by adapting Ido to chemistry. Although unrealized, these hopes were a vital part of the lived experience of being a polyglot scientist at the dawn of the century.

Ostwald, Delegate

In early 1907, all hopeful eyes turned to Couturat's Delegation. After the Delegation was established to explore the question of an auxiliary—their charter, remember, demanded that this not be a living ethnic language—Couturat began enrolling a roster of luminaries to study the various constructed-language projects, not least of them Ostwald. After years of research and the publication of two impressive scholarly monographs on the history and linguistic analysis of a plethora of language projects, on 15 January 1907 Couturat and Léopold Leau formally submitted their materials to the International Association of Academies with a request that this body select which project should be the international auxiliary language. On 29 May, the Association punted, by a vote of 12 to 8 (with one abstention), declaring itself incompetent to resolve the question. According to the charter of the Delegation, it was now up to Couturat to convene a working committee to study the question and issue its own recommendation.

The committee was elected, not appointed by Couturat, although given that of the 253 votes cast by members and member organizations of the Association (out of 351 total), 242 voted for the same twelve names, it seems reasonable to infer that most simply voted for the slate that the Frenchman had submitted for their consideration. That list was impressive: Manuel Barrios, Dean of the Medical School in Lima and president of the Peruvian Senate; Jan Ignatius Baudoin de Courtenay, professor of linguistics at St. Petersburg University; Émile Boirac, rector of the University of Dijon; Charles Bouchard, distinguished physician, member of the French Académie des Sciences; Loránd (better known as Roland) Eötvös, member of the Hungarian Academy of Sciences; Wilhelm Förster, chair of the International Committee of Weights and Measures; Colonel George Harvey, editor of the *North American Review*; Otto Jespersen, the prominent professor of linguistics from

Copenhagen; Spyridon Lambros, former rector of the University of Athens (and later very briefly the prime minister of Greece); Constantin Le Paige, mathematician at Liège and director of the Scientific Section of the Royal Academy of Belgium; and Hugo Schuchardt, a linguist member of the Imperial Academy of Sciences in Vienna. Wilhelm Ostwald joined them as chair. Due to inabilities to attend, resignations, and other perennial banes of large panels, the committee later substituted in another intimidating crew: Gustav Rados, of the Hungarian Academy of Sciences; William Thomas Stead, editor of the London *Review of Reviews*; and mathematician Giuseppe Peano of Turin, the inventor of Latino sine flexione. Since Bouchard, Harvey, and Stead could not attend, they were regularly represented by Paul Rodet (an Esperantist physician from Paris), Paul D. Hugon (lexicographer), and Father Ernest Dimnet (a teacher of modern languages in Paris). At those meetings Boirac could not attend, Belgian pacifist Gaston Moch took his place. Couturat and Leau, as secretaries, raised the total to eighteen.[15] This was an impressive array of scholars from diverse national, linguistic, and disciplinary backgrounds, and hopes were high that by the time they sat for their meeting in mid-October 1907, they would resolve the issue.

Ostwald found it hard going. Until October, most of the work was conducted by correspondence, but in the marathon sessions of October the members interrogated a long series of inventors of constructed languages or their appointed defenders. (Zamenhof could not travel to Paris, so he appointed Louis de Beaufront, the conservative French Esperantist, to represent his language.) As perhaps should not be surprising, the language barrier was a problem, since the native languages of the delegates were Danish, English, French, German, Italian, and Polish. Most of the discussions took place in French, because only Couturat among the French members had a decent command of German. This caused difficulties for Ostwald, since he his understanding of the language was drawn "partly from school memories that had not taken very well, partly from the reading of scientific articles in this language, that as a result of much practice had in fact stuck fluently enough."*[16] Baudoin de Courtenay insisted on speaking German, which was otherwise

*"teils auf den Schulerinnerungen, die nicht sehr eingehend waren, teils auf dem Lesen wissenschaftlicher Abhandlungen in dieser Sprache, das sich zufolge vieler Übung allerdings geläufig genug vollzog."

heard only when interviewing the inventor of the language Parla, Karl Ludwig Spitzer. Peano, characteristically, sometimes spoke in Latino sine flexione.[17] (Apparently no one complained.) Otto Jespersen, whose recollections—translated from the Danish by David Stoner—provide some of our best information about the deliberations, noted that despite the challenges Ostwald "presided over the negotiations with superb skill."[18]

Despite entertaining many language projects, Jespersen recalled that the real decision came down to Esperanto or Idiom Neutral. Esperanto seems an obvious candidate, given the vigor of its movement and its arguable success in serving as precisely the kind of auxiliary the Delegation was supposed to endorse. But Idiom Neutral was also a serious contender in early 1907. Jespersen himself "was rather inclined to vote for Idiom Neutral with a number of amendments."[19] The language was the unlikely descendant of Volapük, issued in 1902 by the International Academy of the Universal Language (Akademi Internasional dc Lingu Universal), the phoenix that rose from the ashes of the defunct Volapük Academy. Idiom Neutral had a more "naturalistic" feel than Esperanto—that is, it tended to resemble ethnic languages, particularly Romance ones, in appearance and use—but that also meant it displayed apparent irregularities.[20] For Jespersen, Idiom Neutral was "the first to carry out this principle [of maximum internationality] scientifically for the whole language."[21] But Neutral also lacked broad support: in 1907, there were only four groups of Neutralists, located in St. Petersburg, Nuremberg, Brussels, and San Antonio, Texas.[22] The rest of the Delegation was inclined to Esperanto.

But they were not that happy about it. (Except for Boirac, who was a militant stalwart for the language, but missed the final deliberations and was thus represented by Moch.) As Couturat wrote to Ostwald on 20 November 1906, almost a year before the final vote, he expected that the decision of the committee would be "to adopt Esperanto in principle, on the condition that important corrections and improvements for science and for practical utility be introduced."* For his part, he "would gladly accept present-day Esperanto if only certain peculiarities in syntax and word-formation were cleaned up and certain poorly

*"das Esperanto im Prinzip anzunehmen, mit der Auflage, die für die Wissenschaft und für die Praxis nötigen Korrekturen und Verbesserungen einzufügen."

chosen word roots were changed."*[23] The objections that circled around Esperanto were by now familiar: the unusual letters, the accusative, and Couturat's worries about word-formation (about which more in a moment).

Ostwald was beginning to chafe at Esperanto for other reasons. In May 1907, the chemist traveled to Dresden for the second congress of German Esperantists, accompanied by his second daughter, who had also learned Esperanto. The event left him "a little shocked, by what kind of shipmates I had undertaken the voyage with."† It was one thing that the participants doggedly defended the ridiculous alphabet. The real clincher was how they treated the *Fundamento*:

> And an older woman, who played a leading role as a very early adherent to the cause, led me into an adjoining room to make visible to me the Esperantists' attitude on this question. There one found against the wall a table with a festive green velvet tablecloth (green is the heraldic color of Esperanto); in the middle there lay a magnificent copy of the 'Fundamento' bound in green leather with a handwritten dedication by the Master [Zamenhof] and two silver lamps with burning candles stood on both sides. The whole thing was an altar dedicated to the cult of the Fundamento's untouchability.
>
> This religious veneration, combined with the blind fanaticism so often attached to religious movements, is very widespread among the adherents of Esperanto.‡[24]

*"ich würde gern das aktuelle Esperanto akzeptieren, wenn es nur von einigen Eigentümlichkeiten in der Syntax und der Wortbildung gereinigt und einige schlecht gewählte Wortwurzeln verändert würden."

†"ein wenig erschrocken, mit welchen Schiffsgenossen ich die Reise unternommen hatte."

‡"Und ein ältere Dame, welche als sehr frühzeitige Anhängerin der Sache eine führende Rolle spielte, führte mich in ein angrenzendes Zimmer, um mir die Einstellung der Esperantisten zu dieser Frage anschaulich zu machen. Dort befand sich an der Wand ein Tisch mit einer feierlichen grünen Sammetdecke (Grün ist die Wappenfarbe des Esperanto); in der Mitte lag darauf ein in grünes Leder gebundenes Prachtexemplar des 'Fundamento' mit eigenhändiger Widmung des Meisters und an beiden Seiten standen zwei silberne Leuchter mit brennden Kerzen. Das ganze war ein Altar, geweiht dem Kultus der Unberührbarkeit des Fundamento.

Diese religiöse Verehrung, verknüpft mit dem blinden Fanatismus, der den religiösen Bewegungen so oft anhaftet, ist unter den Anhängern des Esperanto sehr verbreitet."

By June, Ostwald felt he could no longer defend unaltered Esperanto; it had become "urgently necessary to liberate Esperanto from certain of its greatest imperfections."* [25]

It was in this frame of mind that Ostwald arrived in Paris in October 1907 to attend the meetings of the Delegation committee at the Collège de France, an ordeal that lasted for a total of eighteen lengthy and intellectually taxing sessions. One day late in the month, after a series of especially deadlocked debates, the committee arrived to the meeting room and found, laid out before their places, typewritten copies of a new constructed-language project written by an anonymous inventor who called himself "Ido." Couturat told the committee that he had promised not to reveal the author's identity, but that he guaranteed that it was neither him nor anyone on the committee.[26] What they found in the report was encouraging, even exciting: Ido proposed a systematic language that strongly resembled Esperanto in basic principles, but that took seriously many of the critiques that had been advanced since the 1890s, especially those in Couturat and Leau's 1903 volume on constructed languages. Gone were the circumflexed letters, the compulsory accusative, and several other sins that had given members pause. Jespersen was particularly pleased, thinking that with Ido's "middle course" between Idiom Neutral and Esperanto "we had come close to a solution that might satisfy everyone, even the Esperantists."[27]

Thus, on 24 October 1907, the Delegation voted unanimously to accept Esperanto as the universal auxiliary language "on the condition that certain modifications be made by the permanent Commission in the sense defined by the conclusions of the secretaries' report and the project of Ido, in seeking an agreement with the Esperantist linguistic committee."† [28] The decision was thus *for Esperanto*, but it was not unconditional. The delegates who voted for this proposal left the meeting convinced that Ido had simply proposed a correction within Esperanto. As the Esperanto community began to react to the decision over the next six months, what had been presented as an Esperanto reform began to be seen by both orthodox Esperantists and reformists

*"dringend nötig, Esperanto von einigen seiner grössten Unvollkommenheiten zu befreien."

†"sous la réserve de certaines modifications à exécuter par la Commission permanente dans le sens défini par les conclusions du rapport des secrétaires et par le projet de Ido, en cherchant à s'entendre avec le Comité linguistique espérantiste."

as a separate language. In 1909, Swiss pastor Friedrich Schneeberger, an active advocate, toyed with the names Linguido (meaning "descendant language") and Interlinguo before settling on Ilo—which never quite stuck.[29] Couturat that same year rejected the common moniker "Reform-Esperanto" in favor of the official name: "Linguo internaciona *di la Delegitaro*."[30] Neither had their way, for as an apocalyptic schism fractured the Esperanto world, the name "Ido" morphed from an authorial pseudonym into what even its supporters considered an "absurd" and "stupid name" for a new language.[31]

The Oedipal Language

What was Ido? How did it work? Its advocates were quite sincere when they thought of it (in autumn 1907) as simply a dialect of Esperanto—as Otto Jespersen would have it, a "purified Esperanto"[32]—for the two bore very strong similarities, although as their partisans split socially, they in turn began to separate linguistically. Indeed, some aspects of the language resemble the 1894 reform effort within Esperanto which had been rejected by the readership of *La Esperantisto*.[33] Ido was and remains mutually intelligible to Esperantists, who can learn to speak it fluently in a few hours.[34]

The basic framework does nothing to hide its debt to Esperanto: given a root, all nouns derived from it end in *o*, adjectives in *a*, adverbs in *e*; and the present, past, and future tenses of verbs are marked by the familiar *as*, *is*, and *os*. Visually, however, the language strikes the reader as rather more Latinate. Consider this sentence from the epigraph taken from Jespersen:

> Ma merkez bone, ke omna ta bona qualesi, quin on prizas e laudas en naturala lingui, trovesas nur kande indijeni parolas e skribas oli, ma ne en la boki ed en la plumi di stranjeri.

There are a number of differences beyond slightly variant roots. Plurals are indicated by *i*, not *j* (*stranjeri*). Commands, as in the second word, are indicated by the *ez* suffix, resembling French and differentiating the imperative from the conditional in a way Esperanto does not. Adjectives no longer had to agree in number (see *bona* above, modifying the plural *qualesi*), and the accusative was only to be used in cases where the object preceded the grammatical subject, which was rare, emphatic style. But perhaps most obvious is that the circumflexed letters are gone; the sound that Esperanto expressed with *ĵ* is here written *j*, and the Esperanto *j* is replaced by *y*. Idists argued this made the language more recognizable to the European reader, as well as requiring no special typography. Another significant change was the abandonment of Zamenhof's *a priori* table of correlatives to represent relative pronouns and interrogative words; *ke* is retained for "that," but *quin* and *kande* are added, clearly derived from Latin roots. This is more naturalistic, but also more irregular—one cannot get something for nothing.

If a few morphological and lexical substitutions were all that was involved in Ido, why did the anonymous author bother to construct it? Or, more to the point, what was so attractive about this system to the members of the Delegation committee that they opted for this reform project over the "primitive Esperanto" that thousands were already using? The answer lies in the philosophy built into the grammar and lexicon. Three points stood out for those-who-would-come-to-be-known-as-Idists as evidence of Ido's superiority: univocality, internationality, and reversibility.

As Couturat described the first principle: "The logical rule of the international language is the *principle of univocality*, formulated by Mr. Ostwald: each notion or element of a notion should be expressed once, and only once, and always by the same 'morpheme' (word element); in other words, there should be a univocal correspondence between the elements of ideas and the word elements."* [35] This was often cited as Ostwald's major theoretical contribution to constructed languages, and the chemist invoked it frequently. For example, in a 1911 German article on the application of Ido to chemistry (to which we shall return), he declared that "[i]n Ido, where no bad habits exist yet, one can from the beginning deploy the *principle of univocality*, so that every concept corresponds only to a single word, and each word relates only to a single concept."† [36] The notion is appealing enough: it is con-

* "La règle logique de la langue internationale est le *principe d'univocité*, formulé par M. Ostwald: chaque notion ou élément de notion doit être exprimé une fois, et une seule, et toujours par le même 'morphème' (élément de mot); en d'autres termes, il doit y avoir une correspondance univoque entre les éléments d'idées et les éléments de mots."

† "Im Ido, wo noch keine schlechten Gewohnheiten bestehen, kann man von vornherein das *Prinzip der Eindeutigkeit* durchführen, so dass jedem Begriff nur ein einziges Wort zugeordnet wird, und dass jedes Wort nur einen einzigen Begriff bezeichnet."

fusing in English that the single noun "cast" can refer to a medical treatment for a broken arm, a roll of dice, or a group of actors in a play.

How should one build this univocal vocabulary? Ido proclaimed a systematic method based on the principle of internationality, by which, as Jespersen explained, Ido "is nothing but a systematic turning to account of everything that is already international, that root being chosen in each case which will be most readily understood by the greatest number of civilized people."[37] ("Civilized" was a key word in these discussions. Chinese, Hindustani, and Arabic were not part of the root stock from which these "international" roots were drawn. This exclusion was so self-evident to participants as to go mostly unmentioned.[38]) Jespersen elevated this principle into an easily stated maxim: "*That international language is best which offers the greatest facility to the greatest number*.[. . .] The choice of the words for our neutral language is, therefore, a pure question of arithmetic."[39] But one must not simply add up the number of speakers of a language (in Western and Central Europe, German would be the winner every time), but instead count the *roots* and *divergent forms* of the word for the number of "civilized" speakers, which meant that Spanish, Italian, French, and the French elements in English began to weigh very heavily, and endowed Ido with a more sharply Romance, or even simply French, character than Esperanto—a fact that some Idists felt needed defending but which Couturat (a Frenchman) saw as entirely unproblematic.[40] For him, despite its resemblance to French, it was "nothing other than a purified and idealized extract, a quintessence of the European languages."*[41] Zamenhof selected Esperanto's root vocabulary in a rough approximation to this idea but sometimes arbitrarily. (Why is "bird" in Esperanto *birdo*? In Ido it is *ucelo*, derived from French and Italian.) Esperantists would have to learn a reasonable quantity of new words.

The final principle, that of reversibility, is most clearly attributable to Couturat's influence, dating from his 1903 critique of the arbitrary way Esperanto derived words. *Kroni*, for example, means "to crown," but what does the simple derived noun *krono* mean? Is it "crown," or is it "coronation"? How would one derive "corona"? And could you go backward from a given noun to derive the root verb? This offended Couturat's logician sensibilities, and he championed Ido's alternative: "Every derivation should be *reversible*, that is to say, if one moves from

*"n'est pas autre chose qu'un extrait purifié et idéalisé, une quintessence des langues européennes."

one word to another (in the same family) by means of a certain rule, one should inversely move from the second to the first by means of a rule the exact inverse of the former."* [42] The impact of this principle on the suffix system of Esperanto offered perhaps the widest divergence with its progenitor. Recall that the suffix *id* means "descendant, derived from," and so *Ido* was indeed "that which was derived" from Esperanto—the question of how far it had evolved (or devolved) from the parent was the fundamental issue in the schism.

That a schism within the ranks of Esperanto was even possible was unthinkable in the middle of 1907, and a contemporary manifesto preened that "Esperanto itself is admirably organized, and there are no factions or symptoms of dissension."[43] Almost immediately upon the publication of the Delegation's decision on 25 October this turned out to be manifestly untrue. Ostwald, as chair of the group and widely acknowledged as nonpartisan, assumed the task of negotiating with Zamenhof himself, and he wrote to the "Majstro" on 2 November, trying to win him over to reforms by arguing that Volapük had perished because of failure to reform itself.[44] Zamenhof's response, two days later, was blistering: how could someone as intelligent as Ostwald not understand "that Volapük failed precisely through *reforms*."† [45] The memory of Volapük's rapid disintegration haunted the schism, and in almost every article in the sustained polemic one finds references to Schleyer's doomed experiment.

In 1903, years before he became "the infallible pope of a small schismatic church"[46]—that is, Ido—Louis Couturat offered this balanced judgment of Volapük's rise and fall: "Thus *Volapük* succeded because it seemed to respond to a very sharply felt need, above all in the commercial world; and it failed due to its intrinsic vices, the inflexible dogmatism of its inventor, and the disunion of its adherents."‡ [47] Here were two explanations: internal flaws and social disunion. Idists tended to finger the first, Esperantists the second. If Volapük died because of its failure to reform, then the Esperantists should cling to the olive branch

* "Toute dérivation doit être *réversible*, c'est-à-dire, si l'on passe d'un mot à un autre (d'une même famille), en vertu d'une certaine règle, on doit passer inversement du second au premier en vertu d'une règle exactement inverse de la précédente."

† "dass Volapük gerade durch die *Reformen* zu Grunde gegangen ist."

‡ "Ainsi le *Volapük* a réussi, parce qu'il paraissait répondre à un besoin très vivement ressenti, surtout dans le monde commercial; et il a échoué à cause de ses vices intrinsèques, du dogmatisme inflexible de son inventeur, et de la désunion de ses adhérents."

the Delegation handed to them—and this was an opinion that Ostwald himself espoused.[48] But there was strong evidence that Volapük was functioning fine, as at the Paris conference of 1889, until reformers such as Auguste Kerckhoffs proposed messing with it, leading to the catastrophe.

"[S]ooner or later you will, however, unfortunately be convinced," Zamenhof responded to Ostwald on 4 November, "that your work has achieved nothing positive, but instead unforseeably much that is negative."* Zamenhof blamed Couturat, who "has presented the voices of all 'unsatisfied' Esperantists to you, and you naturally do not hear the voices of all the others.—You thereby naturally have a certain 'optical illusion' and find yourselves under the impression as if all Esperantists were reform-minded."† [49] Ostwald replied on 12 November by questioning Zamenhof's assumptions: "The most essential point of our difference of opinion is that you consider the present Esperantists as a people, a complete organism with its own will. I on the contrary[. . .] am certain that in the present state of the matter *everything depends on individual leaders*."‡ [50] Ostwald's implied analogy was to the structure of the scientific community, which Ostwald saw as decisively guided by the wisdom of leading members (such as himself). He continued his conciliatory approach with an open appeal, dated 21 December and published in *Internacia Scienca Revuo*: "But I cannot accept the supposition that there exists an opposition between the body of Esperantists and the Delegation. By the fundamental decision to choose Esperanto, we were made Esperantists, and there exist many differences of opinion *within the body of Esperantists*, among which we represent a progressive part."§ [51]

* "früher oder später werden Sie sich aber leider überzeugen, dass Ihre Arbeit nichts Positives, dafür aber unabsehbar viel Negatives geschaffen hat."

† "Er hat Ihnen die Stimmen aller 'unzufriedenen' Esperantisten vorgestellt, und die Stimmen aller anderen hören Sie natürlich nicht.—Sie haben daher natürlich eine gewisse 'optische Täuschung' und befinden sich unter dem Eindruck als wären alle Esperantisten reformistisch gesinnt."

‡ "Der wesentlichste Punkt unserer Meinungsverschiedenheit ist, dass Sie die gegenwärtigen Esperantisten als ein Volk, einen geschlossenen Organismus mit eigenem Willen betrachten. Ich dagegen,[. . .] bin sicher, dass im gegenwärtigen Stadium der Sache *alles von einzelnen Führern abhängt*."

§ "Sed mi ne povas akcepti la supozon, ke ekzistas kontraŭeco inter Esperantistaro kaj Delegitaro. Per la fundamenta decido elekti Esperanton, ni fariĝis Esperantistoj kaj tiel ekzistas pleje diferencoj de opinio *interne de la Esperantistaro*, kies progreseman partion ni reprezentas."

At first, the editors of the *Revuo* considered compromise with the Delegation, publishing a declaration in the December issue signed by thirty-three academics (including the editor René de Saussure and Th. Renard of the chemical nomenclature discussed in the last chapter), which defended Zamenhof's language but ended on a conciliatory note.[52] But by January 1908, the same set of scholars had come to think of Esperanto not as just an international auxiliary, but as something close to an ethnic language. "Firstly, Esperanto belongs to the Esperantists," they announced, "in the same way as the English language belongs to the Englishmen. Consequently nobody will be able to impose reforms upon us against our will."* [53] On 18 January Zamenhof wrote to Ostwald that there was no room for compromise—the *Fundamento* would remain in force. Eperantists mounted their own critiques of Ido's grammar and lexicon.[54] In 1911, the president of the Esperantist Academy Maurice Rollet de l'Isle issued a bon mot reminiscent of his countryman Voltaire: "if Ido did not exist, it would have been necessary to invent it in order to show that Esperanto is preferable to it."† [55] Esperantist groups expelled members interested in the new language, cementing the social rift and constructing Ido as a competing language.[56] Later, it became the policy of the global Esperanto community to avoid mention of Ido.[57]

The anathematization of Ido had a great deal to do with the circumstances of its birth, which Esperantists saw as rooted in betrayal. Just who was the inventor "Ido"? Ostwald wrote Couturat on 14 November, within three weeks of the decision, demanding to know Ido's identity.[58] Even earlier, Zamenhof—noting how Louis de Beaufront, his chosen representative and well-known arch-conservative on reformist questions, had quickly endorsed the Delegation's decision—wrote to leading French Esperantist Hippolyte Sébert on 27 October in equal befuddlement (as translated by Marjorie Boulton, Zamenhof's biographer):

> I know nothing about the person of "Ido" and have never seen his grammar. I have not received any kind of letter from Couturat for three weeks. The behaviour of M. de Beaufront seems to me very sus-

* "Unue, Esperanto apartenas al la Esperantistoj tute same kiel la lingvo angla apartenas al la Angloj. Konsekvence neniu povos trudi al ni reformojn kontraŭ nia volo."

† "si l'Ido n'existait pas, il eût fallu l'inventer pour montrer que l'espéranto lui est préférable."

> picious; to show my trust in him, I chose him as my representative before the Delegation, and he, not asking me at all, suddenly and too startlingly went over to the reformers and wrote a letter to me, saying that Esperanto must certainly die, that, after five years, only the memory of Esperanto will remain, and so on.[59]

Zamenhof's suspicions turned out to be on the mark, as Otto Jespersen soon discovered. One day, he received a letter from Couturat addressed to "*mon cher ami*," and he read on, surprised by the uncharacteristic intimacy of the salutation. From the context, he realized that the letter was actually intended for de Beaufront, and it revealed the latter to be the author of Ido! Jespersen was shocked: "At one stroke it changed my view both of him and of Couturat and dismayed me to such a degree that in the first sleepless nights I seriously considered completely severing my connection with the idea of an auxiliary language."[60] He informed Ostwald, who shared his outrage but was somewhat calmer. After Couturat confirmed it, both urged him to persuade de Beaufront to reveal himself.

In May 1908, an article penned by "Ido" appeared in de Beaufront's bilingual journal *L'Espérantiste*, the flagship of the French movement. Ido claimed that he had submitted his proposal to the committee because he was afraid that Esperanto would face "rejection pure and simple."* De Beaufront appended a commentary of equal length which began with the confession: "The declaration that you have just read is mine."† [61] The Esperantist community was apoplectic that Zamenhof's hand-picked representative to *defend* Esperanto before the Delegation would turn out to be Brutus, even Judas Iscariot.[62] (For the record, Jespersen noted that de Beaufront had "really defended Zamenhof's language with great eloquence and skill."[63]) Until the 1930s, most of the Esperanto intelligentsia believed de Beaufront's self-unmasking, claiming that Ido strongly resembled Adjuvanto, the language that the Frenchman had abandoned in the 1890s in favor of Esperanto.[64] (It is unclear how they came to this assessment, as no one seems to have seen a copy of Adjuvanto, which de Beaufront claimed he had destroyed.) He was unseated from his presidency of the French Esperanto society

*"le rejet pur et simple/la puran kaj simplan forĵeton."

†"La déclaration qu'on vient de lire est de moi./ La deklaro, kiun oni ĵus legis, estas mia."

on 8 September 1908. To be an Idist meant by definition not to be an Esperantist.

Today, however, Esperantist opinion is almost universal that de Beaufront had lied even in his confession. In the 1930s, Ric Berger, a devotee of Occidental (a constructed auxiliary published in 1922 by Edgar de Wahl), argued that in fact Couturat himself—the secretary of the Delegation and therefore proscribed from submitting his own plan for consideration—was the author of Ido, and de Beaufront was his cover.[65] Later Esperantist histories consider de Beaufront's assumption of authorship "almost certainly bogus," and assume that Couturat was the author, based on Berger's relatively weak evidence.[66] Ostwald continued to believe that de Beaufront was Ido, as did most Idists, in part because de Beaufront's assumption of authorship was a signal to Esperantists that conversion to reform would be sensible.[67] Perhaps the strangest aspect of this disputed authorship is why de Beaufront might have agreed to protect Couturat. In any event, it is clear that no matter who invented the language, Couturat adopted it as his own; he left his imprint on almost every page of the chief Idist journal *Progreso*, while penning a host of articles viciously attacking Esperanto and Esperantists.[68]

It would take us too far afield from the question of a scientific auxiliary to explore every charge and countercharge of the assembled vitriol. To give but a single example of Couturat's approach, he routinely denounced Zamenhof's resistance to Ido as a commercial ploy, stating that the Majstro was "a *person bound* by contract to a publishing firm, which has acted and now acts to monopolize Esperanto"* through maintaining the circumflexes.[69] With language closely bordering on anti-Semitic, Couturat again and again insinuated that Esperanto was a money-making proposition, and that the movement was "more and more dominated by men of action, that is to say by men of business and of intrigue, politicians and shopkeepers."† [70] Ido's origins in the scholarly Delegation, on the other hand, "excludes any hypothesis and any mercantile intention."‡ [71] (Zamenhof denied these allegations in a letter

*"esas *persone ligita* per kontrato a la librista firmo, qua penis i penas monopoligar Esperanto."

†"de plus en plus dominé par les hommes d'action, c'est-à-dire par les hommes d'affaires et d'intrigue, les politiciens et les boutiquiers."

‡"exclut toute hypothèse et toute intention mercantile."

to Ostwald; his days were spent treating eye diseases in the poor neighborhoods of Warsaw, and his involvement in the Esperanto movement took up his very limited spare time.[72])

As it became increasingly clear that the Esperantists would not reform their language along the lines of Ido, Couturat erected his own infrastructure for his language. In February 1909 the Uniono di l'Amiki di la Lingva Internaciona was established in Zurich to propagandize for Ido, with Ostwald elected as honorary president on 24 May. Ido soon began to prosper. It has been estimated that roughly 20% of the leading figures in the Esperanto movement—journalists, intellectuals, public figures—adopted the new language, but only 3%–4% of the rank-and-file.[73] The new language spread across the ocean as well. The *Internationalist* was quickly released by the Interlinguo publishing company as an American Ido quarterly based in Seattle and edited by A. H. Mackinnon. By February 1910 it transformed into a monthly issued from Philadelphia, with its production values noticeably improved. In August 1910, it changed identity yet again, being absorbed into *The International Language*, edited by Gerald Moore, Esq., out of London. By 1912, the Idists boasted they had 150 societies worldwide (although, since the Uniono had only 600 subscribers at that moment, the claim requires a grain of salt).[74] Ido had ceased to be a reform movement within Esperanto, and became instead a competitor from without.

Ido for Science!

The Delegation at first couched their approach to the Esperantists' Lingva Komitato as a reform program along linguistic lines, and they stressed grammatical critiques: the difficult-to-parse mutations of the roots, the *a priori* table of correlatives, the orthography, and so on. But behind these points was a more general objection to the *social* structure of Esperanto, and in particular the *Fundamento* as the "unchangeable" core of the language. I discussed in the previous chapter how the *Fundamento* was established as the bedrock of the language at the 1905 congress in Boulogne-sur-Mer in order to keep the language coherent against the natural mutating forces of ordinary use. Yet this constraint chafed at some Esperantists, especially those who were linguistically curious and drawn to the Delegation's project.

Idists regarded the *Fundamento*, which many Esperantists had come to see as the core of the movement, as objectionable for three reasons, all linked to their vision of Ido as a scientific auxiliary. The first argument was empirical: no ethnic language had a single text to ground their entire grammar and vocabulary, and so Ido, scientifically derived from such languages both grammatically and sociologically, would have no need of one.[75] The next argument was visible already in Ostwald's revulsion to the altar at the Dresden Esperantist's home: it smacked too much of religious fanaticism, not the sobriety of modernity. Ido "does not have a holy book; it does not have any other *Fundamento*, as Mr. Jespersen excellently said, than its scientific principles, on the one hand, and on the other the collection of European languages from which it draws its material, and which constitute for it the largest and most stable objective base."*[76] Finally, Darwinian theory indicated that the surest path to stability was through evolutionary pressure, as Richard Lorenz explained: "There is, therefore, only one adequate criterion of the stability of an international language, namely, that of suitability or adaptation to its purpose, and we maintain that it is only by means of continuous reforms and improvements that it will succeed in satisfying this criterion and so finally attain to stability."[77]

The Idists not only argued that Esperanto's devotion to the *Fundamento* was antiscientific, they countered that Ido was a uniquely scientific language, and that argument too was rooted in three planks: method, technology, and scientific use. Couturat, Jespersen, and other leading Idists devoted many pages to claiming that the intrinsic logic of Ido's foundational principles guaranteed that it "possesses the advantage over other languages that it is based on rational scientific principles and, therefore, [one] need not fear that some fine day it will be replaced by another and sensibly different language."[78] Or, as Couturat would have it: "In a word, the work of the Committee is the substitution of the scientific, critical, and progressive method for the empirical method of invention, more or less genial but always arbitrary."†[79]

Such claims remained relatively abstract. It was when thinking of language as a *technology*—that is, a tool to accomplish certain ends—that the Idists more concretely linked their linguistic project to the sci-

*"n'a pas de livre saint; il n'a pas d'autre *Fundamento*, comme l'a dit excellemment M. Jespersen, que ses principes scientifiques, d'une part, et d'autre part l'ensemble des langues européennes auxquelles il emprunte ses matériaux, et qui constituent pour lui la base objective la plus large et la plus stable."

†"En un mot, l'œuvre du Comité est la substitution de la méthode scientifique, critique et progressive, à la méthode d'invention empirique, plus ou moins géniale, mais toujours arbitraire."

entific developments of the day, such as the metric system.[80] The argument went further than standardization. Much as Ostwald used the telegraph and the railroad to argue, on energetic principles, that a constructed auxiliary would be best suited to the modern world, so Idists invoked the constant improvements in technology to demonstrate, by analogy, that Ido was the latest model of an artificial language. "Ido is to Esperanto," wrote Couturat, "as today's bicycle is to an old bicycle."* [81] Ido had evolved from Esperanto, becoming better adapted to its environment. "No great invention, no great scientific discovery, ever sprang into the world full-fledged," wrote Otto Jespersen in 1909. "[B]y setting to work on scientific principles it is possible to devise a much better language of a much more truly international character, 'not perfect,' perhaps, 'but always perfectible.'"[82] Indeed, despite Couturat's evident interest in Lamarckian inheritance of acquired characteristics, American Idists pushed for a strictly Darwinian understanding: "Volapuk [sic], Idiom Neutral, Esperanto and Ido are but progressive steps toward the solution of the problem. As with everything else so with international language schemes, only the fittest will survive!"[83]

A language would prove that it was fittest to survive in the competitive world of science when scientists actually used it. Couturat crowed in 1910 that "Science has spoken: *the international language can be none other than Ido*, because that is the only scientific language, in a double sense: because it is the work of the science of linguistics and it is the only one founded on fixed and precise principles; and because it is the only one which is appropriate for scientific use, and which has adapted to the international terminology of the sciences."† [84] To a certain extent, this was true. One could find some American articles written in Ido on refrigeration technology and chiropractics, and Couturat labored diligently over a 1910 mathematical lexicon that translated English, French, German, and Italian terminology into Ido.[85] But the strongest argument for the scientific utility of Ido came from the central science of the day, chemistry, and the man who would make chemistry Idist was none other than Wilhelm Ostwald.

* "l'Ido est à l'Esperanto ce que la bicyclette est au vieux bicycle."

† "La science a prononcé: *la langue internationale ne peut être que l'ido*, parce que c'est la seule langue scientifique, et cela en double sens: parce qu'elle est l'œuvre de la science linguistique et est la seule fondée sur des principes fixes et précis; et parce qu'elle est la seule qui soit appropriée aux usages scientifiques, et qui s'adapte à la terminologie déjà internationale des sciences."

On 20 April 1910, after the schism was complete, Ostwald penned yet another letter to van't Hoff, noting that his "interests had moved ever further away from chemistry. Internationalism, pacifism, and cultural energetics are my problems now."* [86] As determined as he was to support constructed languages, and in particular Ido, Ostwald had spent much of 1908 wavering in his commitment, disillusioned by the behavior of Zamenhof and the Esperantists in resisting reform, but also that of Couturat in promoting it. He wrote in frustration to Jespersen on 25 February 1908: "The individual who is driving me to a resignation is Couturat. He has the makeup of vulcanized rubber: he absorbs every hit, but as soon as these leave off, he assumes his earlier shape."† [87] Couturat wore down opponents through constant correspondence, and Ostwald claimed he could no longer stomach it. Jespersen talked him down.

And just in time, too. On 11 December 1909, Ostwald was awarded the Nobel Prize in Chemistry, and he promptly donated at least $40,000 of the prize money to the Uniono and other Ido projects. (This was in 1909 dollars; in 2014, this comes to roughly $1 million.) Ostwald insisted he had "not one penny for Esperanto and its adherents. My cooperation will be given exclusively to Ido[. . .]."[88] If he had lost patience with some of the organizational work for Ido, he came to appreciate the language itself. He translated some of his energeticist writings into it and told a reviewer that he "found it of great benefit in giving clarity and definiteness to his thought."[89]

Looking for other translation projects of similar worth, he naturally came upon chemical nomenclature.[90] Ostwald was in a unique position to do something about the fin-de-siècle Babel. He was one of the world's most famous chemists, he had leisure on his country estate, and he was the founder and one of the editors of the *Zeitschrift für physikalische Chemie*. In early 1911, he wrote again to his coeditor van't Hoff demanding that the *Zeitschrift* publish abstracts in Ido: "Thereby many who cannot read German would be able to become acquainted with the content of the journal, since Ido is very easily understandable to every

*"Interessen immer weiter von der Chemie fortwandern. Internationalismus, Pacifismus und kulturelle Energetik sind jetzt meine Probleme."

†"Die Persönlichkeit, welche mich zum Ausscheiden zwingt, ist Couturat. Er hat die Beschaffenheit von vulkanisiertem Kautschuk: er weicht jedem Druck, aber sobald dieser nachlässt, nimmt er seine frühere Form an."

Frenchman, Englishman, Italian, Spaniard, etc."*[91] Ostwald would subsidize the annual cost of 100 Marks for the first year.

Van't Hoff had grimly tolerated Ostwald's enthusiasm up to now, but this was the last straw. He responded three days later that he could "not go along" with the abstract idea, and that "upon this change I would no longer wish to support the journal with my name."†[92] Ostwald was rather taken aback, especially since van't Hoff seemed to be allying with the publishing house of Wilhelm Engelmann, which also resisted any introduction of Ido into the journal, a position that Ostwald considered an intrusion of commerce into editorial decisions about scientific content.[93] Ostwald was particularly outraged about Engelmann's resistance to publishing his own German-language article about Ido chemical nomenclature, and van't Hoff provided no support at all. This initiated a huge breach between the friends, and on 21 January 1911 Ostwald penned an angry missive berating van't Hoff. While the letter was en route, van't Hoff fell deathly ill and withdrew from all correspondence. (He died of tuberculosis just over a month later.) Full of remorse, Ostwald wrote to Johanna van't Hoff to apologize (after a fashion) three days after her husband's demise: "I have set aside for the time being the question of the Ido abstracts, concerning which your husband had placed himself in determined opposition to my wishes and plans."‡[94] Mercy seemed to be in the air: the next day, Engelmann agreed to publish Ostwald's article, but only in a supplemental volume, and on the condition that no polemics about constructed languages be allowed to enter the journal in the future.[95]

Ostwald had been laboring over a nomenclature for inorganic chemistry in Ido by scouring chemistry textbooks and translating index items into Ido.[96] The biggest challenge, however, was a complete lexicon of the chemical elements consistent with the three principles of Ido (univocality, internationality, and reversibility) of which internationality was the central factor. In May, July, and December 1910, Ostwald had serialized his chemical nomenclature in *Progreso*, but of course the main

*"Dadurch würden Viele, die nicht deutsch lesen können, doch den Inhalt der Ztschr. kennen lernen können, da Ido für jeden Franzosen, Engländer, Italiener, Spanier etc. sehr leicht verständlich ist."

†"nicht mitmachen"/"daß ich die Zeitschrift bei dieser Änderung nicht mehr mit meinem Namen stützen möchte."

‡"Ich habe die Frage der Ido-Referate, bezüglich deren Ihr Mann sich in bestimmten Gegensatz zu meinen Wünschen und Planen gestellt hatte, einstweilen zurückgesetzt."

audience he hoped to reach—international chemists—did not read Ido, and would not do so until Ostwald persuaded them in German to take it seriously. And so Ostwald's "Chemische Weltliteratur" article was published in the first supplemental issue of the *Zeitschrift* on 28 February 1911, adorned with the Engelmann-imposed footnote: "The author assumes sole responsibility for the content of this article."* [97]

Ostwald began where most such efforts did: by lamenting the growth of "smaller" (*kleinere*) languages alongside the three "big" (*grossen*) ones—already too many—but taking solace in the fact that chemistry was an international science that already had a large body of international nomenclature. Since Latin was no longer a plausible solution, one was compelled to move to a constructed language; hence, Ido:

> In the auxiliary language *Ido* (an improved Esperanto and organized for continuing future improvement) a means of communication presently offers itself whose utility has already been proven many times, and whose continual adaptation to its goal is already secured for the future through an international organization.
>
> This language comprises in general the forms deployed in the European languages, simplified throughout however in the sense of simplicity and univocality.† [98]

At last, here was a possibility to develop a uniform nomenclature building on the shared concepts and substances that chemistry enjoyed. Ostwald turned his attention to inorganic chemistry—essentially, the elements of the periodic table and their basic compounds—rather than the much more complex field of organic chemistry, where the conventions for naming isomers were themselves in flux at precisely this moment in each of the principal ethnic languages in which science was published. Ostwald attempted to construct a name with the maximum internationality for each substance. This task was complicated by occa-

*"Für den Inhalt dieses Aufsatzes übernimmt der Verfasser die alleinige Verantwortung."

†"In der Hilfssprache *Ido* (ein verbessertes und für dauernde künftige Verbesserung organisiertes Esperanto) liegt gegenwärtig ein Verkehrsmittel vor, dessen Anwendbarkeit bereits vielfach bewährt worden ist, und dessen dauernde Anpassung an seine Zwecke auch für die Zukunft durch eine internationale Organisation gesichert ist.

Diese Sprache schliesst sich im allgemeinen den gebräuchlichen Formen der europäischen Sprachen an, vereinfacht diese aber durchaus im Sinne der Einfachheit und Eindeutigkeit."

sional tensions between the international *symbol* and the international *word*. This was most marked for symbols that begin with *C*, as in *carbon*, because the *c* in Ido (as in Esperanto) was pronounced *ts*, and the Ido name for *carbon* would be *karbo*. (Ostwald believed a later generation would change the symbol to K.) Consider *silver*:

> *Ag* argento. The general dictionary has arjento for silver, which we could adopt without further ado if the letter *g* did not occur in the international symbol, which should where possible also be contained in the name. Thus it is suggested that the form argento be adopted for the chemical substance, while arjento remain for general use. The silver moon would be translated with "arjenta luna" [sic: luno—MG], while the sentence: "Silver is soluble in nitric acid," in Ido reads: "argento esas solvebla en nitratocido."*[99]

And so on through several problematic cases. Interestingly, he left without comment the peculiar case of iodine. Considering that Ostwald was the person identified with the principle of univocality—one word, one meaning—it is somewhat disconcerting to observe in Figure 5.1, his table of element names, the Ido word for iodine. It's "ido."

Ostwald remained very proud of the result, as he told the American journal *Science* in 1914: "I showed that a chemic nomenclature in a plastic, artificial language is better, more consistent and more comprehensible than in any natural language."[100] Esperantists responded with their own nomenclature in 1912, which Couturat accused them of shamelessly lifting from Ostwald. Esperantists were in a bind, however, because after the schism it was all the more important to adhere to the *Fundamento* to differentiate their language from Ido. Alexander Batek in late 1909 had attempted to publish yet another nomenclature program in the *Internacia Scienca Revuo*, contending that "the Fundamento is not the foundation (*fundamento*) for 'specialty' naming, and untouchability (*netuŝebleco*) applies only for words of universal

*"*Ag* argento. Das allgemeine Wörterburch hat für Silber arjento, das wir ohne weiteres annehmen könnten, wenn nicht im internationalen Zeichen der Buchstabe *g* vorkäme, den man womöglich auch im Namen erhalten sollte. So ist vorgeschlagen worden, die Form argento für die chemischen Stoffe anzunehmen, während arjento für den allgemeinen Gebrauch bleibt. Der silberne Mond würde mit arjenta luna [sic: luno—MG] übersetzt werden, während der Satz: Silber ist in Salpetersäure löslich, in Ido hiesse: argento esas solvebla en nitratocido."

Ac Aktino (*Ak*).
Ag Agento.
Al Alumino.
Ar Argono.
As Arseno.
Au Auro.
B Boro.
Ba Bario.
Be Berilo.
Bi Bismuto
Br Bromo.
C Karbo (*K*).
Ca Kalco (*Kc*).
Cd Kadmo (*Kd*).
Ce Cero.
Cl Kloro (*Kl*).
Cr Kromo (*Kr*).
Co Kobalto (*Ko*).
Cs Cesio.
Cu Kupro (*Ku*).
Dy Disprozo (*Ds*).
Er Erbo.
Eu Europo.
F Fluoro.
Fe Fero.
Ga Galio.
Gd Gadolinio.
Ge Germanio.
H Hido.
He Helo.
Hg Merkuro (*Mr*).
In Indo.
I Ido.
Io Ionio.
Ir Irido.
K Kalio (*Ka*).
Kr Kripto.
La Lantano.
Li Litio.
Lu Luteto.
Mg Magnezio.
Mn Mangano.
Mo Molibdo.
N Nitro.
Na Natro.
Nd Neodimo.
Ne Neono.
Ni Nikelo.
O Oxo.
Os Osmo.
P Fosfo (*Fo*).
Pb Plumbo.
Pl Palado.
Po Polono.
Pr Praseodimo.
Pt Platino.
Ra Radiumo.
Rb Rubido.
Rh Rodio.
Ru Ruteno.
S Sulfo.
Sa Samaro.
Sb Stibo.
Sc Skando (*Sk*).
Se Seleno.
Si Siliko.
Sn Stano.
Sr Stronco.
Ta Tantalo.
Tb Terbo.
Te Teluro.
Th Torio (*To*).
Ti Titano.
Tl Talio.
Tu Tulio.
U Urano.
V Vanado.
W Wolframo.
X Xenono.
Y Yitro.
Y Yiterbo.
Zn Zinko.
Zr Zirkono.

FIGURE 5.1. Wilhelm Ostwald's nomenclature for the chemical elements in Ido, as published in German in early 1911. Observe that the name for *Ag* (silver), discussed in the text, is mistyped in the figure. The parenthetical symbols were Ostwald's suggestions for better aligning the international system with Ido. Wilhelm Ostwald, "Chemische Weltliteratur," *Zeitschrift für physikalische Chemie* 76 (January 1911): 1–20, on 8.

meaning,"* but his remained a minority voice.[101] Esperantists took the challenge of Ido as a signal to better organize themselves for technical nomenclature, noticeably weak before 1907.[102]

Ostwald continued to organize for international science. In Munich on 12 June 1911, Ostwald established an organization called the "Bridge" (*Brücke*), intended to promote universal standardization among the

*"Sed la Fundamento ne estas fundamento por la nomigado 'faka,' kaj la netuŝebleco atendas nur sur la vortoj de universala signifo."

sciences. Besides making nomenclature uniform within chemistry, he advocated unifying the terminologies of each of the sciences to coordinate with each other. Even page sizes for scientific journals should be of uniform size. And, of course, a crucial plank in his program was "[t]he preparation of an international auxiliary language for publications of universal interest." The time was now, he insisted: "We need only choose one of the artificial systems already at hand. Because Ido is the only one in which a systematic chemic nomenclature has been worked out, we should turn our attention first to that scientifically perfected idiom."[103] In February 1912, Ostwald adopted Ido as the Bridge's official language.

The End of the Experiment

Things were looking good for Ido. The movement was still smaller than Esperanto, but Idists had attracted much of the leadership, and the support of such luminaries as Ostwald granted the language enormous visibility in a short period. But, like many blessings, this defection of the cream of Esperanto's crop had brought its own curse. Part of the reason those individuals had been so active in Esperanto, and why they were attracted to Ido, was their devotion to linguistic experimentation and their search for a *perfect* (not merely "good enough") international auxiliary. The *Fundamento* had constrained them, but now that they used *Fundamento*-free Ido, experimentation could run wild—even amok.

By July 1910, Jespersen began to panic. "I believe that it is now absolutely necessary," he wrote in *Progreso*, "to have certain rules which, without changing the principle of liberality which is our strength and concerning which we are justifiably proud, could slightly restrict it in order to ease the task of the Academy and to lead us as quickly as possible to sufficient stability."* He suggested that any proposed reform come accompanied by a seconding motion from a speaker of a different native language. Even Couturat, the arch-advocate of Ido, appended to the article his agreement that "*some* such rule is absolutely necessary, in order to avoid the 'flood' and the interminable discussion of the same

* "me opinionas ke esas nun absolute necesa havar certa reguli qui, sen chanjar la principo di liberaleso qua esas nia forto e pri qua ni esas juste fiera, povas poke restriktar ol e per to faciligar la tasko di la Akademio e duktar ni max balde posible a suficanta stabileso."

questions."*[104] At a meeting of Idists in Solothurn, Switzerland, from late August to early September 1911, ten years of "stability" were declared for Ido. For a decade, the language would stay fixed in the form it had acquired as of 1 July 1913, in order to give the improvements in the language since 1907 time to solidify, at which point experimentation would be opened again. (The period was then extended for 1924–1926, and again in 1934–1938.) This nakedly ad hoc solution to the problem of stability was precisely why the *Fundamento* had been enshrined at Boulogne-sur-Mer in 1905, and the Idists now had to confront the necessity of a similar move.[105]

But the real threat to Ido was not the innovative zeal of its partisans, or the perfidy of Esperantists—who took the schism seriously enough to found the Universala Esperanta Asocio (UEA) in 1908, an umbrella organization that exists to this day—but from the catastrophe looming over Europe. On 28 June 1914, Archduke Franz Ferdinand, heir to the Habsburg throne, was assassinated by a nineteen-year-old Serbian named Gavrilo Princip in the Balkan city of Sarajevo. The geopolitical machinations and miscalculations that followed set off over four years of generalized massacre that devastated the continent. Patriotism surged, and as Frenchmen, Britons, and Russians slaughtered Germans, Italians, Austrians, and Hungarians—and vice versa—the spirit of global comity that had powered the movement for a universal auxiliary dissipated like a deflated zeppelin. The tenth annual Esperanto congress, scheduled for Paris in 1914, was canceled. The eleventh congress in San Francisco, in the neutral United States, was but a shadow of former meetings. The number of Esperanto periodicals collapsed from a high of over a hundred to under thirty.[106]

Ido was hit worse. On 3 August 1914, while driving in the French countryside, a mobilization truck headed for the front plowed into an automobile out for a quiet drive, killing a passenger. His name was Louis Couturat. Given how much the Ido movement was driven by the indefatigable energy of the Paris logician, it never quite recovered. The first Ido congress, due to take place that September in Luxembourg, was delayed, and the movement revived after the war but never regained its earlier dynamism. Esperanto, in turn, attracted attention from pacifists, bypassing Ido's determined avoidance of those kinds of political issues

*"*ula* tala regulo esas absolute necesa, por evitar l' 'inundo' e la senfina diskutado di la sama questioni."

in favor of a focus on science; and new language projects like Occidental siphoned off enthusiasts. Even Otto Jespersen, member of the Delegation, would abandon Ido in favor of his own universal auxiliary, Novial, in 1928. By 1923 there were only roughly a hundred Idists in all of Soviet Russia (the revolution there had wrought havoc on internationalist movements). Only forty individuals, Soviets included, attended the 1929 Ido world congress.[107] For all their hopes, the war eviscerated the quest for a scientific auxiliary.

In 1914, Ostwald shut down the Bridge. There was a war on.

CHAPTER 6

The Linguistic Shadow of the Great War

> Kranke Völker—und wie wenige sind heute noch gesund!—haben notwendig kranke Sprachen. Ihre Heilung bedeutet zugleich die Heilung des Volkskörpers selbst: ihre internationale Ordnung ist nichts anderes als die Ordnung derer, die sie sprechen. Das Sprachenproblem der Nationen trägt also ein doppeltes Gesicht: ein nach innen gewandtes, das jedes Volk allein und auf seine Weise, ein nach außen gewandtes, das kein Volk für sich, sondern nur in Gemeinschaft mit allen anderen lösen kann.*
>
> FRANZ THIERFELDER[1]

Wilhelm Ostwald was strongly committed to international comity and the power of science and language to unify the world—except when he wasn't. In those early years of the second decade of the twentieth century (which no one would yet dream of as "prewar" since battle among such civilized nations would surely never come to pass), he had crusaded for European uniformity in the guise of scientific standardization. Yet in 1915, he lionized a rather different project. "The breakthrough of our united armies . . . is only the military prelude to a peaceful advance of Germany to the southeast, through which the greatest contiguous mainland complex on Earth's surface, namely the European-Asian land mass, will enter into a new epoch of its history and thereby of world history as a whole . . . ,"† Ostwald thundered.[2] This future was going to be German.

* "Sick peoples—and how few today are still healthy!—necessarily have sick languages. Their healing means at the same time the healing of the body of the people itself: their international order is nothing other than the order of those that speak them. The language problem of nations thus bears a double face: one turned inward, that can be resolved by each people alone and in its own fashion, and one turned outward, that no people can resolve on its own but rather only in communion with all others."

† "Der Durchbruch unserer vereinigten Armeen . . . ist nur das kriegerische Vorspiel zu einem friedlichen Vordringen Deutschlands nach Südosten, durch welches der

Ostwald knew from his study of history that the road to progress was always paved by "*Verkehrsmittel*"—means of communication—ranging from the money that greased the axles of commerce to the roads and trains that transported peoples. The greatest means of communication was, of course, language. Echoing arguments he had deployed just a few years earlier for Louis Couturat's constructed languages, he noted that "[o]nly through language can the isolated personal existence, the individual or the linguistically demarcated people, be set into a fruitful relationship with other persons and peoples."* As German troops advanced, they needed more than ammunition, troops, and fuel: "Everywhere we arrive together with our allies in our push through the world and where we want to protect our mutual interests, the implementation of a common spiritual means of communication, a common language, is the definitely necessary prerequisite."† [3] Ido for world conquest? Hardly. The common language should be, quite obviously, German.

But there, alas, was the rub, for German was "still in a comparatively primitive state"‡; that is to say, it was too complicated. Citing Otto Jespersen's linguistic research, Ostwald insisted that as languages progressed they simplified, shedding inflections, genders, and aspects as they streamlined themselves to charge into modernity. "That is the natural development of every language that is found among all the other means of communication and that is nothing but an expression of the most general law of development," declared the Nobelist who introduced thermodynamics into chemistry, "namely of the *energetic imperative*, whereby the pointless squandering of energy that lies in the multiplicity and irregularity of older linguistic forms is increasingly constrained."§ [4] The success of German armies was due to the

---

größte zusammenhängende Komplex des Festlandes auf der Erdoberfläche, nämlich das europäisch-asiatische Landgebiet in eine neue Epoche seiner Geschichte und damit der gesamten Weltgeschichte eintreten wird. . . ."

* "Nur durch die Sprache kann sich die isolierte menschliche Existenz, das Individuum oder das sprachlich begrenzte Volk, mit anderen Personen und Völkern in fruchtbringende Beziehung setzen."

† "Überall, wo wir im Welttreiben mit unseren Mitmenschen zusammenkommen und unsere gemeinsamen Interessen pflegen wollen, ist die unbedingt notwendige Voraussetzung die Handhabung eines gemeinsamen geistigen Verkehrsmittels, einer gemeinsamen Sprache."

‡ "ein noch verhältnismäßig primitives Gebilde."

§ "Das ist die natürliche Entwicklung jeder Sprache, die sich auch bei allen anderen Verkehrsmitteln wiederfindet und die nicht als ein Ausdruck des allgemeinsten Entwicklungsgesetzes, nämlich des *energetischen Imperativs* ist, wonach die zwecklose

vigor of the German people, who had not been stultified by Frenchified civilization, but this same vigor blocked the language from achieving a higher simplicity. Fortunately, one could learn from the practices of other colonial powers, especially the British, whose advance into Africa was facilitated by the easy communication of pidgin English. Hence Ostwald's modest proposal: "I suggest producing a *simplified German* on scientific-technical principles for practical employment at first in those [occupied] regions. All the avoidable diversity, all that 'richness' of the language so charming for aesthetics, which so tremendously complicates its learning, will be set aside here so that this new means of communication—for which I suggest the name *Weltdeutsch*—can be learned and used by everyone with ease of effort."* [5] Simplify the genders, toss out a few umlauts (following many dialects), and German would be good to go for global domination.

The Idists, understandably, were horrified. Leopold Pfaundler, Ostwald's erstwhile comrade-in-arms, penned an irate missive to the Leipzig chemist on the penultimate day of 1915. After the Ido community had just absorbed the shock of Couturat's untimely death in the summer of 1914 and his widow was forging ahead with a full dictionary of the new language, all of a sudden the world's most famous Ido-speaker had jumped ship—in the middle of a continental conflagration, no less! "[W]e now more than ever need a neutral ground, which is what Ido is,"† Pfaundler wrote. He implored Ostwald to desist: "Therefore I risk an appeal to you not to pursue this plan any further, granting us moreover in this besieged time your exceedingly valuable continued cooperation as well. I remain despite the war in contact with Swedish and Danish Idists and find everywhere the greatest willingness to cooperate. We must advance the work from these neutral states and Switzerland, and not let it slumber."‡ [6]

Energievergeudung, die in der Mannigfaltigkeit und Unregelmäßigkeit der älteren sprachlichen Formen liegt, zunehmend eingeschränkt wird."

*"Ich schlage vor, für den praktischen Gebrauch zunächst in jenen Gebieten ein *vereinfachtes Deutsch* auf wissenschaftlich-technischer Grundlage herzustellen. In diesem müßten alle entbehrlichen Mannigfaltigkeiten, all jener für die Ästhetik so reizvolle 'Reichtum' der Sprache, welche ihr Erlernen so ungeheuer erschwert, beseitigt werden, so daß dieses neue Verkehrsmittel, für welches ich den Namen *Weltdeutsch* vorschlage, von jedermann mit leichter Mühe erlernt und gebraucht werden kann."

†"brauchen wir jetzt mehr als je eine neutrale Grundlage, wie es das Ido ist."

‡"Darum wage ich den Appell an Sie, diesen Plan nicht weiter zu verfolgen, uns viel-

Ostwald was neither surprised nor cowed by Pfaundler's rebuke. "I was very conscious that my suggestion of Weltdeutsch would arouse displeasure and even also protest among my Ido friends," he wrote back on 12 January 1916, "and I naturally owe an accounting in response to your friendly and detailed letter."* His accounting consisted of a reprise of German war aims, including an accusation that Couturat's machinations were intended "to centralize, to monopolize, and at the same time to give Ido a pronounced French character."† With German culture triumphant, that was now an anachronism. Besides, Weltdeutsch was not the same kind of project as Ido: "It is not an international language in the earlier sense which I propose, but rather a language that should serve for an entirely defined goal of at least a reasonably national character, whereupon it might then be seen whether it will be used for general communication around the entire world or not." Ostwald "will not publicly turn my back on Ido, since it represents a very significant improvement over Esperanto under all circumstances, but from the above articulated reasons I can also not any longer expend any special effort on this, in my opinion, hopeless labor."‡ [7] Ido might suffer, but Esperanto would suffer worse. The wartime spirit in action!

From the beginning, the Great War boded ill for both science and language. In December 1914, a despondent Otto Jespersen, in neutral Denmark, wrote to the German émigré (and founding father of cultural anthropology) Franz Boas at Columbia University in New York

---

mehr in dieser ohnehin bedrängten Zeit Ihre uns so überaus wertvolle Mithilfe auch weiterhin zu gewähren. Ich stehe trotz des Krieges mit schwedischen und dänischen Idisten in Verbindung und finde überall grösste Geneigtheit zur Mitarbeit. Von diesen neutralen Staaten und der Schweiz aus müssen wir das Werk weiter fördern und es nicht einschlafen lassen."

*"Ich war mir wohl bewußt, daß mein Vorschlag des Weltdeutsch bei meinen Idofreunden Befremden und wohl auch Widerspruch erregen würde, und ich bin Ihrem freundlichen und ausführlichen Brief natürlich Rechenschaft schuldig."

†"zu zentralisieren, zu monopolisieren und gleichzeitig dem Ido einen ausgeprägt französischen Charakter zu geben."

‡"Es ist nicht eine internationale Sprache im früheren Sinne, welche ich vorschlage, sondern eine Sprache, die für einen ganz bestimmten Zweck von wenigstens halbwegs nationalem Charakter dienen soll, wobei sich dann herausstellen mag, ob sie für den allgemeinen Verkehr auf der ganzen Welt benutzt werden wird oder nicht. [...] Vom ido werde ich mich öffentlich nicht abwenden, da es dem Esperanto gegenüber unter allen Umständen einen sehr bedeutenden Fortschritt darstellt, aber aus den eben dargelegten Gründen kann ich auch fernerhin keine besondere Arbeit auf diese, meines Erachtens aussichtlose Arbeit verwenden."

City, in the still-neutral United States: "When is that dreadful war to end? It poisons everything, not only the minds of the fighting nations, but also to a great extent those of the neutrals and makes all peaceful 'kulturarbeit' more or less impossible—and for a long time to come!"[8] Jespersen was more right than he suspected. The story of scientific languages in the wake of World War I focuses very sharply where Ostwald did: on German, which had become in the latter years of the nineteenth century the fastest growing language of all sciences, especially chemistry, and which appeared set to take over the world in 1915 following the footsteps of German troops. But as those stormy advances bogged down into trench warfare, so did the reputation of scientific German, and the aftermath of the Armistice in November 1918 and then the punishing Treaty of Versailles in 1919 saw linked developments that hobbled, even criminalized, the language that Ostwald was convinced would rise to supremacy.

German for Science, and Vice Versa

By the final year of the war, mocking the German tongue had become something of a national pastime in the countries of the Entente (France, the United Kingdom, and—belatedly—the United States; Russia was already *hors de combat*). A French scholar, mapping the geography of European speech as the continent was tearing itself apart, allowed himself a day-pass from objectivity when it came to the Teutonic tongue. "German is not a seductive language. The pronunciation is rude, hammered by a violent accent at the beginning of each word. The grammar is encumbered with useless archaisms: the nouns for example have multiple case forms different from each other which do not even have the merit of being found in all words, and which serve no purpose since the order of words is fixed and suffices to indicate the sense," he proclaimed. He went further: "The adjective has several uselessly complicated forms. Sentences are constructed in a rigid, monotonic manner. The vocabulary is entirely idiosyncratic, so that neither a Slav, nor a Romance speaker, nor even an Englishman or a Scandinavian can understand it easily. The appearance of the whole lacks finesse, nimbleness, suppleness, elegance."*[9] Thus Antoine Meillet, the most distinguished

*"L'allemand n'est pas une langue séduisante. La prononciation en est rude, martelée par un accent violent sur le commencement de chaque mot. La grammaire en est encombrée d'archaïsmes inutiles: les noms par exemple ont des formes casuelles mul-

French linguist of his age, wrote in the tradition of Gallic attacks on their Eastern neighbors that had percolated even into the rarified discourse of science. As a compatriot biologist noted in 1915, "What characterizes German scholars is patience, prolixity, and obscurity,"* and the same was true of their language.[10]

This is an old story, but a perpetually popular one. German is, to put it mildly, frustrating. There are three genders (masculine, feminine, neuter), where Western European languages typically manage with two (French) or even none (English, sort of). At least the Russian genders are almost always easily identifiable from the ending of the noun, but German offers such solace rarely. Nouns are scattered among the three genders willy-nilly, so that freedom (*Freiheit*) is feminine and death (*Tod*) is masculine, but the charming *Fräulein* stays demurely neuter. (The regional qualities of the language only make this worse: plates are masculine in Berlin but neuter in Munich.) And then there are four cases to keep track of—nominative, accusative, genitive, dative—which typically only inflect the direct or indirect articles of the nouns rather than the words themselves. At least the nouns are capitalized, so you can pick them out; the lower-case verbs are where the trouble truly begins. Whatever verb you conjugate has to come "second" in a sentence (although the second position could be the tenth word, if a lengthy adverbial phrase knocks it back), and then the remaining infinitives, participles, and other verbal detritus are jammed in reverse order at the end of the sentence—which might be a very long way off. (In subordinate clauses the verb order is different yet again.) The forms of verbs can be devilishly irregular, and the abundant prefixes, which often determine the meaning of the word, can leap to the end of the sentence at the drop of a conjugation. (It is probably best not to dwell on the gargantuan compound words.)

To make matters more complex, German has always been a pluricentric language, with no centralized arbiter of pronunciation or even vocabulary and syntax.[11] Despite the different national variants of the

---

tiples, différentes les unes des autres, qui n'ont même pas la mérite de se trouver dans tous les mots, et qui ne servent à rien puisque l'ordre des mots est fixe et suffit à indiquer le sens. L'adjectif a des formes inutilement compliqueés. Les phrases sont construites d'une manière raide, monotone. Le vocabulaire est tout particulier, tel que ni un Slave, ni un Roman, ni même un Anglais ou un Scandinave ne peut l'apprendre aisément. L'aspect d'ensemble manque de finesse, de légèreté, de souplesse, d'élégance."

* "Ce qui caractérise les savants allemands, c'est la patience, la prolixité et l'obscurité."

language—German, Austrian, and Swiss, for example—each region has its own *Dialekt* which serves to shut out even native speakers from other parts of the same country. The written language (somewhat oxymoronically known in German as *Schriftsprache*) is more standardized but contains peculiarities that even natives get wrong—*two* subjunctives!—and the colloquial language (*Umgangssprache*), a kind of averaging of pronunciation and lexicon for the sake of communal interchange, has to make do for oral "non-dialect" communication.[12]

The history of German can be understood as a series of overlapping attempts at standardizing the language. An Indo-European tongue, modern German descended from Proto-Germanic, spoken by tribes who poured into Central Europe at some undefined point before the birth of Jesus Christ. (Julius Caesar fought Celtic-speaking Gauls; his successors had their hands full with bellicose Germans.) Eventually, the languages of the Germanic family split into the progenitors of the Scandinavian tongues, Dutch (from Low German), and of course English. Neither Old High German—"High" referring to location up the Rhine—nor its successor Middle High German (dating from roughly 1150) was standardized, and the massive population explosion among "German" speakers in the high middle ages produced ever greater diversification of regional speech.[13] In the thirteenth century, courtly society in the German states did achieve a kind of unified poetic language known as the *Dichtersprache*, but it was sharply confined by class, while the following century saw the emergence of six main types of "chancellery German" for bureaucratic communication. A (more or less) single learned language for writing really manifested only in the wake of Martin Luther's Protestant Reformation of the sixteenth century, as Luther's translated Bible and writings provided a seed around which standardization could crystallize.[14]

Science, as an activity conducted mostly by learned men (in this period, and for a long time afterward, almost entirely men), tracked alongside these developments. (Throughout the following, it is important to recall that the German word *Wissenschaft*, which I will generally translate as "science," is rather general, referring to "scholarship" or systematized knowledge. In most of the instances I quote, the term refers more narrowly to knowledge of nature.) Latin remained dominant for scientific communication in the early modern period, yet attempts to render Greek or Latin technical terms into German roots began at least as early as the Renaissance polymath Albrecht Dürer in the sixteenth century.[15] Latin may have been the language of instruction for the German universities that began to sprout up in the fourteenth century, but rebels soon called for a shift to ordinary German (whatever that might be). The alchemist Paracelsus had lectured in German in Basel in 1526–1527, and in 1687 the jurist Christian Thomasius became notorious for insisting on teaching in German at Leipzig. He was run out of town, but German was recognized as the teaching language at Halle in 1700 and at Göttingen in 1733. The vision of Johann Balthasar Schupp from 1663 began to be realized: "Knowledge is bound to no language. Why should I not be able to learn just as well in the German language as in Latin how I should recognize, love, and revere God? Why should I not be able to learn just as well in the German language how I can help a sick person, in German as in Greek or Arabic? The French and Italians teach and learn in all disciplines and free arts in their mother tongue."* [16]

One could not, however, just abandon Latin and use German. Many scholars in the German states were gifted classicists and knew that the German they used every day did not possess the vocabulary or the flexibility to reproduce the richness of the universal language of scholarship. This widespread dissatisfaction with German's quality—voiced by leading natural philosophers such as Gottfried Wilhelm von Leibniz—proved instrumental in motivating German academics to improve their native tongue. During the first half of the eighteenth century, Christian Wolff at Halle worked harder than anyone to develop a lexical storehouse to enable German's capacity to "hold" science, much as we saw the Russians labor in the nineteenth century. The standard(ish) German of the north began to solidify as the language of scholarship, science, and poetry.[17] This was precisely the period when Johann Gottfried Herder articulated his extremely influential notion of the spirit (*Geist*) of a people as expressed through its traditions, folklore, and language, a notion that—often implicitly—underscored much of the enthusiasm for German as a language of science.

The establishment of German as such a language by the early nineteenth century was astonishingly rapid when viewed against the back-

* "Es ist die Weisheit an keine Sprache gebunden. Warum sollte ich nicht in Teutscher Sprache ebensowohl lernen können, wie ich Gott erkennen, lieben und ehren solle, als in lateinischer? Warum sollte ich nicht ebensowohl in Teutscher Sprache lernen können, wie ich einem Kranken helfen könne, auf Teutsch als auf Griechisch oder Arabisch? Die Franzosen und Italiener lehren und lernen alle Facultäten und freien Künste in ihrer Muttersprache."

drop of Latin's place in Central European scholarship. Latin first arrived in the area with legions of Roman troops (witness the foundation of such cities as Colonia Claudia Ara Agrippinensium in 40 AD, aka Cologne, aka Köln, aka—in *Dialekt*—Kölle), but it spread beyond the furthest military outposts as Gaulish clerks brought bureaucratic order to the unconquered Germanic tribes. Conversion to Christianity extended the reach of Latin even farther.[18] With the development of moveable-type printing in the early fifteenth century by Johannes Gutenberg—a German whose first publication was a Latin Bible—the emerging *Schriftssprache* began to compete with the venerable ancient language. From records of the Frankfurt book fair, we can judge the proportions of Latin versus German books, but we should take care to note that these documents enumerate not only books for domestic consumption but also those for pan-European use, skewing the results toward Latin. Nonetheless, the numbers are striking: in the sixteenth century, only 10% of publications in Germany appeared in German (of whatever stripe); by 1800, 95% were. The dramatic change happened in the seventeenth and especially the eighteenth centuries. As late as 1570, the percentage of German printed books in Latin was 70%, and German-language books first outnumbered Latin in 1681, and then permanently eclipsed it after 1692. In 1754 Latin production was still at a healthy 25%, but by the eve of the French Revolution in 1787 had dwindled to a tenth. After 1752 German works were dominant in all fields, philosophy and medicine having shifted away from Latin in the early eighteenth century.[19] Latin remained important, but now the Latinity of German scholars could no longer be assumed. At least six different Latin-German botanical dictionaries were published between 1780 and 1820 to assist the grammatically challenged, even as the great mathematician Carl Friedrich Gauss kept his private mathematical journal in the traditional language of scholarship.[20]

Besides Latin, French had enormous impact on the speech of educated Germans from the seventeenth century onward. French's dominance of diplomacy was assured by the very treaties that kept the German states sequestered from each other, and in polite circles it was ubiquitous. Frederick II of Prussia (known as "the Great," reigned 1740–1786) famously ran his Berlin court—and the associated Academy of Sciences—in French. When Voltaire visited in 1750, he wrote to the Marquis de Thibouville that "I find myself here in France. One speaks only our language. German is for the soldiers and for the horses; it is only necessary on the road."*[21] Ten percent of the books published in German lands from 1750 to 1780 were actually in French, and most were consumed internally as well as exported to Paris. French continued as a vehicular language of communication during the German Confederation (1815–1866), a necessity considering the mutual unintelligibility of some German dialects, and was only displaced after the unification of Germany into the *Kaiserreich* under Prussia's aegis in 1871.[22] German became the language of a proud new nation.

By that point, it was already a vital language of science and absolutely indispensible for chemistry. The amazing boom of German chemistry was one of the great dramas of the nineteenth century. There were essentially two related components to this upsurge of Teutonic chemists, one academic and one industrial. With the creation of Justus von Liebig's first large-scale chemical laboratory in the sleepy university town of Giessen in the 1830s, it became increasingly common—first in the German states and then, in explicit imitation, abroad—to require practical laboratory instruction in the training of chemists. Having chemists at one's beck and call in the laboratory meant they could be put to work producing data for one's own research projects and publishing in one's own journal, and Liebig took advantage of both, creating an empire of students and students-of-students far afield. After the abortive revolutions of 1848, various German regional governments beefed up the chemistry facilities at their own universities in the hope, advertised by Liebig himself, that more chemistry would mean better agriculture (staving off famine) and more industry (ditto for unemployment).[23] The academic boom fueled an industrial surge, and vice versa, and the model was widely copied, not least in the United States by Ira Remsen at Johns Hopkins University.[24] The collaboration between industry and academy continued into the twentieth century in the newly unified Germany, and a central figure in facilitating the coordination was none other than Wilhelm Ostwald.[25] Obviously, the language of this cutting-edge research was German.

The stunning successes of German science (synthetic dyestuffs, new pharmaceuticals) inspired consternation abroad, especially in Paris. French scientists looked on enviously at the resources and status of German chemistry while French statesmen—still smarting from the German victory in the Franco-Prussian War of 1870–1871, popularly attrib-

*"Je me trouve ici en France. On ne parle que notre langue. L'allemand est pour les soldats et pour les chevaux; il n'est nécessaire que pour la route."

uted to the superiority of German technology—hatched plans to boost French science back to its previous pinnacle.[26] Particularly worrisome to an observer in 1915, from the point of view of scientific languages, was that "young nations from a scientific point of view,"* such as Italy, Romania, the United States, Japan, and the South American republics, all sent their students to study in Germany rather than France.[27] They brought German home with them, and tended to refer more frequently to German literature.

Consider one particularly virulent eruption of nationalist furor between the Germans and the French. It all began in 1869, when Charles Adolphe Wurtz, distinguished professor of chemistry at the Sorbonne—whom we have already met as one of the instigators of the Mendeleev-Meyer dispute from chapter 2—published his *Dictionary of Pure and Applied Chemistry* in French. In the wake of the tremendous discoveries by German chemists of the preceding thirty years, the first sentence was bound to deliver a shock East of the Rhine: "Chemistry is a French science. It was constituted by Lavoisier, of immortal memory."† [28] Given the increasingly militarized tensions between France and Prussia, certain German chemists were not going to take this provocation lying down. Building on a long tradition after Lavoisier of German chemists refining "Frenchness" out of anti-phlogiston doctrines, Hermann Kolbe of Leipzig printed two pieces in response to Wurtz in his *Journal für praktische Chemie*. The first, by Jakob Volhard, argued that Germanic chemists and not Lavoisier had really developed the new chemistry.[29] The lead editorial, by Kolbe himself, attacking French chemists in general and Wurtz in particular—"a born Alsatian, who is fully knowledgeable about the German language and relations,"‡ and who therefore ought to have known better—in shockingly aggressive terms, triggered an extensive controversy in the chemical world.[30]

One of the most interesting responses to the Wurtz-Kolbe dispute hailed not from Paris or Berlin but from St. Petersburg: an October 1870 editorial signed by four chemists, including D. I. Mendeleev, in the Russian capital's leading German-language newspaper. These guiding lights of the newly formed Russian Chemical Society claimed that they had read Wurtz's original French statement with surprise, but re-

*"nations jeunes au point de vue scientifique"

†"La chimie est une science française. Elle fut constituée par Lavoisier, d'immortelle mémoire."

‡"geborener Elsasser, der deutschen Sprach und Verhältnisse völlig kundig."

ceived Kolbe's hostility with shock. Kolbe, they alleged, was intoxicated by the violent triumph of his nation's armies: "Two great nations stand facing each other in bloody battle; treasures of civilization, of science, and of art—the legacy of centuries—fall now into oblivion in a few days. One of these nations has finally nearly achieved victory, and it has gone, drunk on victory, ever further. Now however it turns out that this drug of victory unhappily is powerful enough to bewitch into battle even peaceful men of science, men of usually sober thought."* Many Germans as well as Frenchmen had thought the same upon reading Kolbe's editorial. (Rudolf Fittig, one of the editors of the ill-fated *Zeitschrift für Chemie* that had published Mendeleev's non-periodic abstract, quipped that "[t]he Journal für practische Chemie should change its title and call itself the Journal für polizeiliche Chemie or the chemical police."† [31]) The reasoning behind the Russian intervention, however, was distinctive. "We however—observing from our neutral position, free of blood and the drug of victory—believe entirely differently," they observed with a soupçon of grandiosity. "In that we stand freely on the sidelines, we are granted the opportunity to observe the performance objectively. The conclusion that one draws from this observation is full of significance: even men of the exact sciences, men of a nation which stands at the pinnacle of civilization, can forfeit their fine humane feelings as soon as their country is overcome by a passionate arousal."‡ Therefore, the very same chemists who had taken to publishing everything first in Russian to bolster their national scientific litera-

*"Zwei grosse Nationen stehen im blutigen Kampfe einander gegenüber; Schätze der Civilisation, der Wissenschaft und der Kunst—der Erwerb von Jahrhunderten—fallen nun in wenigen Tagen der Vernichtung anheim. Die eine dieser Nationen hat beinahe schon endgültig den Sieg gewonnen, geht aber, siegestrunken, noch immer weiter. Nun ergiebt es sich aber, das[s] dieser Siegesrausch unglücklicherweise mächtig genug ist, auch friedliche Männer der Wissenschaft, Männer des gesunden nüchternen Denkens zum Kampfe zu begeistern."

†"Das Journal für practische Chemie sollte seinen Titel ändern u. sich Journal für polizeiliche Chemie oder chemische Polizei nennen."

‡"Wir aber—von unserem neutralen, von Blut und Siegesrausche freien Standpunkte zuschauend—glauben ganz anders.[. . .] Indem wir frei bei Seite stehen, ist uns die Möglichkeit geboten, dass sich Vollziehende objektiv zu beobachten. Der Schluß, den man aus dieser Beobachtung zieht, ist bedeutungsvoll: Sogar Männer der exakten Wissenschaft, Männer einer Nation, welche an der Spitze der Civilization steht—können die feinen humanen Gefühle, einbüßen, sobald ihr Land von einer leidenschaftlichen Erregung bewältigt ist."

ture had chosen to write in German, "so that this can be brought directly to the attention of the nation to which it is addressed."* [32]

Wurtz wrote to his friend Aleksandr Butlerov, one of the authors, in appreciation: "I also have other thanks for you: I was very aware at the time of the protest of Russian chemists against the grotesque and absurd polemic of Kolbe, if you can even call it a polemic."† [33] On the other hand, the completely marginal Russians, to Volhard's mind, were butting in where they were not wanted; he sarcastically noted that "it is understandable that the Russian chemists find no insult in this phrase, since in Lavoisier's time Russian chemistry had not yet played any role in history."‡ [34] Liebig himself wrote to Kolbe that "[t]hese Slavs are full of malice toward the Germans."§ [35] As historian Alan Rocke has documented, all of this was missing the point: Wurtz wrote his declaration not to antagonize the Germans but to shame the French into adopting the theoretical perspectives recently propounded by German chemists, which he insisted (with good reason) had their origins in French intellectual achievements of the 1840s and 1850s.[36] It backfired, but it did so in part because Wurtz could not write in French and hope that only Frenchmen would see it. The implication of one's native language being an international language of science was that what you wrote was open to all literate in it. The Russians could shield themselves behind Cyrillic characters and come out in German when they chose; Wurtz—and Kolbe—were exposed as soon as paper rolled off the presses.

For it was not the case, as one might assume by the present situation of languages (or, one should say, "language," in the singular), that even then-dominant German scientists believed that they needed only their native tongue. German had joined French and English in crowding out other languages of science like Italian and Dutch; it was by no means a monopolist. A case in point was Hermann von Helmholtz, the titan of German physics. Helmholtz was routinely confronted by scholarly developments that manifested at least as often in French and English as in

* "das dieselben direkt zur Kenntniss der Nation bringt, an deren Adresse sie gerichtet sind."

† "J'ai aussi d'autres remerciements à vous faire: j'ai été très sensible dans le temps à la protestation des chimistes russes contre la polémique grossière et absurde de Kolbe si on peut appeler cela de la polémique."

‡ "es ist verständlich, dass die russischen Chemiker in jener Phrase keine Beleidigung finden, da zur Zeit Lavoisier's die russische Chemie noch keine Rolle in der Geschichte gespielt hat."

§ "Diese Slaven sind voller Bosheit auf die Deutschen."

his native German. As a matter of course, he learned to read those languages and soon, as was usual at the time, to do more than read. While traveling to the meeting of the British Association for the Advancement of Science in Edinburgh in 1853, Helmholtz wrote repeatedly to his young wife about his fears that English would trip him up. He needn't have worried. By the end of his journey, he beamed with pride, "I have become entirely so accustomed to English that I understood the better speakers without problems, and among those that I didn't understand, my English colleagues also usually had difficulties. Dr. Cooper said to me that during my visit he learned again how to speak English properly, because I don't understand all the careless usages and provincialisms to which he had become accustomed."* [37] His French was likewise more than serviceable, and he was occasionally mistaken for a native speaker.[38] Helmholtz was exceptional only in the skill he displayed, not in the multilingual pressures he confronted. (Even the jingoist Kolbe, for example, had lived in England for a while, used English sources, and published in it in his youth. He avoided reading French whenever possible, but he clearly knew how.[39])

As the Franco-Prussian War receded in people's memories and the unified German Reich turned into a reality straddling the plains of Central Europe, resentment of German as a language of science receded. It had to, for there was no avoiding it, even as the language itself swelled under the massive lexical expansion engendered by the sciences.[40] British students who had done their tour of German universities returned to their green islands and translated the German books of their advisors into English for those who seemed unable to master German word order.[41] Even Meillet, who scorned the language so poetically at the opening of this section, concluded that "Not knowing German is almost always to renounce being current in the science and technology of the times."† [42] For the moment, those times were stable despite the nationalist swagger. In the summer of 1914, however, old grudges erupted with renewed vigor.

* "Ich habe mich allmälig so an das Englische gewohnt, daß ich die besseren Sprecher ohne Mühe verstand, und bei denen, die ich nicht verstand, hatten auch gewöhnlich meine englischen Bekannten Schwierigkeiten gehabt. Dr. Cooper sagte mir, durch meinen Aufenthalt lerne er wieder richtig englisch sprechen, weil ich alle Nachlässigkeiten und Provincialismen, die er sich angewöhnt hätte, nicht verstände."

† "Ne pas savoir l'allemand, c'est presque toujours renoncer à être au niveau de la science et de la technique de son temps."

Boykott

Although nationalist tensions were ominously growing during the last decades of the nineteenth century, one could also view the era as demonstrating the international character of science, with the possibility that science could provide the glue that would knit nations together. International meetings proliferated. When a group of chemists spanning the entire European continent gathered in the southern German town of Karlsruhe in September 1860 to discuss the standardization of atomic weights, the organization of such conferences was, if not unheard-of, relatively rare. By 1900 it had become commonplace. During the decade before Karlsruhe, international scientific meetings (of a limited geographical scope) took place at a rate of about two a year, between 1870 and 1880 the annual average hovered around a dozen, and in the decade before the turn of the century it rose to roughly thirty each year. International organizations also mushroomed: 25 new bodies created in the decade after 1870, 40 additional in the decade following, and 68 more still between 1890 and 1900. By the time the war broke out, there were roughly 300 such international scholarly bodies, most of them concentrated in the natural sciences, and their "international" character would play a starring role in the reaction to the Great War.[43]

Scientific conflicts share certain features with diplomatic and military ones. One might engage in a spirited debate about who was responsible for initiating World War I—the Triple Alliance headlined by Germany and Austria-Hungary, or the Triple Entente of the United Kingdom, France, and Imperial Russia (the Americans joined in just before the Bolshevik Revolution of 1917 bowed the Russians out)—and likewise fingers can be pointed in multiple directions over whom to blame for the perceived "collapse" of scientific internationalism during and after the war. Since the victors typically write history and the Entente unquestionably won, we can begin with their version. For scientists in France and the United Kingdom, the German "Aufruf," the "Manifesto of the Ninety-Three," cast the first stone.

The War began well, or badly, depending on who you were. If you were Belgian, it began very badly indeed. German troops flooded into Belgium in a flanking attack directed at Paris, bypassing the reinforced Maginot Line on the border between the Reich and France, making amazing speed before the advent of trench warfare plunged the conflict into a standstill. The invasion of neutral Belgium triggered British entry as a belligerent, and also produced reams of hostile propaganda against

German rapaciousness and reported atrocities. In response, a popular German playwright (and incidentally a celebrated translator from the French) named Ludwig Fulda persuaded a stellar array of ninety-two other intellectuals to attach their names to a declaration (*Aufruf*) "To the Cultured World," which began with these outraged words: "We as representatives of German science and art raise a protest before the entire cultured world against the lies and slanders with which our enemies strive to besmirch Germany's pure cause in the hard battle for existence imposed upon it."*[44] The rest of its two pages proceeds pretty much as you might expect.

The "representatives of German science and art" comprised over ten pages of signatories, and their credentials stunned intellectuals in the Entente nations. Six were chemists of international renown: Adolf von Baeyer, Karl Engler, Emil Fischer, Fritz Haber, Richard Willstätter, and—*et tu, Brute?*—Wilhelm Ostwald. A host of other scientists, many laureates of the recently established Nobel Prize, followed, including Philipp Lenard, Ernst Haeckel, Wilhelm Förster, Konrad Röntgen, Walther Nernst, and Max Planck, among others.[45] The contrast between the dignified status of the signatories and their tone of righteous umbrage on the one hand, and the horrific reports of massacres of civilians from the Belgian countryside on the other, combined a violation of international law (the invasion of neutral lands) with a violation of the neutrality of science. This sin against the internationalism of science would be repaid with interest by nationalist partisanship after the Armistice on 11 November 1918.

Propaganda flowed both ways, and you might consider that the first transgression was the vilification of the "brutish Krauts" who had "raped Belgium." The second scientific sin, however, was clearly German in origin, and it was terrible beyond imagination. On 22 April 1915, as belligerent forces continued the interminable conflict outside the Belgian hamlet of Ypres, a greenish-yellow cloud wafted from German trenches over to Entente lines: chlorine gas. Thus German forces birthed chemical warfare, and the man who orchestrated the incredible mobilization of materiel and personnel to militarize the chemical industry was none other than Fritz Haber, the brilliant chemist lauded

*"Wir als Vertreter deutscher Wissenschaft und Kunst erheben vor der gesamten Kulturwelt Protest gegen die Lügen und Verleumdungen, mit denen unsere Feinde Deutschlands reine Sache in dem ihm aufgezwungenen schweren Daseinkampfe zu beschmutzen trachten."

for his world-changing discovery of how to fix atmospheric nitrogen (and berated for his signature on the Manifesto).[46] After the war, the introduction of poison gas was decried as a war crime; during the war, all sides quickly piled on, adding phosgene, mustard gas, and Lewisite to arsenals in every army. This, then, was the state of international science when the guns went silent in 1918.

Planning for postwar retribution to be inflicted on Central European scientists, and especially those of the *Kaiserreich* (which would, as it happened, pass out of existence after the war, ceding to the Weimar Republic, Germany's first democratic government), began months before the end of the war.[47] The penalty seemed obvious: Germans had forfeited their right to participate in international science, and should be excluded from the new postwar scientific order. This was, as a French biologist declared, an obvious consequence of the Manifesto:

> The signatories of the manifesto have disqualified themselves as far as being men of science, and, in making common cause with Teutonic militarism, they are placed at the same level as the ferocious brutes accomplishing the most monstrous crimes under orders. They are much more culpable than these brutes because one cannot deny them intelligence. We can no longer have any confidence in their scientific productions and we can accept as correct only that which we have submitted to a severe critique and verified for ourselves; we must consider all their publications as suspect as any object *made in Germany*.* [48]

The legal force to exclude German scientists was rooted in the humiliating Treaty of Versailles. (This was the first major international accord whose English text was declared equally official to the French, ending almost two centuries of French dominance in diplomacy. American President Woodrow Wilson insisted on the change; the French naturally at first objected, but then acceded when it was clear the Italians and other

*"Les signataires du manifeste se sont disqualifiés, en tant qu'hommes de science, et, en se solidarisant avec le militarisme teuton, ils se sont mis au même niveau que les brutes féroces accomplissant par ordre les crimes les plus monstrueux. Ils sont beaucoup plus coupables que ces brutes, car on ne peut leur dénier l'intelligence. Nous ne pouvons plus avoir aucune confiance dans leurs productions scientifiques et nous ne pourrons accepter comme exact que ce que nous aurons soumis à une sévère critique et vérifié par nous-mêmes; nous devrons considérer comme suspectes toutes leurs publications comme tout objet *made in Germany*."

parties wanted to open the floodgates to a diplomatic Babel.[49]) Articles 282 and 289 of the Treaty allowed for intellectual penalties to be imposed on the defeated powers, and a group of entrepreneurial scientists seized the opportunity in two phases.[50]

The first, and most notorious, was the Boycott.[51] In 1919, British, French, and Belgian scientists created a new scientific organization in Brussels, the International Research Council (IRC, to use the English acronym) to replace the International Association of Academies—the institution that had refused to rule on the Delegation's proposals for an artificial auxiliary language. The IRC served as the umbrella organization for a series of "international unions" replacing prewar international scientific organizations, many of which had been based in Germany. The Executive Committee of the IRC consisted mostly of hardline anti-Germans, especially Emile Picard, Georges Lecointe, Vito Volterra, and the German-born British physicist Sir Arthur Schuster. Only the American representative, George Ellery Hale, was lukewarm. The Central Powers—that is, Germany and Austria, for Austria-Hungary was no more—were excluded from membership until at least 1931 by statute, and the victors stacked the deck against amendments by requiring that even former *neutrals* could only be included by a three-fourths supermajority.[52] Naturally, the official languages of such an organization would be French and English (regardless of Swiss grumbling).[53]

It was a bad time to be a German scientist. Much of the process of actually doing science in this period was conducted at international conferences, and this was the chief target of the Boycott. There were fourteen international conferences in 1919; not a single German was invited. The next year saw twenty such events, and the Germans were excluded from 17 (85%). The intensity of the Boycott declined over the next five years, but only slightly: Germans were excluded from 22 of 36 meetings (60%) in 1921, and 86 of 106 (81%) from 1922 to 1924. The only silver lining was in the neutral countries, where only one of 21 conferences banned German and Austrian participation. Of the 275 international science conferences of the Boycott period, Germans and Austrians were locked out of more than 60%.[54] (An exception was made for Albert Einstein, recognized as a pacifist and "good German" ever since his public opposition to the Manifesto. Einstein hated the Boycott, however, and often served as an intermediary to assist his German colleagues to publish abroad.[55])

German scientists responded with outrage and a counter-boycott. For example, in 1922, the International Union for Theoretical and Applied Limnology held its congress in Kiel, in northern Germany, hosting scholars from twenty countries. Participants from Entente nations were, however, banned. The original call for papers went out in English, French, and German, but the conference itself was held in German.[56] The counter-offensive was largely successful in swaying public opinion, and was surely instrumental in the eventual reversal of the ban. "Such a boycotting of a specific cultural group, here the Central European one, is until now historically unprecedented," Frankfurt's newspaper declared in 1926. "The absurdity of such a decision is most obvious in the area of medicine and the sciences. They are not national, at least not in the chauvinistic sense, but international. A violent sundering of this scientific group is a transgression against science itself."* [57]

At the same time, the leaders of German science, especially Fritz Haber, erected the Notgemeinschaft (Emergency Committee) on 30 October 1920 in Berlin, unifying five academies of science, two other learned societies, and a host of universities and technical, veterinary, agricultural, forestry, and mining schools. The former Prussian Minister of Culture, Friedrich Schmidt-Ott, helmed the self-governing body from October 1920 until 23 July 1934, as it disbursed government and industry money in grants to German scientists. Along with the Helmholtz Society for the Advancement of Physical-Technical Research, it was the most important source of funding for university professors throughout the Weimar Republic.[58] In this manner, the Germans fashioned their own science unhinged from international interchange.

Neutral Sweden helped by granting beleaguered German scientists an enormous share of world recognition. The Swedish Academy of Sciences played an outsized role through its ability to award Nobel Prizes in the sciences, and strongly German-oriented Swedish academics—many had studied in Germany and most Swedish scientists preferred to publish in German—obliged during Germany's time of need. The first postwar Nobel in Chemistry was awarded to none other than Fritz Haber, at that moment roundly denounced for his role in the onset of chemical warfare. He was just one in the German sweep of the 1919

*"Eine derartige Boykottierung eines bestimmten Kulturkreises, hier des zentraleuropäischen, ist bisher in der Geschichte ohne Beispiel. Die Widersinnigkeit eines solchen Beschlusses liegt vor allem für das Gebiet der Medizin und der Naturwissenschaften auf der Hand. Sie sind nicht national, wengistens nicht im chauvinistischen Sinne, sondern international. Eine gewaltsame Zerreißung dieser Wissenschaftskreise ist ein Vergehen an der Wissenschaft selbst."

prizes. Other Nobels to Germans followed during the 1920s (many to deserving scientists, to be sure), although none got the backs of the former Entente scientists up as much as Haber's.[59] The Nobel prizes had skewed pro-German from the beginning of the war; Ostwald himself traveled north during the conflict to argue that prizes for Germans would be valuable in demonstrating the superiority of German culture.[60] As the leading historian of Nobel science prizes put it, the Swedish Academy's "bias toward Germany was never disguised."[61] Sweden ignored the Boycott.

Solidarity with Germans also came from a more surprising source: the newly established Soviet Union. Despite long-standing tensions between German and Russian academics, Russophone scientists preferentially published in German and collaborated with German colleagues before the war, not least because of a sizable Baltic-German and Russo-German community at home. Imperial Russia was Imperial Germany's foe during the war, but after the abortive February Revolution dethroned the Tsar but kept Russia in the conflict, Vladimir Lenin's Bolsheviks deployed their long-standing opposition to the military venture as a justification for their October 1917 (November in the Gregorian calendar) coup against the Provisional Government. The Communists became instant pariahs; in this misfortune they were soon joined by the Germans. In May 1921 Germany and the Soviet Union signed a provisional trade agreement; the following April, in the midst of a 34-nation economic summit in Genoa—one of the few international meetings either party was invited to join—the Russian and German representatives absconded to nearby Rapallo and signed a treaty resuming full diplomatic and economic relations. A host of collaborations followed. In 1925, the *Deutsch-Russische Medizinische Zeitschrift* was founded, providing a Western-language outlet for Soviet scientists, and joint ventures abounded: a research expedition to Siberia here, a Brain Research Institute there (actually, in two places: Berlin and Moscow).[62]

If the Boycott of German scholars from conferences was the short-term punishment for perceived misdeeds during the war, the IRC's second action would have more lasting consequences for the fate of German as a scientific language. The IRC's "international unions" provided umbrellas for postwar scientific governance, and three were erected immediately: the International Astronomical Union, the International Geodesic and Geophysical Union, and, in July 1919, the International Union for Pure and Applied Chemistry (IUPAC). To this day, IUPAC governs global chemistry, serving as the court of final recourse to adjudicate discovery claims of new elements (and the right to name them, thus creating the internationally recognized standard nomenclature that had been noticeably lacking in the nineteenth century). Like many of these organizations, IUPAC was actually a reactivation of a prewar institution—in this case, the International Association of Chemical Societies, proposed in 1910 by Wilhelm Ostwald and Albin Haller, president of the French Chemical Society—but now with the Germans excluded.[63]

Cutting out the Germans implied cutting out German. German had been an official language, with English and French, of the International Association; it was just as obvious to the IRC's movers and shakers that it would not be permitted at IUPAC. Concern over the dominance of German, especially within chemistry, had been simmering for some time. Four days before the Armistice, *Science*, the journal of the American Association for the Advancement of Science, published an editorial entitled "Insidious Scientific Control" by Edwin Bidwell Wilson, which noted that "it has been the feeling of many teachers and of many students that the German language was more essential for scientific uses than any other, and that the German training was the one to which our graduates who were not satisfied with what they found in this country should turn. This American feeling was undoubtedly expressly fostered by the German government[. . .]."[64] Insidious indeed, and only compounded by the universal recognition that Germans had cornered the market on indispensible reference works.[65] Even international organizations like the League of Nations (which also excluded Germany and Austria) and philanthropies like the Rockefeller Foundation used medical and scientific grants with the explicit goal of trying to curb the German language.[66] In all these international venues, German was proscribed, and only (alongside Italian) granted a subsidiary status in IUPAC in 1929.[67]

The exclusion of German as a language of science in an international scientific body might sound like a minor affront. Who would want to go to those boring meetings anyway? But precisely such standardization bodies, which set the ground rules for scientific governance around the globe, have enormous long-term impacts that amounted to an almost irreversible lock-out of German—albeit with a time delay. There were different ways the official disapproval cascaded down to the mundane decisions everyday scientists made about which journals to submit to, or which languages to speak. As in most cases, the effects were not most strongly seen among the Germans themselves, who continued to use their native language, nor among native Francophones and Anglophones, who used theirs. Rather, individuals who had once used a variety of vehicular languages—the Dutch, the Norwegians, the Portuguese—might now choose differently. The official languages allowed at conferences constrained the options. In 1932, for example, French was permitted as an official language at 351 (98.5%) of the international conferences that year, and English at 298 (83.5%). The Boycott being over, German was officially permitted at 60.5%—nothing to sneeze at, but a far cry from the parity one would have expected in the prewar years.[68] Germany also never regained its leading position as a host country of international scientific conferences; from roughly 20% on the eve of the war, this number crashed to about 3% interwar.[69]

Foreigners also submitted to German journals rather less after the Great War than before, which entailed a measurable linguistic shift differentiated by discipline. Foreign contributions to German-language medical journals in 1920, for example, sank 50% (to a total of 23% of all submissions) compared with the level from 1913. The German journals in physics and chemistry witnessed a similar effect: in 1920, 13% of the contributions were by foreigners, compared with 37% in 1913. Astronomy was even more drastic. Under 5% of the articles printed in *Astronomische Nachrichten* after the war came from British and American contributors, down from 15%–20% in 1910, and *zero* Belgians and French submitted. This had the paradoxical effect of increasing the percentage of German-language contributions from 60% in 1910 to over 95% in 1920.[70] The flip side, of course, was that foreign astronomers had moved elsewhere.

The United States of English Speakers

Another major reason World War I was a turning point in the history of scientific languages was the stunning eradication of knowledge of German as a foreign language among members of Entente nations, especially in the United States. I focus here on the United States for two reasons: first, the visceral reaction there against German was more pronounced, more violent, and more prolonged than in the other victorious countries; and second, because the tremendous growth of the American chemical industry during and after the war soon transformed the distant trans-Atlantic outpost into the most productive scientific country in the world. That community was clearly largely Anglophone; the legacy of World War I made it also often non-Germanophone.

Today, Americans are famous for being bad at foreign languages. It was not always this way. German was a dominant language of immigrants to the American continent from before the Revolution, and the Continental Congress published translations of many of its proclamations into German as well as French.[71] Formal foreign language education was rather slow in establishing itself in the early Republic—the first licensed German teacher at Harvard College was Meno Poehls in 1816, although lackluster French had been taught there to supplement the ancient languages as early as 1733.[72] But foreign-language enrollments grew across the nineteenth century, and no modern language appeared to be more popular in the sprawling, industrializing country than German. By 1900, German had a firm foothold in the school curriculum at all levels. As many as 38% of accredited high schools in California offered at least two years of German in that year, and by 1908 that had risen to a stunning 98%, with many offering up to four years. In 1913 72% of *all* high schools in the state, accredited or not, were teaching German, outstripping French in a pattern visible across the country. Universities followed suit. In 1910, of 340 institutions of higher education, all but three taught German, and 101 required some French or German to graduate.[73] Of course, in the Midwest, German was commonly heard on the streets from the children and grandchildren of Central European immigrants, and parochial schools (and some public schools) sometimes taught entirely in the language. When the war arrived in 1914, the Americans were neutral; the figures for 1915 were as robust as before the conflict.

The Americans entered the fray on 6 April 1917 after years of lobbying by President Woodrow Wilson, and the nationwide reaction against the German language was swift and furious. In preparation for the impending conflict, Congress had established a National Council for Defense in late August 1916, and local replicas with a good deal of autonomy proliferated at the state, county, and town level. The local Councils of Defense provided the mechanism for a populist assault on the German language: the Victoria City Council in Texas banned German in 1918; the city council in Findlay, Ohio, fined citizens $25 for speaking German on the streets; the select and common councils in Philadelphia—a city where German was almost as prevalent as English—appealed to the House of Representatives for a ban on the language in public meetings. In May 1918, Governor Warren S. Harding of Iowa (soon to become Woodrow Wilson's short-lived successor) issued an order prohibiting the use of any language but English in public places, over the telephone, and on trains. (Though targeted at German, Harding's reach was too broad, and Iowan Czechs and Danes protested.) Newspapers of every language were regulated, but German-language ones more so. In Collinsville, Illinois, Robert Prager, a German-born socialist, was lynched in April 1918. By the end of the war, 16 states had banned German, a move endorsed by former President Theodore Roosevelt, and *after* the war six more had joined them, bringing the total to 22.[74]

The laws did not last for long. Already after the Treaty of Versailles, a German teacher in Hamilton County, Nebraska, was convicted for teaching ten-year-old Raymond Parpart "the subject of reading in the German language," thereby violating a Nebraska law passed on 9 April 1919. The legislation was explicit:

> Section 1. No person, individually or as a teacher, shall, in any private, denominational, parochial or public school, teach any subject to any person in any language than the English language.
> Sec. 2. Languages, other than the English language, may be taught as languages only after a pupil shall have attained and successfully passed the eighth grade as evidenced by a certificate of graduation issued by the county superintendent of the county in which the child resides.

The case reached the United States Supreme Court as *Meyer v. Nebraska*, which declared such prohibitions unconstitutional in a 7–2 decision. Justice James Clark McReynolds observed in the decision that while "[t]he obvious purpose of this statute was that the English language should be and become the mother tongue of all children reared in this state," nonetheless "[m]ere knowledge of the German language cannot reasonably be regarded as harmful." Nebraska, and other states with similar laws on the books, had exceeded their rights under the fourteenth amendment, McReynolds concluded, for "[t]he protection of the Constitution extends to all, to those who speak other languages as well as to those born with English on the tongue."[75] The dissenters consisted of the arch-conservative George Sutherland (a Harding appointee) and the vaunted liberal lion Oliver Wendell Holmes (who ventured in *Bartels v. Iowa*, an analogous case, that "I think I appreciate the objection to the law but it appears to me to present a question upon which men reasonably might differ and therefore I am unable to say that the Constitution of the United States prevents the experiment being tried"[76]).

The laws had already wrought enormous damage. Even before the tremendous growth of high-school attendance that arrived later, 315,884 students, 28% of all Americans enrolled in secondary school, were studying German in 1915; in 1922, before *Meyer* and therefore at the height of the proscription, there were fewer than 14,000 students of the language, scarcely over 0.5% of the national enrollment of 2.5 million. Even in Ohio, with its heavily Germanic population base, only five high schools offered German in 1925. By 1949, when the high-school population had more than doubled to 5.4 million, the absolute number of students in German had tripled to 43,000, a share of 0.8%. It never recovered. French (15.5% of students in 1922) and Spanish (10%) leapt into the breach, but not for long.[77] One lasting lesson of the temporary criminalization of the German language is that when one foreign language suffers, they all do. Not only did the absolute number of students drop off, thus thinning those who might grow up into polyglot scientists, but the cadres of foreign-language teachers collapsed as well, beginning a vicious cycle that America would confront at the dawn of the Cold War.

The Great Restructuring

When the war was still raging, an American commentator lamented the fate of postwar American scientists, forced to deal with the legacy of scientific Teutons: "Our students should not have to feel that the great majority of the best expository works relating to their subject are to be found only in the language of a people of low ideals imbued with a morbid desire to dominate the world at any cost."[78] The author tacitly pointed to a subtle restructuring in the interrelationships among scientific languages that would start to ripple through the scientific elite in the interwar years. Americans still studied in Germany and the reputation of German scientists rebounded in the postwar years, but something had changed for the Americans. While a scientist of Helmholtz's stature was proud, even happy, to speak in multiple languages, young Americans who entered this brave new world of interwar chemistry imbibed something of the ambient hostility toward foreign tongues. American politics became increasingly isolationist and American education increasingly monoglot. As the American scientific juggernaut climbed to ever-greater heights, it brought with it a generalized reluctance to language study—after all, wasn't the world's important science already appearing in English? The consequences of the shifts born of World War I are with us now.

But they were more or less invisible at the time. Europeans noticed the spasm of Anglomania striking the American heartland, but they saw it against a backdrop of anti-Germanism that bloomed after the war and just as quickly seemed to dwindle away. Much of the retreat was due to the diplomatic breach of the Boycott, led by prominent German scientists such as Fritz Haber, who in 1924 attended the centenary of Philadelphia's Franklin Institute as a German delegate of the Prussian Academy of Sciences and Berlin University. As historian Fritz Stern, whose family was friendly with the chemist, noted in his memoirs, Haber "argued that scientific achievement was the only physical pillar left of German strength, hoping as well to restore the international ties that the war had virtually destroyed."[79] Small countries and neutral countries, especially those in Eastern Europe (former domains of the largely-Germanophone Habsburg Empire) still preferred German, and they pushed against the IRC's strictures.[80]

At the sixth IUPAC conference in Bucharest, Romania, the Boycott was finally breached. Dutch chemist Ernst Cohen, who had earlier organized a small meeting to which he had invited Germans and Austrians, was elected president, a sign that the end of German exclusion was nigh. In June 1926, five years before schedule, the Boycott was lifted and Germany, Austria, Hungary, and Bulgaria were invited to join. Delegates came as guests to a 1928 chemistry conference in The Hague in 1928, and Cohen happily greeted them in German. In the end, Germany only agreed to join IUPAC after its statutes had been modified to grant it full autonomy from the IRC. Progress on that front was much slower. Hungary acceded in 1927, and Bulgaria in 1934, but the Germanophone powers bided their time. Austria only joined the International Council of Scientific Unions (the successor to the IRC) in 1949, West Germany waited until 1952, and international outcast East Germany remained in the cold until 1961.[81] The language rift continued. German was still excluded from the official languages of IUPAC and the IRC. The issue was raised at The Hague, but the threat of opening the floodgates even to so-called "minor languages" meant it was quickly tabled, even though the Germans had originally insisted on the inclusion of German as a precondition of joining the IRC.

And what of Ostwald in this new international moment, the time of the League of Nations and a vanquished Boycott? He lived amid a surprising efflorescence of Esperanto. The League debated allowing Esperanto as an official language, and both the British Association for the Advancement of Science and the French Academy of Sciences in 1921 discussed favorably resolutions endorsing Esperanto as a solution to the language barrier. These were pleasant dreams, but they were no more viable in the long run than Weltdeutsch had been in the short run. Ostwald spent increasing amounts of time working on his new theory of colors and devoted almost none to propagandizing for constructed languages. In October 1931 he agreed, for old time's sake, to being named the honorary president of a new Ido Academy. He died the following year.

CHAPTER 7

Unspeakable

Soviel und welche Sprache einer spricht, soviel und solche Sache, Welt oder Natur ist ihm erschlossen. Und jedes Wort, das er redet, wandelt die Welt, worin er sich bewegt, wandelt ihn selbst und seinen Ort in dieser Welt. Darum ist nichts gleichgültig an der Sprache, und nichts so wesentlich wie die façon de parler. Der Verderb der Sprache ist der Verderb des Menschen. Seien wir auf der Hut! Worte und Sätze können ebensowohl Gärten wie Kerker sein, in die wir, redend, uns selbst einsperren, und die Bestimmung, Sprache sei allein die Gabe des Menschen oder eine menschliche Gabe, bietet keine Sicherheit.*

DOLF STERNBERGER, GERHARD STORZ, AND WILHELM E. SÜSKIND[1]

German science entered the 1930s triumphant. The Boycott had been lifted in 1926, and Germans now attended and hosted international conferences, bathed in the glow of self-righteousness as victims of an unheard-of transgression against scientific internationalism. German scientists raked in Nobel Prize after Nobel Prize, and foreign students flocked to German universities to study at the feet of the titans of the newly emergent quantum physics, then taking the physical sciences by storm. The slow-acting poison pill of the exclusion of the German language from international organizations, the looming threat of competition from American science, and the teeming youngsters of the United States who would grow up without significant exposure to foreign

*"As many and which languages a person speaks, so many and such things, world, or nature is accessible to him. And each word that he speaks changes the world in which he moves, changes himself and his place in this world. Thus nothing is indifferent to language, and nothing so essential as the *façon de parler*. The woes of language are the woes of persons. We are on guard! Words and sentences can be just as much gardens as dungeons in that we, speaking, lock ourselves up, and the definition that language is alone a gift of persons or a personal gift offers no security."

languages—all of these were invisible threats to the dominance of German as a language of science. German still shared the stage with French and English, but the former was evidently in a process of slow decline, and as for English—well, while it was clear that the Americans had overtaken the British as the leaders of Anglophone science and were flooding journals with publications, one could always debate about quality.

Yet, by the end of that decade, the position of German had noticeably changed. Scientists, linguists, and historians ever since have converged on a single point of blame: Adolf Hitler.[2] The timeline accords well with such an account. In September 1930, the National Socialist, or Nazi, Party (NSDAP) won 107 seats in the Reichstag, the parliament of the Weimar Republic, bespeaking the growing appeal of aggressive right-wing populism. True, war hero Paul von Hindenburg defeated Hitler in the presidential elections of March 1932, but it was only a temporary setback; in parliamentary elections four months later, the Nazis bagged 230 seats. It seemed only a matter of time before Hitler, the charismatic Austrian-born leader, would be appointed chancellor, which indeed happened on 30 January 1933. Germany was still a democratic republic, but not for long. On 27 February 1933 the Reichstag caught fire in an act of arson whose ultimate origins remain murky, and even in the wake of the crisis the NSDAP failed to gain an absolute majority in the elections of 5 March, due to continued electoral success by the communists. Nonetheless, Hitler pushed through the so-called Enabling Bill (*Ermächtigungsgesetz*) on 23 March, which he then used to exclude communists from local government. From there, the story is sadly familiar: an expansion of German armed forces in violation of the Treaty of Versailles, increasing persecution of Jews within the newly dubbed "Third Reich" (third, that is, after the Holy Roman Empire and the *Kaiserreich*), the sacrifice of increasing swaths of Central Europe to Hitler, and then the invasion of Poland on 1 September 1939, igniting World War II.

Observed from a distance, it seems obvious that the Third Reich—which wrecked Germany's economy, cities, and moral reputation, and committed the horrific atrocities of the Holocaust of European Jews and the slaughter of countless other innocents—was the great caesura of European history, and the history of the German scientific language should be expected to track. A close look at the graph presented in the introduction tells a somewhat different story. German continued to be an important language of science in the 1930s; in fact, its percentage share in some cases *grew*. Instead of a precipitous drop during the Third

Reich, one rather observes a gradual diminishing of the language's place in the scientific literature, the unfurling of a process that had begun with the Great War. If we want to see the Nazis' impact on German as a language of science, then publications are not the best place to search. Much of this book has focused on written communication among scientists, for good reasons, but there are other ways in which languages matter to science, and this chapter stresses these usually tacit aspects of language, made visible by the unique trauma that beset German science.

Those who emphasize the politics promulgated by the Nazi state point to an important facet of the history, for the *manner* in which Germanophone science was conducted during the Third Reich had an enormous bearing on the shape and the rate of the ensuing decline, if not on its onset. Aside from publishing, scientists teach, and the effort of putting together lectures is substantial even in one's native tongue, let alone a foreign one. Scientists also collaborate with colleagues, mediated through oral and written communication. Last, but far from least, they live as human beings in a milieu saturated by the words of others, both political speech and everyday excursions to the store or to work. This chapter explores what it meant to live in surroundings where the German language—one's native language—became politicized, laden with hitherto unexpected burdens. Therefore, before relating the ways individual Germanophone scientists spoke and wrote both inside and outside Germany, we must step back and examine structural issues (such as employment and unemployment), restrictions on travel, and commentary by intellectuals, linguists, and ideologues about the meaning of German. People live in language in these registers as well, and sometimes they are kicked out of them.

The first part of this chapter follows policy changes in the sciences, continuing the historical explanation begun in the previous chapter, but to end with those developments would only give the outline of the story. The reaction against scientific German was not only a direct consequence of enrollments and academic careers, it was also about human lives crushed by hatred and violence. Language is perhaps our most personal possession, and the particular language each of us considers his or her "mother tongue"—to use the term of art most often deployed in the 1930s, with all its gendered and (as it happens) National Socialist connotations—expresses something deeply intimate for each of us. Beginning outward but then focusing in on scientists in the second half of the chapter, we will see the way these personal valences of language were deployed by scientists, both to express their sincere dismay but also to dramatize to others (sometimes sensationally) the rupture the Nazi-imposed emigration placed on long-standing relationships.

The Great Purge

Hitler's Reichstag wasted no time implementing the racist and anti-Semitic agenda that he had been broadcasting across Germany for a decade. For scientists, many of whom were employed in higher education and thus were civil servants, one of the most momentous acts was also among the earliest: the Law for the Restoration of the Professional Civil Service, passed on 7 April 1933. In its final—strange to imagine, *softened*—form, the law dismissed three categories of employees from the civil service: those of "non-Aryan" descent (mostly Jews), members of socialist or communist parties, and political appointees of the Weimar Republic. (The softening, at the insistence of von Hindenburg, exempted veterans of the Great War and those who had lost a father or son in combat.) Almost immediately, German universities were rocked with firings and resignations.

There is no question the impact of the Civil Service Law was severe. The hardest-hit discipline appears to be physics, especially theoretical physics, which had over the years acquired—especially in the north of Germany as opposed to the predominantly Catholic south—a substantial proportion of scientists of Jewish origin. Some estimate as many as 25% of physicists across the entire country were fired, and at certain centers, most prominently Göttingen, almost the entire department of physics and mathematics was gutted. Later estimates place the impact lower, factoring in other sciences, but not much lower: about one in five, or 20%, of scientists had been driven from their jobs by 1935, followed by another wave when Austria was annexed in 1938 and its institutions (and its citizens working within the Third Reich) were subjected to Nazi laws.[3] Some disciplines, such as biology, got off "lighter," but only as a result of there being relatively fewer Jews appointed: approximately 13% of biologists were fired between the Law's passage and 1938, and four-fifths were for racial reasons. Three-quarters, in a pattern that we will return to shortly, emigrated, never to return.[4]

Chemistry was also badly damaged. Fritz Haber, the architect of chemical warfare, the discoverer of the eponymous "Haber process" to fix atmospheric nitrogen, and the 1918 Nobel Laureate in chemistry, was stripped of his directorship of the Kaiser-Wilhelm Institute for Chemistry in November 1933, whereupon he emigrated in despair to London. All five department directors in the newly Nazified institute were NSDAP members, three of them *alte Kämpfer* ("old fighters") who had joined the Party before Hitler's unsuccessful 1923 Beer Hall Putsch in Munich, a significant indicator of loyalty to the regime and its policies. Twenty-eight Jewish employees were summarily fired after the "coordination" (*Gleichschaltung*) of the institute. The higher status and traditional political conservatism of German chemists meant that fewer Jews had reached positions that would have demanded firing, but the enormous size of the German Chemical Society and the Union of German Chemists—with 40% of their 4,000 members living outside the Reich—meant that the high degree of conformity to the Nazi state had wide reverberations.[5]

The fortunate ones emigrated, their international reputations—gained, of course, by the willingness of foreign scientists to read their German-language publications—securing them positions abroad. Although modest in terms of absolute numbers, the high quality of the émigrés was exceptional. Almost all set sail for the United States. The vast majority of all dismissed German-speaking mathematicians, for example, passed through Ellis Island; more than 100 refugee physicists also arrived between 1933 and 1941. They were not only gifted scientists, they were also overwhelmingly young, with most under 40 and having received their doctorates after 1921. Their youth and quality were not accidental: the ad hoc Emergency Committee in the United States that attempted to find work for displaced scientists focused on scholars over the age of 35 (old enough to have made a substantial name for themselves) but under 58 (to avoid putting a strain on the pension systems of the institutions that hired them).[6] A small subset of scientists, predominantly physicians, headed eastward to the Soviet Union, where learning the new language posed a consequential hurdle.[7]

Although an admittedly significant transformation of German academia, we should be careful before attributing too large an impact to this emigration of the most gifted. Elite scientists represented only a fragment of the very large German knowledge-production system. The emigration did not bleed away most of Germany's talent, but it did inflict a significant threefold harm on German as a scientific language. First, symbolically, foreign scientists began to view the German state with revulsion and expressed reluctance to "collaborate" with the regime in any way. Second, those scientists who left almost all ended up in Anglophone contexts, continuing their high-quality research in a new tongue. The third change was the most immediate and perhaps the one with the longest-lasting consequences: the rupture of the graduate-student and postdoctoral exchange networks.

As we saw in the previous chapter, one of the most salient indications of the importance of German science was the centrality of German universities as the destination of choice for foreign students. Some of America's most famous scientists, such as J. Robert Oppenheimer and Linus Pauling, did their graduate work at German universities. As a side effect, they acquired, and later continued to use, the language. This was also true of Japanese scientists. The modernizing Meiji regime in the late nineteenth century hired dozens of foreign professors (*oyatoi*) to staff new universities, insisting that the academics lecture in their native tongues to encourage the students to learn Western languages. Most *oyatoi* were German. When these professors sent their best students for training to the West, they naturally sent them preferentially to Germany, to the tune of 74% of Japanese students studying abroad in the early twentieth century.[8] The dismissals removed some of the incentive to travel to Germany, even if foreigners—especially Americans in the throes of isolationism and the Great Depression—had been willing to resettle to the Third Reich.[9] These networks did not reassemble until after the war, and they reassembled with the United States as the hub.

Travel by scientists to and from Hitler's Germany became much harder. In 1935 the Reich Education Ministry (REM) assumed control of all lectures by foreigners within Germany; it also decided whether a German scholar would make a fitting representative abroad. The torrent of international exchange dried up, and Germans had to change their patterns of collaboration. (Cooperation with Italian mathematicians flourished, for example, as an offshoot of Axis fellowship.) Scientists who lived in zones under German control, such as the Protectorate of Bohemia and Moravia that had been carved out of the former Czechoslovakia, were also not permitted to go abroad except as part of the "German" delegation, and were required to speak German if they wanted to use any language except Czech or Slovak. (They were forbidden from lecturing within the Reich proper.)[10]

These obstacles had the predictable consequence of snapping connections with foreign scientists. As an indication of the extent of the rupture, consider the guests who stayed at Harnack Haus, founded in 1929 in the Dahlem neighborhood of Berlin to house visiting scholars. (Many of the leading Kaiser Wilhelm institutes in the sciences were located nearby.) The numbers from 1930 to 1933 were stable, with about 200 visitors a year, fewer than half of whom were foreign. In 1933, the number of foreigners fell by nearly half, replaced by an uptick of Germans traveling to Berlin from distant parts of the country—from 45% foreigners in 1932–1933, the number the following academic year dropped to 23%. The original ratio of foreigners to Germans did not resume until 1937–1938, and then the mix of nations was rather different: whereas in 1930–1933, Americans had comprised roughly thirty guests a year (about one-third of all foreigners), they now represented fewer than fifteen a year, replaced by South Africans, Romanians, Dutch, and French (the latter two would later fall under military occupation).[11]

But even with all the purging, emigration, and ruptured collaborations, German science appeared to be functioning. Certainly, scientific journals under the Nazis came out regularly and published work of good quality, and most scientists (until the war) felt little disruption in their work. Foreign scientists may not have traveled to Germany as often, but they still submitted to German journals in noticeable numbers (although here, too, the distribution of nations had shifted). There were, however, changes under the surface, as Emil Julius Gumbel—a famously anti-Nazi mathematician who had been forced out of his Heidelberg position as early as 1932 and was living in France—observed:

> Purely externally most of the physical and mathematical journals seem to be unchanged. They have preserved their appearance. Only upon closer examination does one notice the absences in the world-renowned names: the Jews are eliminated. Against this the proportion of foreign collaborators has grown, since the journals exercise a considerable attraction on the basis of their earlier quality. The quality of the domestic collaborators has dropped; the scientific offspring are partially abandoned. The system is proud of the fact that the number of students has dropped by half. Oddly the past often changes. Certain authors are no longer cited; their earlier achievements are ascribed to impeccable Nordic men; their current works are ignored.* [12]

*"Rein äusserlich sehen die meisten physikalischen und mathematischen Zeitschriften unverändert aus. Ihr Gesicht haben sie gewahrt. Erst bei näherer Betrachtung bemerkt man das Fehlen von weltbekannten Namen: die Juden sind ausgemerzt. Dagegen wuchs der Anteil der ausländischen Mitarbeiter, da die Zeitschriften auf Grund ihres früheren Niveaus eine beträchtliche Anziehungskraft ausübten. Das Niveau der inländischen Mitarbeiter senkte sich; der wissenschaftliche Nachwuchs setzt zum Teil aus. Das System ist stolz darauf, dass die Zahl der Studenten auf die Hälfte herunterging. Eigentümlich wandelt sich vielmals die Vergangenheit. Gewisse

Foreign submissions to German journals declined, both because new regulations restricted the percentage of "non-Aryans" who could appear in each issue, but mostly because foreigners had ceased making German journals their outlets of choice.[13] Citations also changed: about 37% of citations in mathematics articles from 1921 to 1925—at the height of the Boycott—had been to German journals, and this number rose to 39% for 1926–1930; but from 1931 to 1935 the number had sunk to 28%. In the meantime, citations to American journals rose from 14% to 25%, picking up German's lost ground.[14] One should not exaggerate: German was still an essential language for science in the Nazi years; it was just, little by little, somewhat less important.

The Browning of German

Some of the luster of the German language may have faded for foreign scientists, but that was more than made up by the assiduous attention the Nazi state paid to the language in general. These broader linguistic contexts are essential in order to understand the ways in which émigré scientists responded to their native language. It is almost impossible to read any statements about German by the regime or regime-friendly scholars without coming across the adulation of the *Muttersprache*, or "mother tongue." The first recorded reference to the term dates as far back as 1119 (in Latin, as it happens), but appeared in Low German in 1424 and High German in 1520, becoming crucial for the latter. There was no notion more central to linguistics under the Third Reich.[15]

"Mother tongue" might today have roughly the same meaning as "native tongue," but that was hardly the case for those living in the Germanophone world in the 1930s and 1940s. The motherness of it all conveyed lineal heritage, birthright, and intimacy, and quickly became wrapped up in the anti-Semitic quagmire of so much Nazi intellectual output. Jews, so it was said, had no mother tongue, having given up ancestral Hebrew—for centuries a subject of great interest to German philologists—in favor of a mongrelized Yiddish, an earlier variant of other Jewish linguistic Trojan horses like Esperanto. Language for Jews, supposedly, was a matter of communication alone, which was why they so glibly assimilated dominant languages like German, though they could not truly, in their core, understand its depth and richness.[16]

Autoren werden nicht mehr zitiert; ihre früheren Leistungen werden einwandfreien Nordmännern zugeschrieben; ihre jetzigen Arbeiten werden ignoriert."

The language, the new consensus among German linguists would have it, was intrinsically for Germans. (No hint here of creating a simplified Weltdeutsch à la Wilhelm Ostwald; no linguistic concessions would be made to the conquered.) "One often hears it said that no people feels itself more tightly and deeply bound to its mother tongue than the Germans,"* wrote Leo Weisgerber, Lorraine-born specialist in Celtic linguistics, in 1941, and perhaps no better illustration can serve for the tenor of this new language ideology than his musings, distinctive only in their clarity and the intellectual reputation of their author.[17] Even in the throes of enthusiasm for his own "mother tongue," Weisgerber was willing to concede that "[o]ther peoples also recognize the connection to their languages, and if the French have worked with persistent care on the structure of their language, or the English as a matter of course (which surprises us) have carried their language throughout the entire world, then those are also forms of expression of a very strongly felt connectedness, and in their immediate success they are very convincing as so much enthusiasm for the mother tongue." His tolerance had limits:

> But despite the fact that each people feels the far-reaching effects of language in its own life, there remains for us Germans an occasion for ever-renewed reflection, namely that *among the peoples of Europe the Germans are the only ones that have named themselves after their mother tongue.* That is an unmistakable demonstration that the mother tongue is involved in the construction of the life of our people to an especial degree[. . .].† [18]

* "Man hört oft sagen, kein Volk fühle sich mit seiner Muttersprache enger und tiefer verbunden als das deutsche."

† "Auch die anderen Völker wissen um die Bindung an ihre Sprachen, und wenn die Franzosen in unentwegtem Mühen an dem Ausbau ihrer Sprache gearbeitet oder die Engländer in einer uns überraschenden Selbstverständlichkeit ihre Sprache durch die ganze Welt getragen haben, so sind das auch Ausdrucksformen einer sehr stark gespürten Verbundenheit, und in ihrem unmittelbaren Erfolg sind sie sogar mancher Schwärmerei für die Muttersprache weit überlegen. Aber ungeachtet der Tatsache, daß jedes Volk die weitreichenden Wirkungen der Sprache in seinem eigenen Leben spürt, bleibt doch uns Deutschen eines als Anlaß zu immer erneutem Nachdenken, daß nämlich *unter den Völkern Europas das deutsche das einzige ist, das sich nach seiner Muttersprache genannt hat.* Das ist ein unverkennbarer Hinweis darauf, daß am Aufbau unseres Volkslebens die Muttersprache in besonderem Maße beteiligt ist[. . .]."

In a familiar pattern, Weisgerber was happy to ascribe the intellectual achievements of the Germans to language: "where we encounter before us the achievements of German technology, we will bump into the thankful preparation of these creations through the application of linguistic means; where German science acts, it will never be able to entirely set itself free of the assumptions of the German language during all its efforts toward truth."* [19] Of all sciences, none was so specifically German as the science of language itself, declared Hermann Flasdieck, professor of English philology at the University of Köln: "The history of German-born linguistics is a facet of the examination of German and English natures and styles of thinking, and it is no coincidence that precisely linguistics as the most German of all sciences finds no nourishing soil on the other side of the Channel."† [20]

Weisgerber and Flasdieck were simply repeating a central plank of dominant ideology, reflecting the obsessive fixation of Nazi leaders on language and its uses.[21] German itself began to shift under the pressure of vituperative editorials and harangues in mass meetings. These linguistic transformations were only rarely legislated from above, although occasionally even that happened—for example, on 13 December 1937, the state "abolished" the word *Völkerbund* (League of Nations), and on the very day that World War II began with the Polish invasion, it decreed that the word *tapfer* (brave) could be collocated only with *deutsch*![22] Most of the demonstrable changes in German usage happened in a more organic manner, as the structure and especially lexicon of private discourse began to mirror public pronouncements.[23]

Critics of the regime—no less Germanophone than Flasdieck and Weisgerber—were shocked at these developments, and carefully documented precisely the shape of this (mostly) unconscious Nazification of German. One obvious change was the growing abundance of military metaphors, but the influences reached beyond content to form. Nazi discourse tended to nominalize verbs and adjectives: nouns were the way thoughts should be expressed. Verbs became more violent, more

*"wo uns Leistungen der deutschen Technik vor Augen treten, werden wir in der gedanklichen Vorbereitung dieser Schöpfungen auf den Einsatz sprachlicher Mittel stoßen; wo deutsche Wissenschaft wirkt, wird sie bei allem Streben nach Wahrheit sich von den Voraussetzungen der deutschen Sprache nie ganz loslösen können."

†"Die Geschichte der deutschgeborenen Sprachwissenschaft ist ein Teilaspekt der Auseinandersetzung deutscher und englischer Wesensform und Denkart, und es ist kein Zufall, daß gerade die Sprachwissenschaft als die vielleicht deutscheste aller Wissenschaften keinen Nährboden jenseits des Kanals findet."

forceful, explained Eugen Seidel and Ingeborn Seidel-Slotty in their 1961 publication of a manuscript they had begun working on, clandestinely, in the 1930s: "The language of Nazism does not want to explain (*darlegen*), it wants to '*hammer*' (*einhämmern*)."* [24] "Further one should also one more time point to the fact," they continued, "that the language of Nazism is not entirely new, but rather is drawn from various styles and directions, and that only the strength and the extent of this influence signifies something new for the language."† [25]

Others, such as Victor Klemperer, a Jewish academic who survived in hiding in Dresden throughout the war and whose diaries provide a penetrating account of life in the Third Reich, disagreed. Something new *was* going on, so novel that he christened it LTI, Lingua Tertii Imperii (the language of the Third Reich). From the moment of his first encounter—the first LTI word he heard, he vividly recalled, was *Strafexpedition* (punishment expedition)—he became attuned to the phenomenon: "everything that was printed and said in Germany was entirely standardized to the Party; that which deviated from the permissible form in any way did not penetrate to the public; book and newspaper and official letter and form for a job opening—everything swam in the same brown sauce, and this absolute uniformity of the written language explains the homogeneity of the form of speaking."‡ [26] Klemperer, along with the other critics, adhered to something akin to the Whorfian hypothesis that language shapes thought:

> But language composes and thinks not only for me, it also guides my feeling, it directs my entire spiritual nature—the more self-evident it is, the more unconsciously I abandon myself to it. And if formulated language is now formed out of poisonous elements or is made into the bearer of poisons? Words can be like tiny doses of

* "Die Sprache des Ns. will nicht darlegen, sie will '*einhämmern*.'"

† "Ferner ist auch noch einmal auf die Tatsache hinzuweisen, daß die Sprache des Ns. nichts vollkommen Neues ist, sondern aus verschiedenen Stilarten und Richtungen entnommen, und daß nur die Stärke und Verbreitung dieses Einflusses etwas Neues für die Sprache bedeutet."

‡ "alles, was in Deutschland gedruckt und geredet wurde, war ja durchaus parteiamtlich genormt; was irgendwie von der einen zugelassenen Form abwich, drang nicht an die Öffentlichkeit; Buch und Zeitung und Behördenzuschrift und Formulare einer Dienststelle—alles schwamm in derselben braunen Soße, und aus dieser absoluten Einheitlichkeit der Schriftsprache erklärte sich denn auch die Gleichheit der Redeform."

> arsenic: they are absorbed unwittingly, they seem to have no effect, and after a certain time the effect of the poison is just there. If for a long enough time a person says for "heroic" and "brave" the word "fanatical," he actually finally believes that a fanatic is a brave hero, and that without fanaticism one cannot be a hero.*[27]

Such observations were common among adherents of the regime as well as critics. Specifically concerning the language used for scientific discourse, Lothar Tirala—an Austrian psychologist and zoologist notorious for his doctrines of race hygiene—observed that the active character of German science was reflected in language, and vice versa: "The preference for a passive construction is for the Latin exactly as characteristic as the preference for the active for the German."†[28] For postwar observers in the United States, the outcome for scientific German was substantially darker: "The Romance element in German which would have made its scientific vocabulary identical with those all over the world has been beaten and kicked by militant boots until its importance has visibly faded."[29] But it was not just in language that there were attempts to partially Nazify German science.

Consider, this time through a linguistic perspective, the abortive project to produce an "Aryan Physics" (*Deutsche Physik*) that has been extensively studied by scholars. Among the luminaries of the German physics community in the interwar years were two "old fighters," Philipp Lenard and Johannes Stark, both Nobel Laureates (the former for his experimental measurements of the photoelectric effect, the latter for his discovery of the splitting of atomic spectral lines in an electric field). As Hitler solidified his hold on power, Lenard and Stark saw an opportunity to bring Nazi ideology into the heart of physics. They petitioned the Education Ministry to replace the highly mathematical, theoretical approach to elite physics dominant in what they saw as "Jewish"

* "Aber Sprache dichtet und denkt nicht nur für mich, sie lenkt auch mein Gefühl, sie steuert mein ganzes seelisches Wesen, je selbstverständlicher, je unbewußter ich mich ihr überlasse. Und wenn nun die gebildete Sprache aus giftigen Elementen gebildet oder zur Trägerin von Giftstoffen gemacht worden ist? Worte können sein wie winzige Arsendosen: sie werden unbemerkt verschluckt, sie scheinen keine Wirkung zu tun, und nach einiger Zeit ist die Giftwirkung doch da. Wenn einer lange genug für heldisch und tugendhaft: fanatisch sagt, glaubt er schließlich wirklich, ein Fanatiker sei ein tugendhafter Held, und ohne Fanatismus könne man kein Held sein."

† "Die Vorliebe für eine passive Konstruktion ist für den Lateiner gerade so kennzeichnend wie für den Deutschen die Vorliebe für das Aktive."

science with a more "*Deutsch*" physics based on experiment and concrete intuition. At the core of Aryan Physics resided a hostility to quantum theory and relativity, both ably represented by its arch-theorist, renowned pacifist and Zionist celebrity Albert Einstein. Rather than lament the sundered ties of international collaboration, Stark virtually exulted in the autarky of science in the Third Reich, as expressed in this 1934 pamphlet:

> The catchphrase has been coined and is broadcast especially from the Jewish side that science is international.[. . .] Against this it must be enunciated with all emphasis from the National Socialist side that in a National Socialist state the obligation toward the nation against all other obligations holds also for scientists; the scientific researcher also has to feel himself as a member and servant of the nation; he is nothing in himself or for the sake of science, but in the first place has to serve the nation with his work. Therefore none who are foreign to the people can stand in the leading scientific positions in a National Socialist state, but rather only nationally conscious German men can.* [30]

The year 1934 was a good one for Aryan Physics, as Bernhard Rust, the education minister, wrested the Department of Culture from its former home in the Reich Interior Ministry, dismissing Friedrich Schmidt-Ott from his decade-long control of the invaluable Notgemeinschaft that had sustained German science during the Boycott. Rust replaced him with Stark.[31] From there, however, Aryan Physics went downhill, as Rust resented interference in his bailiwick and other powerful factions in the Nazi regime became persuaded that the advance of militarily valuable research required both quantum and relativity theories—although they should be taught, naturally, without reference to Einstein. The

*"Es ist das Schlagwort geprägt und besonders von jüdischer Seite verbreitet worden, die Wissenschaft sei international.[. . .] Demgegenüber muß von nationalsozialistischer Seite mit allem Nachdruck betont werden, daß im nationalsozialistischen Staat auch für den Wissenschaftler die Verpflichtung gegenüber der Nation über allen anderen Verpflichtungen steht; auch der wissenschaftliche Forscher hat sich als Glied und Diener der Nation zu fühlen; er ist nicht um seiner selbst oder um der Wissenschaft willen da, sondern hat mit seiner Arbeit in erster Linie der Nation zu dienen. Darum können im nationalsozialistischen Staate an den führenden wissenschaftlichen Stellen nicht volksfremde, sondern nur nationalbewußte deutsche Männer stehen."

established physics journals continued for the most part uninflected by Lenard and Stark's program; ideological articles were relegated to a new journal, *Zeitschrift für die gesamte Naturwissenschaft*.[32] Aryan Physics crumbled without top-down support, but not without causing a lot of damage.

Aryan Physics obviously reveals a good deal about the tensions between science and the Nazi state, but it also exhibits an important feature of language, one which often passes by without comment: silence. Stark did not laud the German language as essential for science, nor was it highlighted in Philipp Lenard's new textbook for Aryan Physics, where race was clearly the dominant category.[33] Explicit commentary about scientific languages appears only when a language seems to be threatened or when the choice of language is not obvious. During the Boycott, both the avenging victors and the besieged Germanophones brought the question up constantly because both felt themselves to be threatened by foreign tongues; likewise in the debates over Esperanto and Ido. In today's science, almost entirely dominated by globalized English, Anglophones almost never raise the question of scientific languages—that is done by native speakers of other languages, especially Germanophones and Francophones who lament the transformation. Latin's eclipse, too, was not bemoaned until it was already a fait accompli; when it was dominant, people rarely discussed the necessity of writing in it.

For Lenard and Stark, it was simply *obvious* that German scientists would write in German, and ideally with less mathematical formalism and therefore more linguistic content. Not only was that the patriotic and ideologically correct thing to do, but it was also the case that writing in German incurred no costs for international communication. As a result, linguistic policing remained a secondary concern: everyone in the Third Reich would maintain Germanophone uniformity by simply following self-interest. To see the change brought about by this confidence, contrast Lenard's vituperative attack on English science at the dawn of World War I. "One notices in the last ten years in the literature of my science something like the following: England gives itself the appearance of solitary leadership; outwardly the results achieved are richly used, however only openly where they play no essential role; otherwise they are annexed with the help of a certain circumvention," Lenard wrote in 1914. "The origin finds itself recognized in these cases somewhere deep in the interior of the publication or only in an ancillary publication that is difficult to obtain; sometimes also helpful material is used through direct historical distortion."*[34] That is, English science was derivative of German originals but refused to cite properly, and it could get away with it by hiding behind the veil of a foreign language.

Just because Nazi enthusiasts were confident about the continued dominance of the German language within science does not mean that the state did not take measures to guarantee its perpetuation. German forces actively imposed the language in occupied Holland and Denmark, for example.[35] Some American Germany-watchers exhorted their own specialists to learn foreign languages so they could assimilate the advances of foreign engineers and scientists the same way the polyglot Germans were incorporating English, French, Italian, and Spanish sources by their simple ability to read foreign publications.[36] More discriminating and careful observers like Carl Ramsauer, head of the German Physical Society, on the other hand, foretold a different future. He wrote a memorandum to Rust in the late 1930s arguing that the growth in American physics publications was eclipsing German-language ones.[37] Rust could bring Stark to heel, but this was a development he was powerless to act against.

Losing One's Tongue

The propaganda of Aryan Physics calmly assumed that the language of science would remain German, at the very least within Germany. Yet as we have seen, a sizable cohort of once German scientists could no longer call Germany their home. In emigration, these scientists faced what became an all-too-common experience of dislocation, of starting a new life. In almost every case that meant learning a new language. For most of the émigrés, the need to adapt to a new ambient language—often, but not always, English, which they usually had some familiarity with due to their scientific research—was simply a fact of life, and they adjusted to it without comment. A tiny minority, an atypical subset that had the double distinction of being too old to comfortably adapt and

*"Man bemerkt da aus den letzten zehn Jahren in der Literatur meiner Wissenschaft etwa das Folgende: England gibt sich den Anschein alleiniger Führung; auswärts erzielte Fortschritte werden reichlich benutzt, offen aber nur, wo sie keine wesentliche Rolle spielen, andernfalls werden sie mit Hilfe einer gewissen Umgehung annektiert; der Ursprung findet sich dann irgendwo an einer versteckten Stelle tief im Innern der Publikation oder nur in irgend einer schwer zugänglichen Nebenpublikation angegeben; manchmal wird auch das Hilfsmittel direkter historischer Verdrehung benutzt."

also famous enough that their private correspondence has survived to be scrutinized by historians, would come to invoke this exile from their native language repeatedly. Although these statements were in all likelihood sincerely believed, my point here is less psychological than instrumental, to articulate how and why the dramatized story of "losing German" manifested among these extremely elite scientists.

The implications of the crimes of the Third Reich for the language of its victims was not, obviously, an issue only for scientists. After the war, some of the most prominent German intellectuals thrust into exile by the regime would return to the topic of alienation from the language, turning it into a sentimental metaphor to describe the rupture in German history that suggested a potential remedy. In one of her most moving interviews (with the journalist Günter Gaus in October 1964, after the publication of her widely read English-language *Eichmann in Jerusalem*), Hannah Arendt confronted directly the possibility of losing one's native language. "[T]here is no substitute for the mother tongue," she responded to one question. "One can forget one's mother tongue. That's true. I have seen it. These people [other émigrés] speak the foreign language better than I do. I always still speak with a very strong accent, and I often don't speak idiomatically. They can all do so. But it will be a language in which the clichés of others play, because precisely the productivity that one has in one's own language will be cut off as this language is forgotten."* [38] So, what remains after the crimes of German-speaking minions?

> The language remains.[. . .] I have always consciously refused to lose the mother tongue. I have always held a certain distance both from French, which I once spoke very well, as well as from English, which I write in today.[. . .] I write in English, but I have never lost the distance. There is an outrageous difference between the mother tongue and another language. For myself I can say it frightfully simply: In German I know a rather large portion of German poems by heart. They are always moving to and fro somewhere in the back of my

*"[E]s gibt keinen Ersatz für die Muttersprache. Man kann die Muttersprache vergessen. Das ist wahr. Ich habe es gesehen. Diese Leute sprechen die fremde Sprache besser als ich. Ich spreche immer noch mit einem sehr starken Akzent, und ich spreche oft nicht idiomatisch. Das können die alle. Aber es wird eine Sprache, in der ein Klischee das andere jagt, weil nämlich die Produktivität, die man in der eigenen Sprache hat, abgeschnitten wurde, als man diese Sprache vergaß."

head—*in the back of my mind*; that is naturally not something one will achieve again. I allow myself things in German which I would not allow myself in English.*[39]

Like Arendt, Frankfurt-School philosopher Theodor Adorno defended his return to Germany after the war with a simple statement: "Also something objective asserted itself. That is the language."†[40] An autodidact in English (from a three-year stay at Oxford before the war), he claimed it was unsuitable for philosophy. "So one sees, you write in a seriously foreign language, whether or not you admit it, under the spell to communicate yourself, so to speak, so that others will also understand you," he continued, expressing the tension between identity and communication we have seen from the beginning of this book. "In one's own language, however, one is allowed also to hope, if one can only state the matter so precisely and uncompromisingly as possible, that one would be understandable through such intransigent effort."‡[41] Both Adorno and Arendt used the trope of lost—and recovered—German to signal a vital link to the pre-Nazi past, a poetic and philosophical culture worth salvaging.

Most of the émigré scientists whose correspondence I have tracked were less willing to forgive the language than Arendt and Adorno, even while they expressed themselves in it. Julius Schaxel, a prominent anti-Nazi biologist, felt it "repugnant for a German with a healthy national feeling to hear the bad German of *Hitler*, Rosenberg, Franz etc."§[42] So

*"Geblieben ist die Sprache.[. . .] Ich habe immer bewußt abgelehnt, die Muttersprache zu verlieren. Ich habe immer eine gewisse Distanz behalten sowohl zum Französischen, das ich damals sehr gut sprach, als auch zum Englischen, das ich ja heute schreibe.[. . .] Ich schreibe in Englisch, aber ich habe die Distanz nie verloren. Es ist ein ungeheuerer Unterschied zwischen Muttersprache und einer andern Sprache. Bei mir kann ich das furchtbar einfach sagen: Im Deutschen kenne ich einen ziemlich großen Teil deutscher Gedichte auswendig. Die bewegen sich da immer irgendwie im Hinterkopf—in the back of my mind—; das ist natürlich nie wieder zu erreichen. Im Deutschen erlaube ich mir Dinge, die ich mir im Englischen nicht erlauben würde."

†"Auch ein Objektives machte sich geltend. Das ist die Sprache."

‡"Schreibt man in einer ernsthaft fremden Sprache, so gerät man, eingestanden oder nicht, unter den Bann, sich mitzuteilen, so es zu sagen, daß die anderen es auch verstehen. In der eigenen Sprache jedoch darf man, wenn man nur die Sache so genau und kompromißlos sagt wie möglich, auch darauf hoffen, durch solche unnachgiebige Anstrengung verständlich zu werden."

§"Es ist für einen Deutschen mit gesundem Nationalgefühl widerwärtig, das schlechte Deutsch der *Hitler*, Rosenberg, Franz usw. zu vernehmen."

don't use that particular register of discourse; but what if, in order to survive, you *had* to use English as the language for expressing your intellectual labor? A job abroad was not simply an office and a paycheck, but also an obligation: one had to live and buy groceries while bending one's tongue to express foreign words, and one had to teach the science one had always contemplated in German in a new idiom. This was often difficult. While the émigrés usually understood some English, yet there were some who faced similar dilemmas as mathematician Issai Schur, who turned down a job at the University of Wisconsin-Madison because he did not feel that he could lecture in English.[43] Sometimes, accommodations were made: psychologists, for example, often either were allowed to postpone their first bouts of lecturing or were allowed to teach in German, precisely because psychology did not yet have the international vocabulary common in the physical sciences.[44] Some scholars were fortunate enough to end up at the Institute for Advanced Study in Princeton, New Jersey, a research institution with no teaching; it functioned almost entirely in German during the war years.[45]

Publication was another matter. By 1940, essentially all of Germanophone Europe (except Switzerland) was under Nazi control, which meant that most German-language outlets for scientific work were compromised by their association with the regime. Nonetheless, some Jewish exiles in the United States continued to submit to Reich journals. At Purdue University, Cornelius Lanczos, a one-time assistant of Albert Einstein from Berlin who practiced a form of mathematical physics alien to the more pragmatic style of most American journals, found himself stymied by journal editors in the United States. Faced with rejection after rejection, he decided to submit papers to the *Zeitschrift für Physik* in Germany. Einstein was outraged. He "can however not understand that you as a Jew still publish in Germany. This is after all a kind of treason. The German intellectuals have as a whole behaved disgracefully concerning all the abominable injustices and have richly deserved to be boycotted. It is already sad enough when non-Jews abroad do not do it."*[46] Lanczos, for his part, refused to blame all Germans for the actions of the state: "Since I consider the *Zeitschrift für Physik* to be thus entirely

*"kann aber nicht begreifen, dass Sie als Jude noch in Deutschland publizieren. Dies ist doch eine Art Verrat. Die deutschen Intellektuellen haben sich im Ganzen bei all den scheusslichen Ungerechtigkeiten schmachvoll benommen und haben es reichlich verdient, boykottiert zu werden. Wenn es die Nichtjuden des Auslands nicht tun, ist es schon traurig genug."

an organ of German physicists and not Germany's journal, I felt no obstacles to placing my work there[. . .]."* He added that he felt himself discriminated against in the United States. Partly, he alleged, that was because he was not as well known as he had been in Germany, but the rejection letters also targeted his kind of science. Editors could not even hide behind the convenience of rejecting his poor English: "Thereby the consequent difficulties do not at all consist in the English formulation, because I have taken precautions against the well known excuse of 'bad language,' since I subject the text to a thorough revision with good friends."† [47] Behind this exchange we can infer a widespread practice of disciplining émigré Germans into particular norms of scientific decorum through the editorial policing of grammar and syntax.

Although Einstein dropped the subject in future correspondence with Lanczos, it is unlikely that these protestations appeased him. Einstein's acquaintance with languages other than German was shaky. He had learned French for his final examinations at the ETH in Zurich, and his parents' removal to Milan in his high school years had given him at least some familiarity with Italian (although he considered his capacities atrocious).[48] His lifelong correspondence with Michele Besso, his close friend from university days, was entirely in German on Einstein's part, but Besso at times wrote in French or Italian, without apparent difficulties for the recipient. In 1913—rather late considering that his international reputation had begun to rise from his 1905 publications on special relativity, the photoelectric effect, and Brownian motion—Einstein confided to Besso that "I am learning English (at Wohlwend's), slowly but thoroughly."‡ [49] It proved quite useful after his emigration to the United States in 1932 and his residence at Princeton, but he was never quite comfortable with it: "I cannot however write English due to its underhanded orthography. If I read, I hear it before me and do not remember how the form of the word appears."§ [50]

Einstein was renowned for his attachment to the German language,

*"Da ich die 'Zeits. f. Phys.' durchaus also ein Organ der deutschen Physiker und nicht als eine Zeitschrift Deutschlands betrachte, empfand ich keinen Hinderungsgrund, meine Arbeit dort zu placieren[. . .]."

†"Dabei sind mir die konsequenten Schwierigkeiten durchaus nicht aus der englischen Formulierung entstanden, denn ich habe der bekannten Ausrede mit der 'schlechten Sprache' immer dadurch vorgebeugt, dass ich den Text mit guten Freunden einer eingehenden Revision unterzog."

‡"Ich lerne Englisch (bei Wohlwend), langsam aber gründlich."

§"Englisch aber kann ich nicht schreiben von wegen der hinterhältigen Orthogra-

and he deployed it with a grace and poetic feeling lacking in his charmingly ungrammatical English. Yet he apparently felt bound to the latter language for his publications and correspondence in latter years, reserving German for speaking to fellow émigrés or Americans. He spurned all attempts to tie him back to the German academic community from the moment he learned of the Holocaust, writing with great volubility to Otto Hahn, the co-discoverer of uranium fission, in 1948:

> The crimes of the Germans are really the most disgusting that the history of the so-called civilized nations has to display. The attitude of the German intellectuals—considered as a class—was not better than that of the rabble. Remorse and an honest will, the least that could be done in order to redeem things that might be redeemed after the enormous murder, have not shown themselves even once. Under these circumstances I feel an irresistible aversion against being associated with any single affair that embodies a piece of German public life, simply out of a need to keep clean.* [51]

Such views generated tension even with his closest friends, such as Max Born, one of the many evicted from the University of Göttingen because of the Civil Service Law. (Born eventually found refuge at the University of Edinburgh in Scotland.) He was raised in Breslau (now Wrocław, in today's Poland) with a classical *gymnasium* education, including Greek and Latin. (He was particularly fond of Greek.[52]) One of the architects of quantum mechanics, he developed an extraordinarily wide range of international contacts, producing what was quite possibly the first textbook on quantum mechanics from a series of lectures he delivered at MIT in Cambridge, Massachusetts, *in English*. (He published a German version almost immediately.)[53] When he suddenly found himself banished from his own institution, he of course consid-

phie. Wenn ich lese, höre ich es vor mir und erinnere mich nicht, wie das Wortbild aussieht."

*"Die Verbrechen der Deutschen sind wirklich das Abscheulichste, was die Geschichte der sogenannten zivilisierten Nationen aufzuweisen hat. Die Haltung der deutschen Intellektuellen—als Klasse betrachtet—war nicht besser als die des Pöbels. Nicht einmal Reue und ein ehrlicher Wille zeigt sich, das Wenige wieder gut zu machen, was nach dem riesenhaften Morden noch gut zu machen wäre. Unter diesen Umstanden fühle ich eine unwiderstehliche Aversion dagegen, an irgend einer Sache beteiligt zu sein, die ein Stück des deutschen öffentlichen Lebens verkörpert, einfach aus Reinlichkeitsbedürfnis."

ered emigration. At first, his options were grim. As he wrote to Einstein in June 1933, he despaired that his best offer might be one from Belgrade. "The scientific wasteland that probably still reigns there scares me, as well as the language. I am extremely ungifted in languages, and it seems to me almost impossible to learn a Slavic one. But if nothing else comes along, then I will undertake it."* He would prefer, he wrote, "to naturalize my children in a Western land, best would be England.[. . .] I also studied in England 26 years ago, know the language and have many friends."† [54] Then Cambridge, England, came through, and he gladly left "since I knew the country and the language."[55]

Throughout these difficult years, he continued his correspondence with Einstein, even at one point—just before the Battle of Britain—penning a letter in English. (Einstein responded in German.) Commenting on this document later, Born observed: "This is the first letter in English, which at that time was barely more familiar, but after the outbreak of war it was more appropriate to my voice than German was."‡ [56] After the war, just before he was awarded the 1954 Nobel Prize in Physics, Born suddenly found himself facing financial hardship and the need to retire. He opted to resettle back in Germany. "Life in Germany is again truly pleasant," he wrote Einstein in 1953, "the people are fundamentally shaken to rights—in any event there are many fine, good people. We have no choice, because there I have a pension, here I don't."§ [57] The sage of Princeton would have none of it, lambasting the stinginess of British bean-counters and Born's blitheness about returning to "the land of the mass murderers of our fellow tribesmen."¶ [58] While for Einstein, adherence to or rejection of German was a matter of moral principle—difficult but obligatory—for Born the question was

*"Mich schreckt die wissenschaftliche Öde, die da vermutlich noch herrscht, und die Sprache. Für Sprachen bin ich äußerst unbegabt, und eine slawische zu lernen, scheint mir fast unmöglich. Aber wenn nichts anderes kommt, so würde ichs unternehmen."

†"Ich möchte meine Kinder in einem westlichen Lande einbürgern, am liebsten in England.[. . .] In England habe ich auch vor 26 Jahren studiert, kenne die Sprache und habe viele Freunde."

‡"Dies ist der erste Brief in englischer Sprache, die mir damals kaum geläufiger, aber nach Kriegsausbruch meiner Stimmung gemäßer war als die deutsche."

§"Das Leben in Deutschland ist wieder recht angenehm, die Leute sind gründlich zurechtgeschüttelt—jedenfalls gibt es viele feine, gute Menschen. Wir haben keine Wahl, weil ich dort eine Pension habe, hier nicht."

¶"das Land der Massenmörder unserer Stammesgenossen."

one of expediency. Both scientists were torn by a larger conflict between pragmatism and rigor, but linguistic choice was one of the major tropes through which they debated it.

One of the most interesting cases is that of Lise Meitner—an Austrian-born Jew (although baptized as an adult) who worked with Otto Hahn on the problems of the uranium nucleus until the moment, after the *Anschluss* absorbing Austria into the Third Reich in 1938, when she was forced to escape the country. Infamously, the Swedish Academy awarded the first postwar Nobel Prize in Chemistry to her collaborator Otto Hahn alone, spurning not only her but his assistant Fritz Strassmann.[59] Both she and Hahn were linguistically dextrous—she had taught French at a girls' school in her youth, and Hahn had spent student years in London—but he was allowed to stay in his homeland; she was sent into exile.[60]

Meitner ended up in Sweden, without any ability to speak the language and grudgingly hosted at the Royal Swedish Academy of Sciences in Stockholm at the laboratory of Manne Siegbahn, who disliked the refugee.[61] With little alternative, she applied herself to studying Swedish, and soon spent long nights reading Swedish literature (but preferring to indulge her passion for ancient Greek classics—also in the original).[62] Max von Laue, the most outspoken anti-Nazi physicist remaining in Germany and one of the few Meitner maintained a correspondence with, was amazed at her facility. "As far as concerns my knowledge of languages, I fear that you overestimate it," she wrote von Laue in 1940. "My general incompetence in life makes itself felt also in my capacity for languages. I learn to read each language very easily and to speak it only with great difficulty. On the other hand you do yourself an injustice with respect to English. I happen to remember that you read the book *Gone with the Wind* in English and with enthusiasm. I have retained that memory because for me that book also made a very strong impression at that time, although in places it seems almost like a pulp novel."* [63]

* "Was meine Sprachkenntnisse anbetrifft, so fürchte ich, Sie überschätzen sie. Meine allgemeine Lebensuntüchtigkeit macht sich auch in meiner Sprachbegabung geltend. Ich lerne jede Sprache sehr leicht lesen und sehr schwer sprechen. Übrigens tun Sie sich selbst Unrecht mit dem Englischen. Ich erinnere mich zufällig, daß Sie das Buch 'Gone with the wind' englisch gelesen haben und mit Begeisterung gelesen haben. Ich habe das in Erinnerung behalten, weil mir dieses Buch seinerzeit auch einen sehr starken Eindruck gemacht hat, obwohl es stellenweise fast wie ein Colportageroman wirkt."

Margaret Mitchell notwithstanding, von Laue considered himself handicapped with foreign languages, despite schoolboy education in Latin, Greek, French, and German. The problem, quite understandably, was oral: "And if I now must speak at all in a foreign language, it sets me immediately into torment and never allows me to come to a fluent and correctly pronounced presentation."* [64] The issue was particularly severe in English, he recalled in an autobiographical memoir: "There was then no instruction [in English] in German gymnasia; I have later felt this to be the most terrible lack in my education. I learned English after my school years from scientific journals and books, that, already for a long time, had presented themselves as so indispensable; I spent months in America and was instructed in English there."† [65] Meitner likewise felt ill at ease with English, but yet "mediate[d] the correspondence for an array of friends and colleagues who have relatives in the belligerent countries, and this means a doubled writing and rewriting, and on top of that partially in the English language, which does not come easily to me."‡ [66]

After the war, Meitner's future employment and residence remained uncertain. Unlike Max Born, she felt she could not return to Germany. (She refused a chair at the University of Mainz.)[67] While Einstein excoriated those Germans who stayed, regardless of their own felt degree of complicity with the regime, Meitner had worked and communicated with certain individuals—like von Laue and Hahn—for so long during the darkest years that she believed she understood some of the pressure they were under. At the same time, she also felt they bore responsibility, and was unstinting in her praise for Max Planck, who had spoken with her honestly about the terrible things Germans were doing; his admission of personal responsibility was a balm to her.[68]

*"Und wenn ich nun gar in einer fremden Sprache reden mußte, so wurde mir dies geradezu zur Qual und erlaubte mir nie, bis zu einem fließenden und ausspracherichtigen Vortrag zu kommen."

†"Es gab damals keinen Unterricht darin auf den deutschen Gymnasien; das habe ich später als den bösesten Mangel meiner Bildung empfunden. Ich habe Englisch nach der Schulzeit aus wissenschaftlichen Zeitschriften und Büchern gelernt, die sich, je länger, als umso unentbehrlicher herausstellten; ich habe Monate in Amerika zugebracht und war dort auf das Englische angewiesen."

‡"Ich vermittle für eine Reihe von Freunden und Kollegen, die Verwandte im kriegsführenden Ausland haben, die Korrespondenz und das bedeutet ja ein doppeltes Hin- und Herschreiben dazu teilweise in englischer Sprache, was mir nicht leicht fällt."

When, after the war, Hahn and von Laue accused the Allies of tormenting Germans, she lost her composure and excoriated both of them for failing to recognize the enormity of the atrocities committed by Hitler and his minions.[69] After von Laue continued to resist seeing how the Allies and the victims of Nazism could feel justified in imposing certain hardships during the postwar occupation, Meitner chose to deploy the framework of losing one's language to make it vivid for him:

> You can I suspect not entirely comprehend how much one must have his natural and uninhibited behavior under control, when one as a 60-year-old person comes to a foreign land whose language one has until then never spoken, and if one is on top of that dependent on the country's hospitality. One never enjoys equal rights and is always internally alone. One always speaks a foreign language—I don't mean the external formulation of language, I mean mentally. One is without a homeland. I wish that you never experience this, and not even that you understand it.* [70]

By using precisely this kind of sentimental imagery, Meitner attempted to elicit a sentimental reaction. If von Laue couldn't understand Meitner's anger and frustration through abstract analysis, the way to make it concrete for him was to describe exile as linguistic alienation. While Born and Lanczos (and Arendt and Adorno) used their comfort with the German language as a proxy for forgiveness and continuity, Meitner, like Einstein, invoked her linguistic position as a fitting analogy to the trauma the Nazis brought into her world.

The Boycott That Never Was

During and after the war, old outrage at Germans, some of it left over from the Great War, resurfaced. If the Germans had been punished with a boycott after the First World War, when their crimes were in-

*"Sie können vermutlich nicht ganz realisieren, wie viel man von seiner natürlichen und unbefangenen Art unter Kontrolle haben muß, wenn man als Mensch von 60 Jahren in ein fremdes Land kommt, dessen Sprache man bis dahin niemals gesprochen hat, und wenn man dazu noch auf die Gastfreundschaft des Landes angewiesen ist. Man ist niemals gleichberechtigt und ist immer innerlich einsam. Man spricht immer eine fremde Sprache, ich meine nicht die äußere Sprachformulierung, ich meine das Gedankliche. Man ist heimatlos. Ich wünsche Ihnen nicht, es zu erleben, und nicht einmal, es zu verstehen."

comparably milder than Hitler's, should not the same reaction follow the Holocaust? Harvard physicist Percy Bridgman had opened the door to this kind of thinking already in 1939, when he advocated a voluntary preventative boycott—more of a self-imposed gag order—against Axis scientists, especially the Germans: "I have decided from now on not to show my apparatus or discuss my experiments with the citizens of any totalitarian state.[. . .] These states have thus annulled the grounds which formerly justified and made a pleasure of the free sharing of scientific knowledge between individuals of different countries."[71] Such measures could be understood as part of the war effort; now that the war was won, what kind of punishment should be exacted on German scientists?

Theodore von Kármán, the aerospace engineer and mathematician of Hungarian origin, fumed to Warren Weaver—director of the National Sciences Division of the Rockefeller Foundation, whom we will meet again in the following chapter—that a boycott was most decisively necessary. "He thinks that at least 80 per cent of all the present German faculties and German students are completely unrepentant and arrogant," Weaver recorded in his diary. "He says that if we do one thing for them, we will simply justify their own opinion of us as fools. When I ask him what he thinks we ought to do with them, he shrugs his shoulders and says: 'Just leave them alone for about fifty years.'"[72] Von Kármán's sentiments found support among many scientists.[73] No German scientists attended the first major conferences after the war, but this was more the result of travel restrictions imposed by the Allies rather than an explicit boycott. Max von Laue, due to his anti-Nazi reputation, was allowed to travel already by July 1946.[74]

So much of World War II seemed to repeat the mistakes of World War I, that Niko Tinbergen, the Dutch-born ethologist, wished not to replicate the botched postwar. Considering his long-standing collaboration with Austrian animal-behavior specialist Konrad Lorenz—whose connections with the Nazi Party were substantially closer than a staunch Resistance fighter like Tinbergen could easily stomach—Tinbergen felt "it is impossible for me to resume contact with him or his fellow-countrymen, I mean it is psychologically impossible. The wounds of our soul must heal, and that will take time.[. . .] In order to avoid the mistakes from 1918-'26, I did not want, as then, to begin cooperation between allied scientists and leaving the Germans out altogether."[75]

Tinbergen reflected a growing consensus, endorsed even by hardliners like Dutch-born physicist Samuel Goudsmit, who had lost his parents in concentration camps and spearheaded the American investigation into the Nazi uranium project. "It would be understandable if many among us were reluctant to converse with our German colleagues again as if nothing had happened," he observed after the war, but such individual resentment must be overcome to avoid poisoning German rehabilitation. "We must again communicate with them as in the days before Hitler. The exchange of scientific literature, now practically at a standstill due to our indifference, should be actively promoted."[76] Only a minority of émigrés considered a boycott a good idea, and Meitner, for one, was relieved that it was unlikely to come to pass.[77]

Yet the absence of a boycott was not enough to restore the state of German science to its interwar heights, let alone its upward trajectory of the first decade of the twentieth century. The damage to German as a scientific language was locked into the governance statutes of international organizations and the educational infrastructure of the United States, whose scientific community—and language—continued to blossom through the Second World War. Educational connections between the former scientific superpower and the newly emergent ones also needed help; as one 1978 analysis observed: "The break between the United States intellectual community and Germany in the thirties was radical and complete. With few exceptions the postwar effort to restore the broken ties started from point zero."[78] There was also a new scientific power on the horizon in the East, speaking a different language and writing with a different alphabet. The Soviet Union posed multiple new challenges to American science, and exacting vengeance on German physicists and chemists did not rank high on the list of priorities.

As Germany began to rebuild, and be rebuilt, after the war, Hannah Arendt allowed herself to feel a bit of optimism about her country, and her language. "And besides that the experience that German is spoken on the streets," she said to Günther Gaus. "That pleased me indescribably."*[79] This, at least, was something.

* "Und außerdem das Erlebnis, daß auf der Straße Deutsch gesprochen wurde. Das hat mich unbeschreiblich gefreut."

CHAPTER 8

The Dostoevsky Machine

Что же, сказали мы,—пора иностранным ученым изучить русский язык.

Сосед покосился недоверчиво,—да всерьез ли это сказано?

Всерьез—и очень всерьез! Без русского языка ужс ссйчас нельзя быть подлинно образованным человеком. . . .*

DAVID ZASLAVSKII[1]

Everyone called it an experiment, but it was more of a demonstration. On 7 January 1954, at the world headquarters of the International Business Machines Corporation (universally known as IBM) at 57th Street and Madison Avenue in midtown Manhattan, members of the press filed into a room dominated by the 701, IBM's first commercially available scientific computer. The 701 was an attraction in its own right: costing roughly $500,000 (over $4.4 million in 2014 dollars), it consisted of eleven separate units and took up as much area as a tennis court. (See Figure 8.1.) Computers were scarce commodities: there were roughly seventy computers in the United States in early 1954. The 701 was especially rare, having been shipped to its first customer—Los Alamos National Laboratory, America's nuclear weapons design facility—just a year earlier. (The Atomic Energy Commission, which oversaw the laboratory, controlled over 25% of the large computers in the country.)[2] They were about to witness this wonder machine perform a feat scarcely imaginable a decade earlier.

The 701 was going to translate Russian into English. Léon Dostert, the director of Georgetown University's Institute of Languages and

* "'All right,' we said, 'it is time for foreign scientists to study the Russian language.'

My neighbor looked askance at me incredulously: 'Is that said seriously?'

'Seriously—and very seriously indeed! Without the Russian language already now it is impossible to be a genuinely educated person. . . .'"

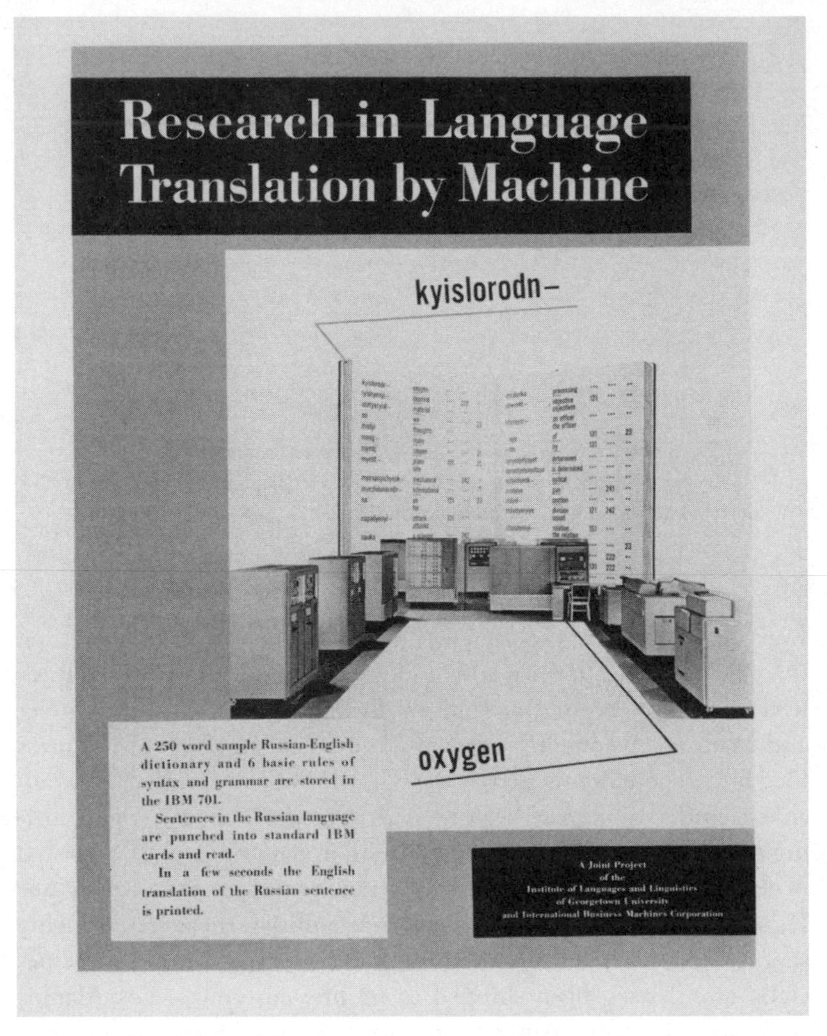

FIGURE 8.1. Publicity still from the 7 January 1954 Georgetown-IBM experiment, depicting the IBM 701. All of the objects in the picture (except the chair) are part of the computer. Courtesy of Georgetown University Archives.

Linguistics, and Cuthbert Hurd, the director of IBM's Applied Science Division, presided over the public unveiling of over a year's worth of work—mostly a collaboration between Georgetown linguist Paul Garvin and IBM mathematician Peter Sheridan—to apply a vocabulary of 250 Russian words and six rules of "operational syntax" to render over sixty Russian sentences into readable, indeed entirely grammatical English, at a rate of one every six to seven seconds. The IBM press release bubbled over with enthusiasm: "A girl who didn't understand a word of the language of the Soviets punched out the Russian messages on IBM cards. The 'brain' dashed off its English translations on an automatic printer at the breakneck speed of two and a half lines per second."[3]

Yet the real star of the day was Dostert. A dapper man sporting his customary well-trimmed mustache, he knew how to play to the crowd. He called the experiment a "Kitty Hawk" of machine translation (MT), meaning there was substantial work remaining on both the linguistic and computing sides of the problem before MT became widely applicable and functionally error-free. Yet he exuded optimism. "Those in charge of this experiment now consider it to be definitely established that meaning conversion through electronic language translation is feasible," he said, and predicted that "five, perhaps three years hence, interlingual meaning conversion by electronic process in important functional areas of several languages may well be an accomplished fact."[4]

Russian was the obvious place to begin, as Thomas Watson, Jr., chairman of IBM, noted: "We chose Russian because we believe that today it is of very great importance to be able to communicate with the Russians in the shortest possible time with the hope that through increased understanding we will be able to make faster progress toward the goal of world peace."[5] Also crucial for Dostert, however, was scientific communication: "The value to research of having current literature in scientific fields readily and promptly available in various idioms is another practical objective."[6] MT of scientific texts would open the door to full automatic translation of not just Russian science into English, but any words in any language into any other. The press was appropriately amazed, and a series of articles flooded popular and scientific media about the "Georgetown-IBM experiment" and the future of MT.[7]

There was very little "experimental" about what happened on that January day. In early October 1953, progress was going so well on the highly constrained language program that Dostert anticipated a public demonstration to be possible in "early November."[8] Complications arose, however, and the public trial run was pushed back. In the archives of Georgetown University, you can find a dot-matrix printout of transliterated Russian sentences accompanied by English translations—similar but not identical to the sentences which appeared at the public trial in January 1954, signed by Dostert and addressed to Father Edward B. Bunn, S.J., the president of Georgetown University, with the exultant aura of a recently accomplished feat:

> On this day [24 November 1953], at 11:45 A.M., in the headquarters of the International Business Machine Corporation, 57th and Madison Ave. in New York City, the first segment of language translation by electronic-mechanical process was achieved on this sheet. This experiment was the result of research conducted jointly by the University's Institute of Languages and Linguistics and the I.B.M. The formula for the testing of the basic principle was prepared by the undersigned, who gratefully presents this paper as a memento to Father Rector.[9]

By late November, Dostert and Hurd knew the experiment would be a success. Dostert arranged for two demonstrations on 7 January; the big public shebang was the second. Earlier, about forty government officials received a separate briefing, "because many of the gov't officials are security minded and do not wish to be publicly associated with the project."[10] Security minded indeed: representatives from the Central Intelligence Agency (CIA), National Security Agency (NSA), and Office of Naval Research (ONR) were among those present.

The birth of MT represents the convergence of several strands of the complicated history of scientific languages in the early years of the Cold War. On the one hand, attention to science, especially Soviet science, was on the rise. Without question the most significant event for American perceptions of their geopolitical rival's science and technology was the Soviet launch of the first artificial satellite, Sputnik, in 1957. Yet anxiety about Soviet achievements stretched back at least to the surprisingly early detonation of the first Soviet nuclear device in 1949. The Cold War was shaping up to be, at least in part, a scientific race, and maintaining good intelligence about the other side was essential. The Soviet Union's science infrastructure became the largest in the world in the 1950s, and in principle one could simply follow the vast published literature and obtain an adequate lay of the land. There was, however, a practical obstacle: Russian. Thus, almost as soon as the guns stopped firing in 1945, the American science establishment came to believe itself deep in the throes of a crisis of scientific language.

Obviously, government power brokers were not especially interested in Soviet findings on the chlorination of benzene; instead, political pressure in the United States was focused on reading Soviet documents for intelligence purposes. The goal was to translate Russian rapidly, full stop. Then what was special about *science*? Over the course of the first postwar decade, we see a sharp redefinition of scientific language as a peculiar subspecies of language—one that was lexically, semantically, and syntactically simpler, and thus more tractable to the limited computer power available. MT was spawned by the interaction of two forces: scientists who wanted to read Soviet publications, and backers who became convinced that scientific language was the key to unlocking the secrets of Russian grammar. To see how all this came about, we have to delve into the mood of the late 1940s concerning Russian. That is, we have to panic.

The Russians Are Writing!

The numbers were bad, and getting worse. According to one estimate in 1948, more than 33% of all technical data published in a foreign language now appeared in Russian; even German had reached only 40% of such data at its peak.[11] The most comprehensive source of information on foreign publications was the *Chemical Abstracts*, a publication of the American Chemical Society that was necessarily limited to that science (broadly construed), but had a lengthier series of figures. In 1913, the number of Russian-language publications abstracted was 2.5% of the total; in 1940, that number had risen to 14.1%, and in some subfields like mineralogic chemistry was a whopping 17.3% of the published papers. Much of this expansion came at German's expense: over the same period, the once-regnant language of chemistry had imploded from 34.4% to 13.4%. By 1958, among the fifty different languages from which *Chemical Abstracts* drew their information, Russian (17%) trailed English (50.5%), but was already greater than German (10%) and French (6%) combined. In 1970, Russian had reached 23%, and the Soviet Union was producing as many publications in chemistry as the United States.[12] Chemistry was not an extreme case, either; the American Geological Institute calculated in the early 1960s that the Soviets were producing 29% of the world's geological literature, and the United States 23%.[13] These numbers must be read alongside the growing baseline: the amount of chemical activity in the world from 1909 to 1939, for example, had *quadrupled*, so Russian's percentages here were vaster slices of an ever expanding pie, a growing mountain of Cyrillic science. As the editor of *Chemical Abstracts* laconically put it in 1944: "The necessity in chemistry of the reading of Russian will increase."[14]

This need not have generated a crisis. After all, Americans had managed to keep track of German-language literature during its dominance. They simply, as a matter of course, learned German. But suddenly, as

World War II receded into the past, American scientists looked around and found their hard-won knowledge of Teutonic word order and compound nouns less and less important. In a 1958 survey, 49% of American scientific and technical personnel claimed they could read at least one foreign language, yet only 1.2% could handle Russian. (The situation was even worse for Japanese, which had as many publications in this period as French and yet only 0.2% of the sample claimed competence in it.)[15] That in itself was an improvement. In 1953, the National Science Foundation had sponsored a sampling of 400,000 scientists and engineers, and found roughly 400 who could read Russian without difficulty—a tenth of a percent.[16] Some scientists dismissed the Cassandras, declaring that Soviet science was biased and overly skewed toward applications and therefore could be safely ignored, but such arguments sounded less and less frequently.[17] The architects of science policy were very worried. From these quite low numbers, by 1962 some 5.6% of American scientists claimed to read Russian. That seems like progress, but it must be compared with the backdrop of 50% who knew German and 35% who knew French.[18] For the newborn National Science Foundation (created in 1950), language was the central problem: "The principal barrier to gaining knowledge of the Russian scientific effort is the Russian language. Very few scientists and engineers are able to read scientific papers in the original Russian."[19]

What could be done? What might be the solution to an incipient (nay, already present!) Babel of insurgent national languages which were overwhelming the delicate ecology of English, German, and French—but, for these scientists, mostly English? We have seen this question before, and a tiny minority of commentators once again proposed the same solution: a constructed language. For some, the answer was not just any constructed language, but the *same* language we encountered in chapter 4: Esperanto. On 10 December 1954, the United Nations Educational, Scientific, and Cultural Organization (UNESCO) passed a resolution encouraging the adoption of Esperanto for international communication. Outside of Europe (always Esperanto's stronghold), the Japanese were principal advocates of the language, and the Brazilian Government Institute of Geography and Statistics officially accepted Esperanto as its auxiliary language even earlier, on 18 July 1939.[20] Esperanto, however, remained marginal to scientific languages.

Interlingua was a somewhat different story, representing an American-centric Cold War approach to Scientific Babel, although with interwar roots. In 1924, chemist Frederick G. Cottrell, Ambassador David Hennen Morris, and his wife Alice (née Vanderbilt Shepard, granddaughter of William Henry Vanderbilt) established the International Auxiliary Language Association (IALA), dedicated to creating a viable auxiliary. Assembling a team of scientists, engineers, journalists, and assorted intellectuals, the IALA selected German-American linguist Alexander Gode to construct the eventual solution.[21] Or, rather, *extract* it. For Gode, there already *was* an international language located within science:

> In interlinguistic terms all this means that even though the "language" of science and technology is not a full-fledged language, even though it can supply us only with a vast number of words and phrases of international validity in various peculiarly national but easily recognizable forms, it does represent a nucleus of a complete language. It does represent fragments of the only international language we have.[22]

That is, when the collected scientific texts of the world were perused, "the result will be a welcome *rapprochement* of the several systems and projects [of constructed languages] which may thus be more clearly recognized as what they really are: variants or dialects of the same interlingua."[23] In 1951, they published their system and produced a series of primers to educate people in reading the language.[24]

Interlingua faded from view by the mid-1960s, but a decade earlier it seemed a potential solution to the cacophony of languages. Explicitly drawn from scientific publications, Interlingua found a natural home among their ranks. Thanks in large part to Gode's exhortations and Alice Morris's financing, a few publications undertook printing abstracts in Interlingua: first the *Quarterly Bulletin of the Sea View Hospital*, followed by the *Journal of Dental Medicine*, the very prestigious *Journal of the American Medical Association*, and the *Danish Medical Bulletin*. In line with this heavily medical theme, the official program of the Second World Congress of Cardiology, which took place in Washington, DC, in 1954, contained summaries in English and Interlingua, the first mass trial of the language.[25] Full periodicals followed, with the journal *Spectroscopia Molecular* appearing in 1952, followed by a newsletter *Scientia International*. As historian of constructed languages Arika Okrent observed: "By attaching itself to science, and refraining from grand claims, Interlingua spread a little further than it otherwise might have."[26]

Reflecting on this brief efflorescence of interest in constructed languages—*Spectroscopia Molecular* folded in 1980, a decade after Gode's death removed the movement's guiding spirit—exposes in miniature this Cold War notion of a "scientific language." Interlingua was built out of the vocabulary that Gode pulled from science, and thus like Ido seemed related to the international scientific project, but unlike Ido was designed to be *read*, not spoken or written. Specialized abstracters would render English, French, Russian, or Malayalam into Interlingua, and then those translations would be open to all. This limited internationality stopped when you moved from the abstract to the article, and it reflected a largely textual scientific community. In the early Cold War, with limited personal contact between Soviet and American scientists, this seemed reasonable, and the attention to abstracts and publication reflects a primarily American understanding of the scientific language barrier, which explains the Western orientation of the Interlingua publications. In the end, the constructed language could not generate enough enthusiasm to appeal to Americans, let alone the Soviets. The Soviets faced an even larger language barrier—they had to deal with the overwhelming quantities of English publications, setting aside German and French—which raises the question of how they coped with what the Americans understood as a linguistic catastrophe.

How the Other Half Read, Especially Science

To understand how Soviet scientists themselves treated the language barrier, it is important to recognize that Russophone space—that is, the Soviet Union—was not monoglot. The United States functioned (and still functions) as primarily an Anglophone society, even though there is sizeable linguistic diversity among immigrants, heritage speakers, and Native American populations. By contrast, the Communist superpower was a better analog to India: enormous linguistic heterogeneity, with a leading cultural language (Russian, Hindi/English) used for bureaucratic unity while tolerating, even encouraging, regional difference. Parts of the former Soviet Union, especially the Caucasus, remain some of the most linguistically diverse regions in the world. Politically, the Soviet leadership exploited its status as a multilingual country—visible in the fifteen official languages of the fifteen constituent republics, or in the many regional languages within each of those units—for propaganda value in the decolonizing world, a source of some concern to American commentators. Language was so central that it was "recognized as the main criterion of nationality in the USSR."[27] There were as many different peoples as there were languages, and there were an awful lot of languages.

That was the image, and during the first two decades of the Soviet regime it was more or less the reality too. Beginning in 1917, V. I. Lenin's official policy toward "minority languages" granted enormous linguistic (but of course not political) autonomy to the peoples who spoke them. In order to extirpate what the regime saw as baleful religious influence, Turkic-speaking peoples (such as the Azeris or Uzbeks) were required in August 1929 to discard their Arabic-derived alphabets, but instead of switching to Cyrillic—which might be reminiscent of Russification policies pursued by the loathed Tsars—alphabets derived from Latin were created for them and for dozens of "pre-literate" minority peoples. This policy changed in 1938, when learning Russian became a compulsory subject in Soviet schools, and in 1939 all of the Latin scripts were discarded and every alphabet for the hundreds of Soviet languages—with eight exceptions—were transformed into Cyrillic.[28] Both measures resulted in significant Russification of these languages, especially with respect to the stock of nouns, and in no field so much as science with its well-developed Russian vocabulary.

The 1938 reform of language policy greatly benefitted native speakers of Russian, and was driven in part by the need for a common language in All-Union institutions like the Red Army. Officially, Russian was still primus inter pares, to be taught to every child in the country starting in the second grade, even though education might still be dominated by the local minority language. In some regions, such as Dagestan, seven languages were used in schools, but this case was exceptional. As children advanced through the grades, Russian became more entrenched and minority languages dropped out; all higher education in Russia and the five Central Asian republics took place in Russian.[29] In science, the dominance of Russian was essentially total. For example, a dissertation had to be written in Russian or the title language of one's republic; but if a scholar took advantage of the latter option, he or she had to have it translated into Russian so that a Higher Attestation Board could approve it.[30] World War II intensified the increasingly Russocentric character of the Soviet Union, and by 1949 it was easy to find statements such as: "The Russian language is great, rich, and powerful, it is a tool of the most progressive culture in the world."* [31]

This overwhelming emphasis on Russian vis-à-vis all the other languages of the multilingual Union did not imply the neglect of foreign languages. On the contrary, Soviet scientists and engineers were compelled to master at least one foreign language so they could navigate through the scientific literature. The development of this policy mirrored the marginalization of minority languages internal to the country. In 1932 foreign languages were officially introduced into Soviet schools with an emphasis on grammar and not on use, apparently with the goal of training translators. By 1948, schools were established in major Soviet cities where the language of instruction itself was a leading foreign language.[32] A fourteen-year-old in the Soviet school system would already have received three years of a foreign language, and with very few exceptions (for fields such as physical education or agriculture) admission to university was contingent on passing an oral test in English, German, or French, with Spanish added in 1955. (There was extensive debate in both the Soviet Union and the United States about the rigor of these examinations in practice.) True, the number of hours of compulsory language study dropped from 270 hours (four hours a week for two years) in 1950–1951 to half that by Joseph Stalin's death in 1953, but in 1954 the Communist Party boosted language requirements back up, specifically citing concern for science. Despite quibbles about quality, an American analyst noted that "the fact remains that *every* student in the Soviet institutions of higher learning studies at least one foreign language."[33]

For chemists, that language was usually English, although some Soviet chemists contended German remained essential for the still important prewar literature.[34] Soviets cited disproportionately more Russian-language studies (according to a 1966 study, 51.6% of the time) compared with its share of the global literature, but this was true for everyone: the French tended to cite French works disproportionately, the Germans German, and of course the British and Americans English. Nonetheless, that meant 48.4% of Soviet citations were to foreign material, and almost half of that was to English-language publications, which roughly 80% of Soviet scientists were ostensibly able to read.[35]

*"Русский язык велик, богат и могуч, он является орудием самой передовой в мире культуры."

Even as early as 1951, a bibliographic survey of the languages of chemistry noted that "the Russian chemist relies on the chemical literature of other countries to a greater extent than the American chemist."[36] A year after Sputnik, Jacob Ornstein, a frequent commentator on scientific Russian, expressed the common perception starkly: "Everything considered, there seems little doubt that the Soviet language effort is the most sizable one of any leading modern nation and that the American program dwarfs by comparison. If one may speak of a 'language race,' all signs indicate that the Soviet Union is well in the leading position."[37]

Raw numbers told only half the story. In the late 1940s, at precisely the moment the Cold War set in and Stalinism entered its final phase, language became intensely politicized within the Soviet Union in a manner that severely diminished Westerners' access to Russian science for the following decade. There were two key episodes in this renewed militancy surrounding the Russian language, one of which was quietly issued by fiat in 1947, and the other presented as a great public debate in 1950. Both had scientific controversies at their roots, and both would frame the development of machine translation.

Consider the latter incident first. Academic disciplines featured as zones of heavy contestation from the early years of Soviet power, but the intensity of conflict waxed and waned in response to internal dynamics within the Kremlin. The postwar moment was one of renewed clashes, perhaps the most vigorous of the entire Soviet period, and the two most visible fields were biology and linguistics. The biology story is a sad one, but it is quickly told. Beginning in the 1930s, a young agronomist named Trofim Denisovich Lysenko challenged the fast-consolidating scientific consensus that heredity was transmitted by genes that remained essentially unaffected by an organism's surrounding environment, proffering instead a modification of Jean-Baptiste Lamarck's early nineteenth-century theory of the inheritance of acquired characteristics. According to Lysenko's doctrine of "Michurinism"—named after Ivan Michurin, a quirky Russian plant breeder—heredity could be manipulated through a series of practices called "vernalization" in order to generate desired qualities (such as greater resistance to cold or higher yields) that would be passed on to future generations. Lysenko effectively packaged his attacks on geneticists in the discourse of Stalinist ideology. The conflict went into abeyance during the war, but erupted with renewed vigor immediately afterward, resulting in an August 1948 declaration by Lysenko that the Central Committee of the Party—that is, Stalin—had approved his theory and condemned the geneticists. Genetics remained an officially forbidden doctrine in the Soviet Union until Lysenko's fall from power in 1965.[38]

Academics of all stripes took note: the Party had intervened in an intellectual dispute and established an orthodoxy around the more Marxist-inflected variant. Most scholars tried to keep their heads down; others scented opportunity. A vocal and idiosyncratic faction of linguists felt that the time had come to impose the linguistic theories of Nikolai Marr (1865–1934) as decisively as Lysenko's "creative Soviet Darwinism." Beginning in the 1920s, Marr aggressively promoted his "New Theory of Language," which rejected the dominant historical-comparative framework of Western linguistics—in which languages were grouped into "families" such as Indo-European or Semitic on the basis of inferred common descent from a proto-language—and instead argued that language had developed independently and repeatedly across the world, with stages of linguistic development connected with socioeconomic conditions. That is, the rise of Indo-European languages such as Greek and Latin in the Mediterranean was not the result of migrations into the region of peoples who spoke languages belonging to this family, but rather the transformation of the underlying "Japhetic" languages. Marr insisted that peoples of similar classes around the globe would speak languages more similar to each other (with respect to certain linguistic features) than peoples of the same region from different social backgrounds.[39] Marr seemed the perfect complement to Michurin for an orthodoxy in Soviet linguistics: a native-grown doctrine, hostile to Western theories, saturated with Marxist rhetoric. Indeed, by April 1950 these linguists seemed poised to take over the field, and had already started eradicating conventional linguistic categories. (For example, the Sector of Comparative Grammars of Indo-European Languages of the Academy of Sciences was renamed "General Linguistics.") Stalin's intervention was all that remained.

Things did not turn out as expected. On 20 June 1950, Stalin published "On Marxism and Linguistics" in *Pravda*, arguing against Marr that language was not part of the "superstructure" to be influenced by socioeconomic relations. The Marrists were routed. There were numerous reasons why Stalin acted as he did: to keep a faction of academics in line, to ameliorate the negative impact Marrist ideas had produced on Soviet language teaching, and to improve relations with satellite countries that looked to those methods to provide unity among the Slavic languages.[40] An important consequence of Stalin's intervention was a renewed emphasis on the Russian language. At the Sixteenth Party Congress in summer 1930, a younger Joseph Stalin had predicted that a future universal language "of course, won't be either the Great Russian language or German, but something new."*[41] It seems his opinions had changed.

The other major development created less fanfare, but was no less consequential. Given how consistently Stalin's Soviet Union projected an aura of autarky, especially toward his death in 1953, it is surprising to realize that at the end of the war scientific collaboration with the West seemed possible—represented, for example, by the lavish 225th anniversary celebration in 1945 of the foundation of the Academy of Sciences.[42] Hopes were soon dashed by a controversy over two biomedical researchers, Nina Kliueva and Grigorii Roskin, who had published two papers in American journals concerning a potential cancer cure. A manufactured scandal about scientific espionage, dubbed the "KR Affair" after the protagonists' initials, enabled cultural elites such as Andrei Zhdanov, Stalin's ideological second-in-command, to assert Soviet particularism. On 14 July 1947, the Party's high command issued a verdict on publications, inspired by Kliueva and Roskin's perceived transgression (as translated by historian Nikolai Krementsov):

> The Central Committee considers that the publication of Soviet scientific journals in foreign languages injures the interests of the Soviet state, [and] provides foreign intelligence services with the results of Soviet scientific achievements. The Academy of Sciences' publication of scientific journals in foreign languages, while no other country publishes a journal in Russian, injures the Soviet Union's self-respect and does not correspond to the task of scientists' reeducation in the spirit of Soviet patriotism.[43]

This decision had momentous consequences for Soviet scientific publications. First, the three "prestige journals" published in foreign languages within the Soviet Union were shuttered: the *Comptes rendus* of the Academy of Sciences, in French, and the *Acta Physicochimica* and the *Journal of Physics of the USSR*, which published in English.[44] (The German-language *Physikalische Zeitschrift der Sowjetunion* perished in 1938, a victim of Nazism.) Even worse, it had been customary for Russian-language publications to include either abstracts in English or

*"конечно, не будет ни великорусским, ни немецким, а чем-то новым."

German (and sometimes French) or at least a translated table of contents in those languages. After September 1947, both were eliminated, making it all but impossible for non-Russian readers to figure out what was in a journal.[45] At the very moment when it had become more important than ever for Americans to follow Soviet science, and when there was more and more of it to read, the vast trove of it was locked behind the bars of the impenetrable Russian tongue.

Inventing "Scientific Russian"

It is not, of course, impossible to learn Russian. Millions of children routinely do so, and many millions of adults have as well (albeit with significantly more effort). The Stalinist transformations of Soviet scientific communication had renewed the pressure for Americans to learn how to read the language, and by the end of World War II the resources were in place to allow them to do so. The question was which *kind* of Russian they ought to learn.

The first Russian classes at an American university date to the last five years of the nineteenth century. Harvard College, spurred by the pressure of its chief librarian, Archibald Cary Coolidge, hired Leo Wiener to teach Russian in 1895, promoted him to assistant professor in 1901 and full professor in 1911, and then guided him to retirement in 1930. Across that career, Wiener seeded the slow bloom of Russian language teaching across the country. His student George Rapall Noyes was hired by Berkeley in 1901, and the University of Chicago and Yale shortly followed suit with appointments of their own.[46] Enrollments were low, but these institutions and the few that joined them persisted through the early decades of the twentieth century.

An awful lot of Americans already knew Russian, but they were not terribly eager to speak it. According to the 1910 census, which counted 92,228,496 Americans, 57,926 people listed Russian as their native language, out of a total of 1,184,382 who named Russia as their land of birth.[47] (Many of the latter were presumably Jews whose native language was Yiddish, but it is likely that quite a few also knew the language of the Empire.) That meant potentially 1.3% of the American population at that time could understand Russian, which is quite a good deal better than how we found our scientists at the beginning of the perceived translation crisis in 1950.

The fact that the capacity for Russian was not even lower among educated professionals at the dawn of the Cold War had a lot to do with World War II. With wartime mobilization on multiple fronts—rationing, the draft, massive armaments buildup, and so on—came a vigorous effort to train American officers to speak the languages they would need to conduct this global war. The Army Specialized Training Program (ASTP), which eventually encompassed some 40 languages, took on the task, transforming American language pedagogy. Before 1914, there had been only five universities in America with regular instruction in Russian, which had grown to 19 in 1939. By the time of the attack on Pearl Harbor, the ASTP had shot that number up to 86 campuses around the country, and after the Soviets' climactic victory at Stalingrad that number rose to 112 by 1946. Cornell, and then Georgetown—where Léon Dostert helmed the language program—were the first to adopt ASTP methods.[48]

Despite 211 universities and colleges offering Russian in 1953, however, enrollments dropped by 25% from the peak of 1947–1948, the opposite of the growth trend in other foreign languages. In 1954–1955, there were 4,000 students enrolled in Russian courses, compared with 70,000 enrolled in German, 95,000 in Spanish, and 110,000 in French.[49] Part of the difficulty was the absence of feeder programs from the high schools. Only 10 American high schools offered Russian in 1957, down from a high of 17 a decade earlier.[50] And then the Soviet Union launched Sputnik, and the whole situation changed, seemingly overnight.

Léon Dostert was paying attention. As he noted in 1960, only a crisis seemed to rock Americans out of their habitual disregard of foreign languages: "Prodded by unexpected and external developments—be it a Pearl Harbor or the orbiting of a Sputnik—we are suddenly brought to a realization that the national efforts in the teaching of foreign languages and related fields have not been adequate to meet our need."[51] Congress had passed the National Defense Education Act in 1958, funneling more than $28 million alone ($230 million in 2014 dollars) for pre-university education in foreign languages, besides even greater sums for science education. By 1959, four hundred American high schools offered at least one course in Russian. Thanks to this infusion, 19.1% of American high school students were enrolled in foreign-language courses—a huge improvement, although still lower than the equivalent 19.5% in 1934, and the whopping 35.9% enrolled in modern languages in 1915. (That same year, 37.3% of all American high schoolers were taking Latin.)[52]

Yet it was not enough. Despite a relative resurgence of training in

Russian among America's future scientists, the absolute numbers were appalling in the face of the juggernaut of Soviet publication—and did nothing to help current scientists who needed to grapple with this scholarly literature. Attitudes needed to change. Alan Waterman, the first director of the NSF, announced in November 1953 that the "problem of languages can be met on a long-term basis only by stiffer language requirements for science students."[53] To meet the challenge, planners redefined the Russian language, so that when scientists learned Russian, they would not be learning what you and I might conventionally understand as "the Russian language."

Instead, they would learn "scientific Russian" or "technical Russian." This was, according to most commentators, a different beast from the tongue of Dostoevsky and Pushkin—a more docile, friendlier beast. As one booster of this idea noted as early as 1944: "Many of the factors that make conversational and literary Russian so forbidding are absent in scientific Russian, and an impressive number of new factors, inherent in Russian scientific writing, come to the aid of the reader."[54] The latter are easiest to fathom: the international vocabulary of science (the same that had inspired Interlingua) and the presence of mathematical and chemical formulas made general orientation easier. But this was not what individuals like George Znamensky, who taught generations of scientists to read Russian at MIT, meant when he declared that "scientific Russian is comparatively simple."[55] They meant that *the Russian itself* was different. Consider V. A. Pertzoff's rather extreme take on pronouns from 1964:

> Let us do a little statistical analysis. Not counting the indefinite pronouns, there are approximately 350 bits of information which you must carry in your head if you wish to locate a particular pronoun in its proper place in the case-gender system.[. . .] In order to spare you unnecessary labor, we undertook the rather arduous task of determining which pronouns are most frequently used in scientific texts. Scientific language is specialized, and, of course, these findings apply only to this type of exposition.[56]

Not having to learn unnecessary pronouns or all the verbal forms? Now *that* made Russian easier. There were also syntactical transformations: a reliance on the passive voice (never mind that Russian has three ways of conveying this), simplicity of clauses, and authorial emphasis on clarity over stylistic virtuosity.[57] "Virtually everything about technical Russian," one advocate insisted, "except the alphabet and pronunciation, differs to some degree from the study of Russian as we normally know it. The aim, the scope, the student population, the teacher, the material, and the teaching method—all are specialized."[58]

The invention by American scientists and Russian teachers of this category of "scientific Russian" implied the need for different kinds of courses for these technically savvy students. Before World War II, American physics departments had typically required doctoral candidates to develop a reading knowledge of French and German, and few bothered with other languages. Throughout the 1950s, however, many graduate programs allowed the substitution of Russian to meet this requirement, and in fall 1958 the Mechanics Department at the Illinois Institute of Technology became the first program to require Russian as one of its two language qualifications.[59] Meanwhile, chemists "constitute[d] the largest group of students studying Russian," and self-teaching guides began to appear in major chemical journals as early as 1944 to meet the demand.[60]

New courses leapt into the breach: "In any institution which offers graduate work at the doctoral level, with its attendant language requirement, a technical Russian course is indeed a necessity."[61] Already in 1942 Znamensky began offering MIT students a yearlong course for three hours a week "enabling a good student by the end of the year to read scientific articles in Russian."[62] Debates raged about how much time one had to invest in learning this reduced, simplified, "scientific" language—or rather, how little. There were three-month courses, 16-week courses (with two one-hour meetings a week), double courses of two hours a week with 18-week semesters, and so on.[63] As of 1951, when general enrollments in Russian were declining, 24% of all schools that taught Russian also offered courses on scientific Russian, and by 1957, of the over 4,000 students taking the language, between 10% and 20% were enrolled in these specialized courses.[64] In order to assuage students' fears of the formidable language, conventional comparisons of the language as "very similar to Technical German with regard to sentence structure and inverted word order,"[65] or—in a more common but completely contradictory refrain—that "[p]erhaps the most important similarity is the word order, which is so nearly the same that, once the corresponding English words have been written under the successive words in a Russian sentence, very often no rearrangement is needed to produce understandable English sentences and minor rearrangement suffices to provide good idiomatic English."[66] (Given that English and

German do not share the same word order, both cannot be right. Personally, I think the English side wins this argument.)

Textbooks and scientific readers proliferated, ranging from James Perry's magisterial *Scientific Russian* to compressed pamphlets that read more like reference manuals than plausible texts for classroom study.[67] People even experimented with teaching by radio or television. The first instance of the latter, "Basic Russian for Technical Reading," was taught by Dr. Irving S. Bengelsdorf, a chemist working for General Electric's main laboratories in Schenectady, New York, for two mornings a week for twelve weeks. Originally meant for 250 scientists in the upstate New York area, it became a runaway success—its final audience reached between ten and twelve thousand.[68] Others that had nothing to do with science followed, which demonstrates a significant point: in matters of language training during the early Cold War, scientific languages often led the way, both by highlighting the language barrier, and—crucially—in specifying the kind of Russian to be mastered. Amid all this tumult, a man from Georgetown emerged as an unlikely messiah.

Mr. Dostert's Wondrous Device

Léon Dostert was the last person one would have imagined working to replace human translators.[69] He was born on 14 May 1904 in Longwy, France, a few kilometers from the Belgian border. When he was ten, he found his village overrun by German troops in the European cataclysm of the Great War. As a schoolchild in occupied territory, he was forced to learn German, which he mastered quickly, and the Germans set him to work as an interpreter. After the Americans liberated Longwy, Dostert began to study English, which he likewise soon commanded. The sickly teenager, weakened by the hardships of wartime, once again translated, and the American soldiers became fond of him. When his health recovered, a few of them sponsored the boy to study in the United States, and he enrolled in Pasadena High School in 1921, and then at Occidental College in Los Angeles three years later. He transferred from Occidental to Georgetown University, and earned his bachelor of science from the School of Foreign Service in 1928 (and an additional bachelor of philosophy and master's degree from Georgetown in 1930 and 1931, respectively). He studied at the Sorbonne for a year, and began work toward a doctorate at Johns Hopkins University (completing the coursework in 1936). He was appointed Professor of

French at Georgetown in 1939 and named chair of the Department of Modern Languages.

Dostert had two indisputable gifts: a facility with languages and a talent for getting people to do him favors.[70] Both would stand him in good stead as Europe—and the world—was once more engulfed by war. In September 1939, France was again at war with Germany, and Dostert (still a French citizen), served his tour in the infantry as an Attaché at the French Embassy in Washington, DC. After the fall of France in July 1940, Dostert spurned the collaborationist Vichy regime and in August 1941 became an American citizen. Relieved of duty, he taught as Professor of French Civilization at Scripps College in California, where he penned a pamphlet to educate the American public about France's recent history in order to mobilize support for intervention in Europe.[71] After Pearl Harbor, Dostert was appointed a Major in the US Army, served as liaison officer to the Free French General Henri Giraud in North Africa, and was General Dwight Eisenhower's French interpreter. He also worked with the Office of Strategic Services (OSS), the wartime intelligence organization that would seed the CIA.

In 1945, having been decorated by the French, Moroccan, and Tunisian governments, Dostert (now a colonel) was assigned the unprecedented task of arranging for simultaneous translation of English, French, German, and Russian at the Nuremberg war crimes trials. As the story goes, while observing how distracting it was to have interpreters whispering all the time—and, even worse, the immense delays of consecutive interpretation—he hit on the idea of sequestering the interpreters in a booth and piping sound to the parties through headsets. He persuaded Thomas Watson, Jr., a prewar acquaintance who would become (in 1952) the second president of IBM, to have his company donate the equipment for the venture.[72] In 1946, Dostert was asked to do the same for the fledgling United Nations in Flushing Meadows, New York, and then ascended the ranks of international translation, moving to Mexico City in 1948 as Secretary General of the International High Frequency Broadcasting Conference under UN auspices. In 1949 he was called back to Georgetown University as the first director of the newly created Institute of Languages and Linguistics.

Dostert published little (essentially nothing in linguistics) but organized a great deal; most of his efforts were directed to either technological or institutional modernization of language instruction. In addition to administering the teaching of 36 languages, he established programs for teaching English in Yugoslavia and later in Turkey, with the goal of giving military officers a chance to familiarize themselves with the language before coming to the United States for training. He was also a fierce advocate of language laboratories—a controversial innovation at the time—and pioneered the "binaural apparatus" to enable students to simultaneously hear native and foreign language versions of the same text (a modification of the Nuremberg technology).[73] This man, a polyglot interpreter with little interest in formal linguistics and no facility with electronics, became the key proponent of MT.

Having lived through two world wars, it stands to reason that preventing a third lay at the root of Dostert's surprising foray into computing. His Institute, located at 1717 Massachusetts Avenue NW in Dupont Circle in the American capital, was geographically embedded among the pressures of the Cold War, and Dostert was convinced that translation was vital to national security. In 1951, in the Army journal *Armor*, Dostert questioned the efficacy of force commitments to the newborn North Atlantic Treaty Organization (NATO) precisely on the grounds of linguistic incommensurability: "This writer believes that unless the problem of multilingualism inherent in the creation of an integrated international force is recognized, properly defined and analyzed, and practical action taken to meet it squarely, we shall fall way short of our potential effectiveness in this important field." Committed contingents of "co-equal sovereign governments" spoke English, French, Dutch, Danish, Norwegian, Italian, Portuguese, and potentially Icelandic—a military debacle in the making.[74] Could the West do better? Within a year of publishing this article, a silver bullet presented itself to slay this nightmarish Babel.

Like the Cold War itself, Dostert's panacea emerged out of the rubble of World War II. Among the canonical technologies developed during the war—including nuclear weapons, radar, and the jet engine—perhaps the last to receive widespread attention was the electronic computer. As is well known, in wartime the calculating machine was turned to a variety of ends (including computing cross-sections for nuclear physics), but the most glamorous was code-breaking, and it was from this context that MT popped into the mind of Warren Weaver, the long-serving director of the Division of Natural Sciences at the Rockefeller Foundation. Drawing from his own wartime experience and a conversation he had in 1947 with British electrical engineer Andrew Donald Booth, Weaver wrote to MIT polymath Norbert Wiener—the son of Leo Wiener, the first professor of the Russian language in the United

States—on 4 March 1947 about the possibility of machine translation, later excerpted in a memorandum on the question he penned on 15 July 1949 and circulated widely:

> Recognizing fully, even though necessarily vaguely, the semantic difficulties because of multiple meanings, etc., I have wondered if it were unthinkable to design a computer which would translate. Even if it would translate only scientific material (where the semantic difficulties are very notably less), and even if it did produce an inelegant (but intelligible) result, it would seem to me worth while.
>
> Also knowing nothing official about, but having guessed and inferred considerable about, powerful new mechanized methods in cryptography—methods which I believe succeed even when one does not know what language has been coded—one naturally wonders if the problem of translation could conceivably be treated as a problem in cryptography. When I look at an article in Russian, I say: "This is really written in English, but it has been coded in some strange symbols. I will now proceed to decode."[75]

There are several points of interest in this passage: the role of the language barrier in complicating the postwar world order; the specific focus on Russian; and the emphasis on scientific texts. All three would become dominant themes of the first decade of MT. Wiener, fluent in several languages, dismissed the idea as computationally and linguistically unworkable: "I frankly am afraid the boundaries of words in different languages are too vague and the emotional and international connotations are too extensive to make any quasimechanical translation scheme very hopeful."[76] Others were more receptive, including universal science-policy maven Vannevar Bush, who responded in October 1949 that "I think the job could be done in a way that would be extraordinarily fascinating."[77]

Weaver could afford to indulge his pet ideas. In 1952 he sponsored the first Conference on Mechanical Translation, held at Wiener's own MIT from 17 to 20 June 1952.[78] At this point, the community of scholars interested in machine translation was rather small, but MIT had already appointed Israeli philosopher Yehoshua Bar-Hillel for a one-year position in this field (in collaboration with the Research Laboratory of Electronics, the postwar successor to the Rad Lab, where radar had been developed during the war), and his early papers on MT's philosophical and methodological problems proved foundational.[79] Bar-

Hillel brought together all of the fledgling field's advocates. (The total bibliography of works related to MT at this point comprised under two dozen research reports and publications.) Léon Dostert, curiously, chose to attend.

It is unclear why the organizers thought to invite Dostert, who had demonstrated no interest in this question before being asked to MIT. His five-year plan for the Institute of Languages and Linguistics, submitted in 1952, made no mention of machine translation, and yet by the end point of that proposed time-frame (1958), Georgetown would have the largest MT program in the country.[80] (The other grand venture at Dostert's Institute during the 1950s was the promotion of spoken Latin, a pet project of some of the Jesuit priests who administered the university.[81]) No written text of his presentation, entitled "Ordinary Translation and Machine Translation," survives, but a participant recalled that Dostert drew on his experiences at Nuremberg and the United Nations to present the perspective of human translation, describing "systems employed in setting up efficient simultaneous translation systems and also rapid printed translations in international gatherings. These systems were remarkably similar in their organization to machine organization for computer application. He confessed that he came to the Conference as a skeptic."[82]

He did not stay one long. "The experience and impressions gained at that conference," he later recalled, "led me to the conclusion that, for a plausible approach to the general problem, one would have to accept as a first postulate that the primary difficulty is really a linguistic one." A second conclusion followed: "[R]ather than attempt to resolve theoretically a rather vast segment of the problem, it would be more fruitful to make an actual experiment, limited in scope but significant in terms of broader implications." It was thus easy to see that "[t]he Georgetown-I.B.M. experiment was, in a sense, a direct result of this meeting."[83] Dostert had pitched a machine demonstration at the conference itself, suggesting, as one participant recalled, "the early creation of a pilot machine or of pilot machines proving to the world not only the possibility, but also the practicality of MT."[84]

The MT community was, from the beginning, "well aware of the linguistic and engineering problems involved," as Erwin Reifler put it, and did not proceed "in blissful ignorance of the manifold difficulties of the task."[85] How did you organize a dictionary given the limited amount of storage available on magnetic drums? Should you insert a dictionary entry for every morphologically different form of a verb ("think"/"thinks"), or create some algorithm to undo the transformations (but then how to account for the past tense "thought")? Could you codify rules for transpositions of word order? What about the omission of features of the source language (like the Russian particles "же" and "ли") that did not have counterparts in English? Or the inclusion of features in the target language absent in the source (such as definite and indefinite articles, which Russian lacks)? Would texts have to be simplified in advance by a native speaker of the source language, to eliminate lexical and syntactic ambiguities ("pre-editing")? Or would you need to rely on a native speaker of the target language to fix the output ("post-editing")? From the very first publication on MT in 1951, these issues were hashed out practically and theoretically, on computers and on paper.[86]

MT as a field was torn by serious debates about almost every assumption and approach, both before the Georgetown experiment and after, but there were two areas of conspicuous agreement. The first was the language of translation. Although MIT stuck with German, and there were ventures in French in Booth's laboratory in Great Britain, the vast majority of Anglophone researchers were interested in Russian, just as Warren Weaver had been in 1949. There were obvious geopolitical reasons for the attention to Russian as a source language, but there were also intellectual ones—namely, the perceived quality of Soviet technical achievements—which brings us to the second point of consensus: focusing on scientific and technological texts. This was a direct consequence of the widespread attention given to the category of "scientific Russian." If you had to focus on the Soviets because the sponsors wanted it, then advocates insisted the only way to handle the task was to target scientific Russian, since it was the only sort that was tractable. When looking for your lost keys at night, it is best to stay under the lamppost, where you have a prayer of seeing them.

Most researchers would echo Kenneth Harper's assertion in 1953 that "[i]t is only within this limited sphere of 'scientific Russian' that our mechanistic and perhaps naive approach is valid."[87] The year after the Georgetown-IBM experiment, he continued along the same vein, claiming that the simplification was not just lexical (limited word choice) but syntactic (how words were put together to generate meaning), for in "scientific writing, Russian sentence structure is definitely close to English—much closer than is normal for other forms of Russian prose." He extended his reasoning to morphology itself—that bugbear of dictionary creation in an age of limited storage:

> The problem of identification of verb forms is less difficult in scientific Russian than in normal Russian prose; scientists very rarely make use of the imperative, of the first person singular, present tense, or of the second person, singular or plural, present tense. In the present (or the future) tense, therefore, we need be concerned only with three forms: third person, singular, and first and third persons, plural. The following also require identification: infinitive, past tense (four forms), and present and past adverbial participle (four possible forms). This gives a total of eleven forms that we must be prepared to distinguish[. . .].[88]

We can see the influence of this conception of scientific Russian throughout the construction of the protocol for the Georgetown-IBM experiment.[89] Before actually approaching a machine, Dostert arranged a human simulation of how a computer might approach language. That is, he tried to break down the process of parsing and translating a sentence without any attention to meaning. The computer would not "understand" the text, so the humans had to approximate that state. The result was the "Card Test":

> This involved giving to individuals who did not know the source language, Russian, sentences in that language written in Romanized script. They were directed in writing to go through a look-up, not only of lexical items but of the syntactic manipulations as well. The look-up was based on instructions reduced to strictly mechanical terms rather than "thinking" operations. The subjects were able to take a sentence presented to them in Romanized Russian and to come up, by going through instructions a machine could follow, with a correct English rendition of the Russian sentences. True, it took them from 10 to 15 minutes to achieve the translation of a 10 to 15 word sentence. But the significant fact is that, without knowing the Russian language, and, therefore, without contributing anything except their ability to look up, which is what the computer is capable of doing, they came out with the correct English version.[90]

By this method, Garvin and Dostert isolated the rules of syntax that were minimally necessary to rendering the Russian as English, eventually settling on six basic operations—what Peter Sheridan, the IBM mathematician who handled the programming, would call "rule-tags."[91]

However, in determining the syntactic patterns that could be handled by the machine, the linguists in turn were constrained to select input sentences that could be processed using those rules and *only* those rules, a tiny subset of the perhaps over one hundred rules they expected to be necessary to handle arbitrary samples of scientific Russian (let alone texts drawn from any sector of the language).

The Georgetown-IBM system was what would later be termed a "direct" translation system.[92] It was designed to move from Russian into English without an "interlingua" to handle semantic features. (As a result, it could not be applied in reverse, to undo the English and yield up the Russian.) The grammar rules were tagged to individual words in the dictionary. Each of the 250 Russian words had up to three numerical codes attached to it: the Program Initiating Diacritic (PID), and the two Choice Determining Diacritics (CDD_1 and CDD_2). Those codes defined a binary decision tree so that the program could select between two dictionary definitions (no word in Dostert's dictionary had more than two definitions, and many if not most had only one), or between retaining the word order or inverting it. For example, if the PID of a word was "121," then the computer should scan the following complete sentence word and see whether its CDD_1 was "221" or "222." If the former, it should select the first English equivalent in the dictionary; if the latter, the second. If the PID was "131," was the CDD_2 "23"? If so, select the second English equivalent in the dictionary and retain the word order; if it was not, select the first equivalent and invert the word order. And so on.[93] There are obvious limitations to this system: the assumption that all choices could be reduced to two; the redundant coding required of every single dictionary term; the absence of negative particles or compound and interrogative sentences; and the ability to scan only one word forward or backward, rendering it unable to deal with complex inversions or rearrangements of adjectival phrases.[94]

Nonetheless, the result was impressive, by any measure. For ease of programming, romanized Russian sentences (using a rather idiosyncratic transliteration system) were rendered on punchcards (Figure 8.2), and then run into the machine. Here is a selection of the sentences with their rendered translations from the 7 January 1954 demonstration:

KRAXMAL VIRABATIVAYETSYA MYEKHANYICHYESKYIM PUTYEM YIZ KARTOFYELYA

Starch is produced by mechanical methods from potatoes.

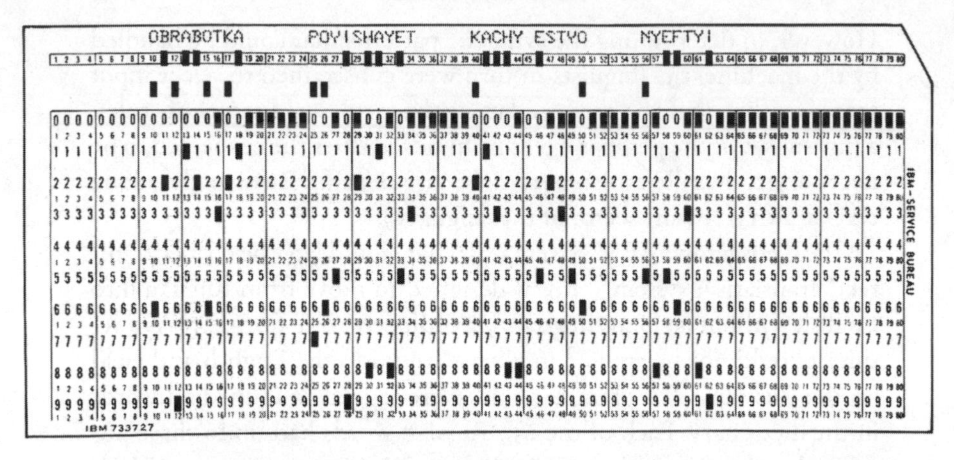

FIGURE 8.2. Punch card from the Georgetown-IBM experiment. This sentence was rendered as "Processing improves the quality of crude oil." Courtesy of Georgetown University Archives.

VYELYICHYINA UGLYA OPRYEDYELYAYETSYA OTNOSHYENYIYEM DLYINI DUGI K RADYIUSU

Magnitude of angle is determined by the relation of length of arc to radius.

MI PYERYEDAYEM MISLYI POSRYEDSTVOM RYECHYI

We transmit thoughts by means of speech.

VOYENNIY SUD PRYIGOVORYIL SYERZHANTA K LYISHYENYIYI GRAZHDANSKYIX PRAV

A military court sentenced a sergeant to deprival of civil rights.

DOROGI STROYATSYA YIZ BYETONA

Roads are constructed from concrete.

DYINAMYIT PRYIGOTOVLYAYETSYA XYIMYICHYESKYIM PROTSYESSOM YIZ NYITROGLYITSYERYINA S PRYIMYESYIYU YINYERTNIX SOYEDYINYENYIY

Dynamite is prepared by chemical process from nitroglycerine with admixture of inert compounds.[95]

In January 1954, and even today, this was amazing.

No Takers

Amazement was precisely what Dostert was banking on. Within five months of the demonstration he was already making the rounds to potential patrons with deep pockets, pitching his system to the Navy, the NSA, and other organizations, several of which showed real interest.[96] Dostert's ambitions were enormous judged on the scale of contemporary linguistics: "The plan, as now conceived, would involve the assignment of four part-time senior research consultants and eight to ten full-time junior research workers to do the linguistic processing; consultation with experts in various related fields, as required; and occasional testing on existing instruments of the language material as it becomes processed." If someone could supply $125,000 ($1.1 million in 2014 dollars), Dostert believed that within 12–18 months "the language data processed would permit the handling of a considerable amount of technical translation and would afford valuable experience for designing of an electronic instrument specifically built to handle language translation."[97] And yet, despite some nibbles, no one bit.

The price tag was only one of the issues. Paul Howerton, a specialist on Soviet chemical bibliography who was the CIA representative at the January demonstration, mentioned to colleagues that the experiment was "rigged" and "premature." Word got back to Dostert, and he was furious. In partial self-exculpation, Howerton explained:

> I did refer to the experiment as "rigged" in the jargon of the laboratory chemist. (To the chemist, a "rigged experiment" is one in which there is no variable as yet untested; i.e., a confirming experiment.) I regard the term "rigged" as synonymous with "controlled." I do not mean, in any sense, to impugn the validity or objectivity of the demonstration in New York.

But he would not back down on the issue of timing: "I felt that the demonstration was premature because of the several years research necessary to bring the instrument to actual routine operation."[98]

Perhaps, but MT seemed to be establishing itself nevertheless. Although later critics would declare that the Georgetown-IBM experiment had "no scientific value," they still credited it with alerting practitioners to the need for closer communications, so they could not be blindsided again.[99] Thus, also in 1954, William Locke and Victor

Yngve of MIT established the first journal for the field, *MT: Mechanical Translation*. A little later that same year Harvard University awarded the first PhD on the topic (in applied mathematics) to Anthony Oettinger. His subject: creating a dictionary to translate Russian to English—technical Russian, of course.[100] These were baby steps, however, and the field continued at a low simmer until Dostert managed to find a suitable patron, which happened in 1956. That was when the CIA finally signed on to Dostert's vision, and it did so because of news from Moscow. Once again, it was time to panic.

CHAPTER 9

All the Russian That's Fit to Print

> Органическая химия уже стерла черты между живой и мертвой материей. Ошибочно разделять людей на живых и мертвых: есть люди живые-мертвые и живые-живые. Живые-мертвые тоже пишут, ходят, говорят, делают. Но они не ошибаются; не ошибаясь—делают также машины, но они делают только мертвое. Живые-живые—в ошибках, в поисках, в вопросах, в муках.*
>
> EVGENII ZAMIATIN[1]

One day in 1954, Aleksei Liapunov, the Soviet Union's leading figure of cybernetics—the science of feedback and control that had been pioneered in the United States by MT-skeptic Norbert Wiener—was leafing through *Referativnyi Zhurnal*, a new journal of scientific abstracts, when he came across an interesting report. Hmmmm. Georgetown. Russian-to-English. Machine translation. Now *here* was an interesting idea. Since he could read English, he obtained the original article, wrote some memoranda, and organized a group of researchers at the Steklov Institute of the Academy of Sciences. Léon Dostert opened the door to experimenting with operational machine-translation devices, and Liapunov would walk right through. He brought company.

There is no question that the Georgetown-IBM experiment—or, more precisely, the reportage about that experiment—drove early Soviet research. Two main groups developed under the auspices of the Academy of Sciences. In addition to Liapunov's own program, which focused mainly on translating French into Russian, Dmitrii Iu. Panov

*"Organic chemistry has already rubbed out the boundary between alive and dead matter. It is erroneous to divide people into the alive and the dead: there are alive-dead people and alive-alive people. The alive-dead also write, walk, speak, act. But they do not make mistakes; machines also don't make mistakes, but they produce only dead things. The alive-alive exist in mistakes, in searches, in questions, in torments."

at the Institute of Precision Mechanics and Computing Technology emphasized English-to-Russian direct translation. Panov even visited IBM headquarters in New York to observe the 701 in action, and commissioned a fairly detailed account by two computer scientists of what was known about the experiment for the major journal of Soviet linguistics.[2] He and his colleagues were even permitted to publish a high-profile review of their work in *Pravda*, the Party's central newspaper.[3]

From these beginnings, it only grew. The first Soviet publications on MT began to appear in late 1955. Just three years later, a conference in Moscow drew 340 representatives from 79 different institutions (21 of these were ensconced within the cavernous domain of the Academy of Sciences) to hear 70 presentations. By 1964, Yehoshua Bar-Hillel—who had been the very first full-time researcher on MT in the world in 1952—declared the Soviet Union "the leading country of MT."[4] Given the relatively limited contact between the Soviet and Western groups in the early years, it is unsurprising that programming strategies began to diverge. The most significant difference was to separate the program "into two fundamental parts—analysis and synthesis,"* that is, parsing the sentence first and then demanding a different protocol to inflect the root stems and endings.[5] The Soviets also pioneered the development of "interlingua" programs, which rendered source material into an abstract code which could then be transformed into several other languages by independent protocols, a strategy indebted both to programming and linguistic traditions as well as the Soviets' need for multilateral translations due to the multilingual nature of their country.[6]

The Soviet Union erected this massive MT establishment out of fear of the Americans; the United States returned the compliment. News of Soviet interest in machine translation jumpstarted Léon Dostert's abortive efforts to obtain a large grant to develop the limited Georgetown-IBM experiment. Dostert noted with great satisfaction (and even greater understatement) in 1957 that publicity of Soviet experiments on their BESM machine "was not unrelated to a renewal of interest and support for work in MT in the United States. In June of 1956 Georgetown University received a substantial grant from the National Science Foundation [NSF] to undertake intensive research for the translation of Russian scientific materials into English. This grant has been renewed for a second year of continued research."[7] The push-me-pull-you character of the "MT race" between American and Soviet programs was an

* "на две основные части—анализ и синтез."

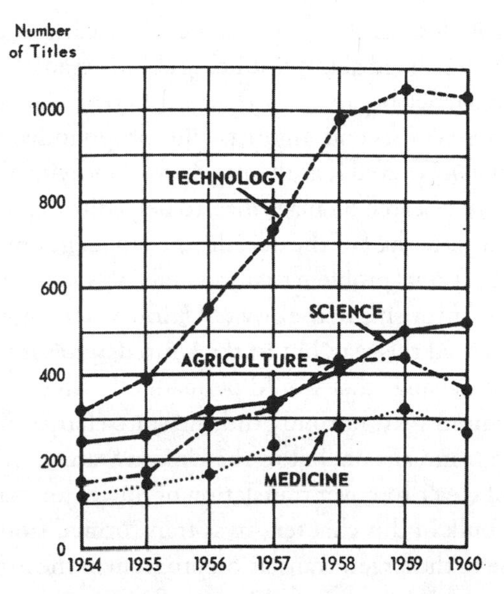

FIGURE 9.1. Number of different journal titles in science and technology produced in the Soviet Union, from 1954 to 1960. Both the sheer quantity and the massive growth were characteristic of Soviet scientific production during the Cold War. Boris I. Gorokhoff, *Providing U.S. Scientists with Soviet Scientific Information*, rev. ed. (Washington, DC: Publications Office of the National Science Foundation, 1962), 4.

open secret among the global community of researchers. Anthony Oettinger, who led Harvard University's program, later described it as "a kind of amiable conspiracy to extract money from their respective governments, playing each other off with various 'experiments' and 'demonstrations' that sometimes bordered on fraud."[8] These accusations evoke some of the negative evaluations of the original Dostert experiment, and that 1954 demonstration would continue to be a touchstone—positive and negative—for the developments that followed.

The Americans still perceived themselves in the throes of a translation crisis, a necessary by-product of the scientific and technological competition that gripped the superpowers. MT was only one of the solutions proposed to deal with the perpetual avalanche of Soviet publications in the natural sciences. The number of journals alone expanded almost exponentially, and each of these cried out for American readers to figure out what "Ivan" was up to. (See Figure 9.1.) Exhortation to learn Russian thus continued as a parallel strategy alongside MT. (And, ironically, one rather desperately needed by MT researchers themselves. One of the consequences of Soviet investment in this area was the production of valuable Russian-language studies that many of the American researchers complained about not being able to read![9])

The language barrier began to assume a different character in the mid-1950s. Those scientists and linguists who propounded the notion of "scientific Russian" viewed it as the province of individual decisions: to learn Russian, to research on machines, to pen editorials that the sky was falling. In the latter half of the decade, the language crisis began to be perceived as a national problem that was amenable to solution by the state. This was less surprising in the Soviet Union, where most scientific problems were viewed this way, but in the United States it represented something of a sea change that would permanently alter the terrain of debate. Thus, around 1955, not only did the state charge into MT on both sides of the Iron Curtain, but in the United States even the fledgling enterprise of cover-to-cover translation of Soviet journals—which will occupy the bulk of this chapter—was transformed from an entrepreneur's gamble to the largest translation program in the history of science. In following this intertwined history of MT and cover-to-cover translation, we see that the "language barrier" comprised three distinct, though related, issues: language of publication, quantity of information published, and access to the material. MT focused on the first to the neglect of the others, an oversight that would be partly responsible for its catastrophic collapse by 1966.

The Great MT Gold Rush

The beginnings of the 1956–1966 boom in machine translation can be traced not so much to the Georgetown-IBM experiment as to its progenitor, Léon Dostert. Dostert, as we have seen, persuaded Thomas Watson of IBM to donate the dictaphone technology that made simultaneous translation a success at Nuremberg and at the United Nations, and then again to underwrite the enormous opportunity cost of time on the IBM 701 that made the 1954 experiment possible. But during the war, while working alongside the Office of Strategic Services, America's intelligence organization, he met his greatest patron: Allen Dulles. In 1956, after the Office of Naval Research and Army Intelligence had declined to fund MT, Dulles came through in a big way. Three years earlier, Dulles had become the head of the Central Intelligence Agency (CIA) in the new administration of President Dwight Eisenhower. Dulles wanted to know what the Soviets were doing, he had far too few

Russophone analysts, and "Léon" claimed he could make this happen through his machines.[10] Dulles was willing to pay.

Although not, at first, entirely openly. Dostert had pitched an apparently very modest goal to the NSF in 1956: "To focus research for the purpose of achieving, mechanically, as complete translation as possible from Russian into English in the field of chemistry, primarily organic." Basing themselves on texts from the Soviet *Journal of General Chemistry* (*Zhurnal obshchei khimii*), the most important chemical publication in the Soviet Union, Dostert's team of researchers "will aim at the presentation of *unedited* Russian texts at the input and strive to achieve *semantically accurate translation* in English at the output, although the output material may require stylistic editing if this is found to simplify the storage problem." He expanded his earlier plan, now proposing to hire seven linguists, eight linguistic research assistants, five Russian-to-English translator-lexicographers, six bilingual clerical assistants, a bilingual secretary, and an administrative secretary, for a budget of $103,850 (over $900,000 in 2014 dollars).[11] Dostert made sure the award received wide publicity, both in Georgetown publications and in Washington newspapers.[12] What the reports did not say was that a good deal of this money was CIA cash, simply funneled through the NSF. The NSF numbers from 1956 through 1958 were $100,000, $125,000, and then $186,000—$305,000 of which was from the CIA—with subsequent direct CIA infusions without the NSF middleman totaling $1,314,869 (over $9.7 million in 2014 dollars). This was by far the largest award of funds for MT to any institution in the United States, and by 1962 even Georgetown's publicity team openly acknowledged CIA sponsorship.[13] When questioned by a Congressional committee about these sums in 1960, both the anonymous CIA witness (almost certainly Paul Howerton, the one-time MT skeptic who became CIA case officer for the project) and Dostert defended these numbers by laconically noting that the Soviet Union was even more heavily invested in MT than the Americans.[14]

Dostert built up an MT program at Georgetown commensurate with these sums, unheard-of for almost any project outside of nuclear physics or public health.[15] He stressed organic chemistry, because, as this 1959 internal report made clear, that science lent itself to MT:

> The theoretical necessity for such a routine lies in the fact that the number of organic compounds is in theory infinite; and in practice

> it is enormous. Therefore it seems wasteful to burden the main dictionary with literally hundreds of thousands of very long items. Secondly, compounds can be and are created in the laboratory for the purpose of studying them. Names are created for them according to established rules, names which can be translated by this type of routine. But after laboratory testing the particular compound may never be made again, and its name never appears in the literature again. Thirdly, the freedom with which carbon can combine with itself over and over and with other elements means that a dictionary which is relatively complete in other areas of the chemical language can never hope to have all the organic compounds in it. Therefore the need for a machine technique to analyze chemical terms.[16]

The corpus of words garnered from the analysis of only a few years of the *Journal of General Chemistry* was enormous (24,000 words by 1957).[17]

The linguistic results were promising. For example, the rules for adding definite articles to plural nouns in chemistry texts applied about 80% of the time—which seemed pretty good—and the rule worked "even for general texts, although to a lesser degree."[18] Coding continued apace. By the end of the decade, 85,000 more terms in organic chemistry had been keypunched, composed of about 8,000 distinct words, which reduced to 3,200 entries (notice the compression characteristic of scientific language). Labor costs became a concern, and Dostert rented commercial space in Frankfurt, Germany, in 1960, recruiting 200 keypunch operators for $80/month, a quarter of the American wage.[19] Georgetown won the coveted contract to translate Russian atomic-energy documents into English for both the Atomic Energy Commission at Oak Ridge, Tennessee, and for EURATOM in Ispra, northern Italy.[20]

Not surprisingly, Dostert's success sparked resentment. When A. D. Booth and William Locke published the proceedings of the 1952 MIT conference in 1955, Dostert's was the only essay to receive a cautionary editorial footnote: "Its inclusion in this book reflects the editors' desire to cover all aspects of the application of machines to translation and should not be taken as indicating their acceptance of all the author's views."[21] Dostert's lack of hard-core linguistic publications was probably the heart of the worry, but the showmanship contributed. Anthony Oettinger would later recollect Dostert as "a great conversationalist[. . .], but as a researcher I was unsure about him, whether he was just a figurehead or whether he was a bit of a fraud—the Georgetown MT demonstrations seemed always to be contrived; they made impressive publicity for the sponsors, but they soured the atmosphere by raising expectations that nobody could possibly fulfil."[22] Booth, for example, gleefully dismissed Dostert as "one of the less esteemed members of the American MT community."[23] MT colleague Winifred Lehmann was overheard describing him as "a wart on the field of linguistics."[24]

Yet Dostert's rising tide lifted all boats: the more he stumped for MT, the more grant money flowed to everyone. By 1960, five separate governmental agencies—NSF, CIA, the Army, Navy, and Air Force—were all funding mechanical translation at a steadily growing rate. The National Defense Education Act, passed on 2 September 1958 as a response to Sputnik, specifically indicated that the NSF and other groups "undertake programs to develop new or improved methods, including mechanized systems for making scientific information available."[25] That same year, the Army and the Navy joined in the funding boom.

Meanwhile Victor Yngve and William Locke at MIT devoted themselves to building a professional community of MT researchers. In 1954, the year of the Georgetown-IBM experiment, they established *MT: Machine Translation*, the first journal exclusively devoted to this topic. The early issues were composed on an electric typewriter in Locke's office, and eventually the journal moved to a commercial compositor. *MT* could only sustain that change with page charges, which granting agencies were initially happy to subsidize. In June 1962, a professional society for MT was founded, at which point the irregular *MT* had already published 52 articles and 187 abstracts in its total of 532 pages. It moved to Chicago with Yngve, and then to University of Chicago Press, but the costs proved too great and the journal foundered in 1970.[26] The story of *MT* is the story of MT in miniature: high hopes in 1954, massive grants, and then, around 1965, a precipitous collapse.

But catastrophe was the furthest thing from the minds of the dozens, and then hundreds, of linguists, programmers, statisticians, and engineers who flocked to machine translation. MT began to reshape linguistics in turn. Some have linked MT to the revival of structuralism—an approach developed decades earlier based upon the teachings of Ferdinand de Saussure (brother of the editor of the Esperantist *Internacia Scienca Revuo*). The tremendous postwar rise of structuralist analysis of language is often identified with the immensely influential work of Noam Chomsky in the late 1950s. Yet the receptive audience for that work was partially conditioned by the flurry of MT publications. In 1963, Dostert noted that the "development of structuralism in contemporary linguistics is at the basis of the concept of machine translation, since, without structuration procedures, the idea of sign-substitutions or automatic transfer of linguistic data would hardly be conceivable," and even his arch-rival, A. D. Booth, considered it axiomatic to assume "that structural linguistics as a science has already progressed to a state in which it is possible to devise adequate rules of procedure for translation from one language to another in terms which can be understood by a computing machine."[27] Structuralism's resurgence in Western thought—in linguistics, in philosophy, in anthropology—fit perfectly into the climate fueled by the intensity and raw financial support that flowed into MT.

This was especially visible across the geopolitical divide. Stalin's 1950 intervention in linguistics had the consequence of firmly establishing historical-comparative linguistics within the Soviet Union, as we saw in the previous chapter. In the Soviet context, that attention to the diachronic evolution and transformation of languages effectively countered structuralism's emphasis on synchronic analysis of linguistic structures. A year after Stalin's death, Liapunov jumpstarted Soviet MT. Given the symbiotic relationship between algorithmic machine-translation processes and structuralist analyses of language, it is no exaggeration to say that Soviet research at this intersection rescued Soviet structuralism, morphing Soviet linguistics into perhaps the most structuralist of any national community in the world.[28]

Abstracting the World

I have described the story of the Russian-English language barrier as principally an American story, one in which scientists and policy makers in the United States confronted a challenge posed by Soviet technical publishing. This lopsided emphasis has something to do with the sources: the Americans simply wrote more about this problem *as* a problem. Yet despite their panic, Americans could afford to be nonchalant about the language barrier. After all, by the 1950s over 50% of world publication in such sciences as chemistry was appearing in English. The Soviets could not just ignore that work, and so their approach to the language barrier tended to be more holistic than that of the Americans, who sometimes casually dismissed Soviet work as substandard. For the Soviets it was not just about *language*, but also about *quantity* of information and *access* to it. The Soviet solution to the latter two problems would in turn challenge American attitudes to the organization and distribution of scientific information.

Consider a seemingly simple problem: how did you learn about scientific findings happening outside your laboratory, whether down the street or across the ocean? (This was, recall, in the days before there was an Internet, let alone online databases or search engines.) One approach was to select the main journals in your field and then regularly thumb through each issue, studying the table of contents, reading many of the abstracts, and focusing on the relevant articles. Of course, if an article cited a significant paper in a journal outside the regular set, that would lead to another article, and crawling down the citation chain could enrich your research. Yet this approach was maddeningly incomplete, essentially guaranteeing missing important articles in your subfield unless some other scientist happened to have a broader bibliographic base and then published about it. There had to be a better way.

There was: the abstract journal. The most comprehensive of these in English was the American *Chemical Abstracts* (which surpassed the German *Chemisches Zentralblatt* in coverage by the interwar period). The editors of *Chemical Abstracts* surveyed a very broad set of journals, in several languages, and then paid a per-abstract fee to an army of chemists to summarize articles from the journals to which they were assigned. You used it like a massive index. Yet there were three difficulties with this system: size, speed, and scope. *Chemical Abstracts* was huge. By the late 1950s, each annual issue produced 100,000 abstracts spread across 10,000 pages of close printing; even the index was 5,000 pages long.[29] Physically handling these volumes, let alone extracting useful information from them, was a chore. Time was a related issue: the more articles there were, and the more journals needed to be covered, the longer it took to abstract the current year. And then there was scope: *Chemical Abstracts* was confined to chemistry (albeit broadly construed). Could the abstract journal be fixed?

By the late 1950s, it seemed to scientists on both sides of the Iron Curtain that the Soviets had done it. Russians had been abstracting for a long time, but partially and incompletely. The first Russian abstract journal, the *Guide of Discoveries in Physics, Chemistry, Natural History and Technology* (*Ukazatel' otkrytii po fizike, khimii, estestvennoi istorii i tekhnologii*) appeared from 1824 to 1831, and despite its title was hardly comprehensive, even for its limited lifespan. Individual subfields developed their own abstract journals in Russia: there was one for medicine from 1874 to 1914 and one in railroad engineering from 1883 to 1916, but the Great War and the Russian Revolution ended those. Very little was done to systematize scientific information during the first decade of Soviet power. On 9 January 1928, the state established a Commission for the Compilation and Publication of Indexes of Scientific Literature, designed to abstract everything published in the Soviet Union, but it soon bogged down under the weight of material and Stalinist upheaval. The 1930s saw a return to abstract journals in several fields, but once again war's advent ended several projects. Medical abstracts returned in 1948, but the other sciences were left uncataloged.[30]

Then, on 19 June 1952, the Soviet Academy of Sciences established an institute specifically to collate and publish information on scientific publications from around the world, responsibility for which was soon shared with the State Committee of the Council of Ministers of the USSR on New Technology (Gostekhnika). The new institute was called the All-Union Institute of Scientific and Technical Information, or VINITI in its Russian acronym. At first, VINITI's central product was *Referativnyi Zhurnal* (*Abstract Journal*)—from which Liapunov had learned about the Georgetown-IBM experiment—which assiduously sifted through the international literature. It quickly eclipsed its Western rivals. Beginning in 1956 VINITI also produced *Ekspress-Informatsiia*, translations of crucial Western articles and pamphlets into Russian, and in 1957 issued a monograph series, *Advances in Science and Technology*. The Institute also put out photo-offset copies of roughly 300 Western journals (such as the American behemoth *Physical Review*), identical to the original except for noticeably poorer paper quality. On 29 November 1966, VINITI assumed control of all the science-information services of the Union republics as well, becoming "the largest scientific information centre in the world," according to a British delegation. At the end of its first decade, VINITI's permanent staff reached 2,500, not including those working at its publishing house, or the 22,000 specialists who produced the over 700,000 abstracts it printed each year.[31]

VINITI was intended to be a solution to all three aspects of the language barrier. By centralizing information and reprinting foreign journals, it could more easily tame the exponentially increasing quantity of global scientific information as well as granting access to foreign scientific periodicals within the Soviet Union. As for the language aspect of the barrier, *Referativnyi Zhurnal* bypassed it. Each abstract in the sixteen subsidiary abstract journals (divided by science, with chemistry being the largest) had the same form: article title in Russian, author's name in Russian transcription, title in the original language, author's name in the original, name of journal, year, volume, issue, page, the name of the language, the abstract in Russian, and the abstractor's initials. All the foreign language text was reproduced in the original typography, whether the writing system was Cyrillic, Latin, Arabic, Devanagari, or Chinese.[32] Soviet scientists were expected to command several foreign languages, and many in fact did, yet VINITI continued to translate 85% of the world's scientific tables of contents into Cyrillic so that everything could appear in standardized form.[33]

As one might expect, it was difficult to keep such an enterprise going indefinitely. By the late 1970s, *Referativnyi Zhurnal* lost its edge while VINITI became increasingly strapped for resources and personnel, stabilizing its coverage at a whopping 1.3 million abstracts a year while the scientific literature mushroomed ever larger.[34] But in the 1950s and into the 1960s, VINITI was the envy of American science planners, ostensibly demonstrating why the Soviet Union had been able to assume the lead in the space race, as well as eclipsing the United States in the training of scientists and engineers. The tremendous American investment in machine translation has to be understood against the backdrop of the total picture of Soviet science-information efforts as obsessively tracked by Western observers. And just as the Soviets were pouring money into automated translation, the Americans believed they needed to do something else to surpass (or at least keep up with) the Soviets. Complete centralization was unlikely in the American political climate, but even a partial intervention to bridge the language gap would be welcome.

Retail, Wholesale, and Welfare Translation

According to the science press in the 1950s, MT played the starring role in the drama of Cold War scientific languages. As the decade progressed, however, a bit player began to assume an ever greater share of the lines: direct human translation of Russian articles into English. While MT focuses on the linguistic aspect of the language barrier, human translation adds access to sources by providing readers with a version of the article they want in a language they can read. (Both approaches are, however, bedeviled by quantity: the more material there is, the more there is to translate, and the harder it is to keep up.) The idea behind training Americans to read scientific Russian was, of course, to turn each scientist into his own translator. Translation journals were supposed to be a stopgap.

If you wanted to read a short story by Anton Chekhov but did not know Russian, you would look for a translation into a language you did know. Therefore it is not surprising that from a very early stage the notion of translating selected articles was seen as a remedy for the Cold War translation crisis. Starting in the late 1930s, a consortium of American petroleum companies employed one of the leading bibliographers of Soviet science, J. G. Tolpin, to edit and privately circulate translated tables of contents, abstracts, and selected Soviet articles on hydrocarbon and petroleum chemistry, a venture that lasted for eight years.[35] The shared interests of the consumers drove the choice of what to translate, and the deep pockets of the industry bankrolled the staggering costs.[36]

For the federal government, the selection problem was more fraught. After World War II, it outsourced the editorial selection of translations to those presumably in the know. For example, the American Mathematical Society in 1948 initiated a program (funded by the Office of Naval Research) to translate the highlights of recent Soviet mathematics, and the newly created Brookhaven National Laboratory on Long Island began to translate the tables of contents of important periodicals but soon found themselves swamped by even this limited quantity of material.[37] To pool privately commissioned translations, the NSF funded a Translations Center at the Library of Congress (as well as a selective article-translation program for atomic energy at Columbia University), building on the Translation Index developed by the Special Libraries Association (SLA) in New York City. In 1953, the SLA Translation Pool moved to the John Crerar Library in Chicago, and in 1956 assumed the duties of the Library of Congress in its entirety, issuing monthly catalogs of the translations deposited with them.[38] The United Kingdom's Department of Scientific and Industrial Research experimented with a hybrid of translation pool and translating service: if two or more researchers independently requested a translation of an article, the state would pay for it and deposit it for general access.[39] These approaches suffered from two intrinsic faults: they were unsystematic, the selections being made arbitrarily at the whim of the editors; and they were untimely, since by the time the translation was deposited and cataloged, others might have already commissioned translations, or the information might simply have turned stale.

Earl Maxwell Coleman, who by his own admission had "no translation skills whatsoever," stumbled into this ramshackle world of technical translation by accident, and founded a publishing operation called Consultants Bureau, Inc., with his wife, Frances, (and the measly capital of $100) in 1946. That year, Coleman learned of a trove of twenty-one tons of captured German technical documentation, and he sensed that someone could make money translating this material into English. Coleman approached the American Petroleum Institute, which had 100 microfilms of German-language technical reports at 1,000 pages per reel, and he made them an offer based on an unheard-of price scheme: $2 per thousand words instead of the industry standard of $12 (Coleman paid his men $10), but with multiple orders the price would drop gradually to a floor of 50 cents per thousand. That is, by pro-rating the translations he gave himself a guaranteed profit only if enough copies were ordered. Coleman kept losing money under this arrangement, while his translator, he claimed, flourished. Then Coleman had his second major insight: "I was paying him at a freelance rate even though I was keeping his lance at full tilt. It was as though I was paying him at a *rate* of $1,000 a week in a world of $50 a week salaries." Coleman ran back to the office and slashed the pay: from $10 to $4—starvation wages for a freelancer. "He ranted and raved and swore that he'd quit. He never did. Where else could he get as *much* work?—the *key*." By turning translation into assembly-line labor, Coleman changed the economics of the profession. He hired more translators, standardizing the job description in 1947 according to rules he maintained until the end of his career:

1) To work for me you had to have English as a mother tongue.
2) You had to have command of the target [sic] language because a) you'd studied it, or b) it was spoken fluently in your family.
3) If you were translating chemistry you had to *be* a chemist, or at a minimum have an advanced degree in chemistry.[. . .]
4) You had to be willing to work for me at $4 a thousand despite the impressiveness of the above demands. Implicit in the notion of so low a rate of pay was the following: You had to be able to translate *fast* or you wouldn't make enough money to keep you interested.[40]

In 1949, the same year Warren Weaver penned his memorandum on MT, Coleman revolutionized scientific publishing. He had developed a new industry but had no market. The difficulty was twofold: he focused on German, and there just was not enough demand; and he produced discrete articles. He decided to change both premises: "*Suppose*, I conjectured, you translated a whole Russian journal."[41] From the Consultants Bureau offices at 153 West 33rd Street in New York City, Coleman decided to translate the entire run of the *Zhurnal obshchei khimii* (the same journal Dmitrii Mendeleev had published in almost a century earlier, albeit under a different name), as *Journal of General Chemistry of the USSR*. Translations of the first issue of the 1949 volume appeared in November 1949, eleven months late, to a total list of thirteen subscribers. Coleman borrowed money to keep his business afloat; within five years he produced five journals and within seven offered twelve entirely in-house. Coleman had become one of the most powerful individuals in scientific publishing.[42]

Coleman modeled Consultants Bureau on a factory template. First, the Russian originals would arrive by air mail and translators were invited to select specific pieces from the tables of contents. The editor then distributed the work, ensuring no piece was left out, getting translations back six weeks later. Those were edited for style and referred back to the translators (and, when he later subcontracted for learned societies, to boards there) for queries, then typeset, and hit the shelves six months after they arrived.[43] With this mode of production on a vastly larger scale, Coleman reduced the expense to 18 cents per thousand words—that is, under *2%* of his 1946 expense (before taking inflation into account).[44] Who were these shockingly underpaid translators? A 1970 study of Coleman's stable found that most had PhDs and translated in their spare time, but there were full-time translators who had been there from the very beginning. This group turned out more than 34,000 pages of English from Russian originals a year.[45]

It wasn't pretty. The volumes initially came out on 8.5x11 sheets of paper—rather larger than the close-printed original—and were little more than bound mimeographed typescripts. The pagination did not match the Soviet originals, although starting in the "September 1949" issue (which appeared sometime in 1950) the table of contents listed both sets of page numbers. Images were crudely mimeographed, with all the annotations in the original Russian, and appended to the end of the articles rather than printed in-line as they came from the Soviet Union. By the second year the translated *Journal* included indices for author, subject, and organic chemical empirical formula. It also wasn't cheap. Coleman charged $7.50 for an individual article, $12 for an issue, and $95 for the whole year (almost $940 in 2014 dollars). Consultants Bureau took "cover-to-cover" seriously: there was no selection of the articles, and the idiosyncrasies of the Soviet original (lavish attention to the periodic table, nationalist priority claims, obituaries, historical pieces) were reproduced without comment.

The real question had to be: was it any good *as translation*? All too few American chemists could follow the Russian originals, so this was the best they had—but would an American chemist get *accurate* translations from Coleman's product? Consider a spot analysis on a randomly selected article by G. I. Braz on the reactions of ethylene sulfide with amines, taken from the third year of publication (1952). Early in the translation, one comes across this sentence:

> As we might expect, ethylene sulfide behaves similarly with diethylamine. When freshly distilled ethylene sulfide is added at room temperature to a solution of diethylamine in methanol, a white precipitate of polymeric compounds containing no nitrogen begins to settle out within a few minutes, the precipitation being complete within a few hours.[46]

Here is my rather literal rendition of the original Russian:

> As one would expect, ethylene sulfide behaves analogously also in relation to diethylamine. If one were to add at room temperature freshly distilled ethylene sulfide to a solution of diethylamine in methanol, then already in several minutes there begins the separation of a white precipitate of polymeric compounds which do not contain nitrogen, which ends in several hours.* [47]

Hardly any cause to complain. Toward the middle of the article, however, in a description of its core experimental procedure, I came across the following: "Fractionation of such a solution in a current of nitrogen after it had stood for 5 days at room temperature produced a yield of 60% of β-phenylaminoethanethiol."[48] Here is what the Russian actually says, in my translation: "Letting such a solution stand at room temperature for five days after fractionation in a stream of nitrogen produced β–phenylaminoethylmercaptan with a yield of 60%."† [49] The chemical product is listed differently. This is in fact the same compound, but the

*"Как и следовало ожидать, аналогично ведет себя этиленсульфид и по отношению к диэтиламину. Если к раствору диэтиламина в метаноле прибавить при комнатной температуре свежеперегнанный этиленсульфид, то уже через несколько минут начинается выделение белого осадка не содержащих азота полимерных соединений, заканчивающееся через несколько часов."

†"При стоянии такого раствора при комнатной температуре в течение 5 суток после фракционированной перегонки в токе азота β-фениламиноэтилмеркаптан получается с выходом 60%."

translator was inconsistent throughout the article about how he represented it, sometimes using the modernized name, sometimes an older nomenclature. It could have been a typesetter's error, or the product of rushed translation—either way, to a reader, this article would seem to be at best confused. The problem, as contemporaries indicated, was ineradicable in precisely this area: "Organic nomenclature problems arise from faulty translations.[. . .] This type of error is extremely difficult to catch in editing."[50] Yet what choice did the reader have? He or she could not check the original Russian. Cover-to-cover was all there was.

And soon, it was everywhere. Extant scholarly accounts of the history of cover-to-cover translation emphasize the massive translation initiative of the American Institute of Physics (AIP), supported by the NSF.[51] In 1955, the AIP sent a survey to 300 physicists about their views on providing either complete or selected translations, and 269 replied, with results that "(a) an overwhelming majority favor establishment of a Russian-to-English translation service, (b) appreciably more than half believe complete translations of Soviet journals would be preferable to translation of selected articles, and (c) about 90% are of the opinion that they or their organization would subscribe to such a journal."[52] According to the survey, 79.8% supported cover-to-cover translation "[b]ecause of the technical value of the research now in progress in the USSR," and 72.3% added a caution about "the national danger of underestimating the strength of the USSR, particularly as far as scientific advances are concerned."[53]

While Coleman created the industry almost by accident through calculating profit margins, the AIP debated the rationality of the venture from every angle before committing resources to it. In 1954 Elmer Hutchisson, later the director of the AIP and the man behind the 1955 survey, offered a laundry list of reasons why cover-to-cover was superior to selected translations:

> First, the administration and mechanics of a project in which selections are made is much more complex than one in which journals are translated completely in a regular manner. Second, in many cases it would be necessary to translate abstracts, at least, so that the judges would be able to determine which articles should be translated and which should not. Third, any attempt to make a selection will undoubtedly cause a delay. One would hope that after this process gets established and air-mailed page proofs are coming regularly, the English edition would be available shortly after the Russian edition.

> Further, because of the confidence that everything would be translated, which is available in a given journal, the appeal of the translated edition would be far greater and a more ready market would be found among libraries and industrial organizations.[54]

The reasoning made sense to the NSF, and they underwrote the cost of the first volumes, which helped keep the subscription price down. The procedures instituted by the AIP for their first journal—*Soviet Physics JETP*, a translation of *Zhurnal eksperimental'noi i teoreticheskoi fiziki*—edited by Robert Beyer, professor of physics at Brown University, were remarkably similar to the work flow developed at Consultants Bureau, and the experiences of the physicists who translated in the wee hours to earn diaper money are also reminiscent of Coleman's employees.[55]

As a matter of fact, the AIP was surprised to learn in 1954 that Consultants Bureau had been operating in the field—with five different journals already for sale, and without any government subsidy. "The venture presumably is successful," a liaison from the Library of Congress noted, "since it has been in operation for several years."[56] With NSF support, the economics were even more favorable than they had been for Coleman. A typical Soviet journal contained 500 words per page; a science journal, because of tables, images, and formulas, came in at roughly 300. With a volume comprising 1,000 pages (a reasonable estimate in 1955, a shocking undercount by decade's end), that meant 300,000 words to translate a year. This would cost $6,000–$9,000 for a single organization at industry rates, but with NSF support the production cost was reduced to 2.5 cents a page. That meant with 150–200 subscribers, you could cover the expense of translation; with 600 subscribers, you covered production; and with over 750, you would be turning a profit and could subsidize a new journal.[57] By 1956 *Soviet Physics JETP* had grown to 700 annual subscribers, with 2,600 pages of translated material distributed for the cost of translating 10–12 pages—a real bargain. The NSF promptly agreed to support three new journals: *Journal of Technical Physics,* the physics section of the *Proceedings of the Academy of Sciences of the U.S.S.R. (Doklady)*, and the *Journal of Acoustics*.[58] The pump was now primed, and cover-to-cover ventures proliferated. As each title gained subscribers, the importance of the subsidy diminished and could be moved to start a new journal. The cycle repeated itself, and private firms started to join the gravy train. By 1958, there were 54 cover-to-cover translations of Soviet journals, and 85 in 1961.[59] (See figure 9.2.) Consider what was happening here: each

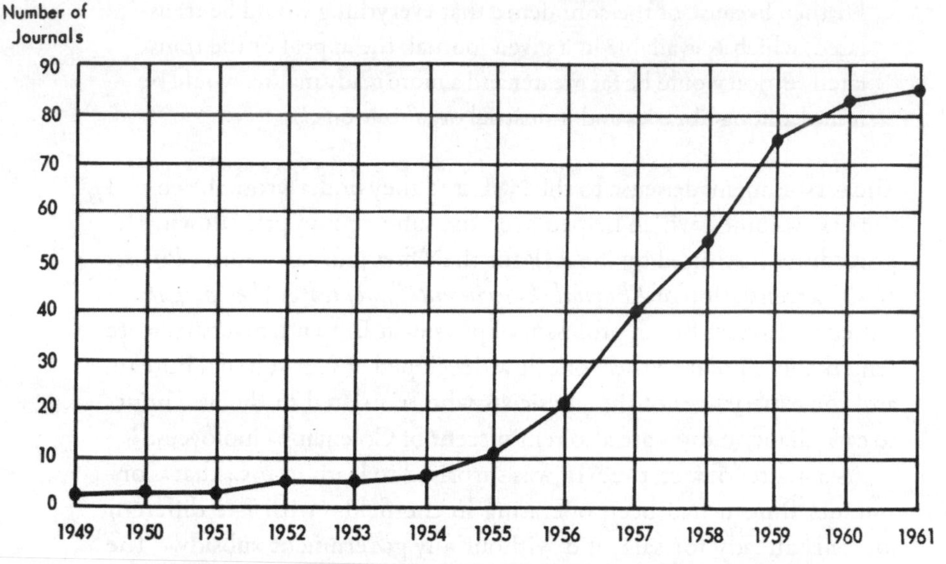

FIGURE 9.2. Number of English cover-to-cover translation journals from Russian, from the inception in 1949 (*Journal of General Chemistry of the USSR*) to 1961. Notice the rise after the AIP/NSF collaboration began in 1955. Boris I. Gorokhoff, *Providing U.S. Scientists with Soviet Scientific Information*, rev. ed. (Washington, DC: Publications Office of the National Science Foundation, 1962), 15.

month a hefty tome would arrive at an office in the United States, be ripped apart, distributed, translated, edited, stitched back together, and printed, all within six months—and this was done for dozens of journals, every month, for decades. It was the largest scientific translation project in the history of the world.

It was also, in the eyes of some, an administrative and bibliographic disaster. The complaints were legion, encompassing every aspect of the enterprise and expressed both behind closed doors at the AIP and openly in the pages of science journals. Did cover-to-cover violate copyright? (Since the Soviet Union did not yet adhere to international copyright conventions, this concern was dismissed.)[60] The leaders at the AIP were also very worried about explicitly propagandistic articles in Soviet science journals. Should those be translated alongside the regular science? Views ranged on both sides of the issue for years, but the AIP settled in the end for omitting such "non-scientific" pieces.[61] (This issue did not bother Coleman; *Journal of General Chemistry of the USSR* translated, without comment, a congratulatory message to Joseph Stalin on his seventieth birthday.[62]) Lack of standardization bedeviled cover-to-cover enterprises, with at least fourteen different transliteration systems from Cyrillic, with different translators, editors, and journals rendering authors' names differently, frustrating indexing and abstracting.[63]

The two most significant problems were the delay and the expense. Time-lags were inevitable, given that one had to wait for the Soviet journal, translate it, and produce an entirely new issue. Robert Beyer noted in January 1957, after two years of experience, that his journal was "appearing 7 months behind official publication date."[64] Politics on both a micro and macro scale were partially responsible. In 1948, the Soviets temporarily held up all shipments of scientific journals to the United States because an ill-considered regulation required all foreign mail to be cleared through Foreign Minister Viacheslav Molotov's office, generating an incredible backlog. When journals arrived in America, on the other hand, the New York post office had been known to quarantine the material lest it contain dangerous propaganda.[65] Those hiccups were cleared (relatively) quickly, but they masked a deeper problem. The intrinsic delays in producing cover-to-cover journals and the vagaries of the Cold War meant that scientists who wanted their translations right away continued to commission their own, which generated wasteful duplication and further taxed the limited corps of technical translators.[66]

Cost was even more serious. *Soviet Physics JETP* contained 1,500 pages in 1955, which yielded a net price of $30 for an annual subscription (converting to an astonishingly low translation cost of 2 cents a page). By 1965, however, the Soviet original had bloated *threefold*, which meant the price had to skyrocket to $90 (not counting inflation) to break even, which in turn depressed demand.[67] And this was for the AIP's flagship translation journal, with the greatest subsidy and the greatest reader demand. For more boutique periodicals, like *Soviet Aeronautics*, the cost of the cover-to-cover translation reached 28 times the price of the Russian-language original, and the average circulation of the journals was 200–300, well below break-even.[68] Eugene Garfield, the information scientist who pioneered the *Science Citation Index* and transformed scientific bibliography, was scathing about the practice—and the government's intervention—in 1972, when cover-to-cover had taken over as America's chief strategy for following Soviet science:

> Since the government did not allow the demonstrated needs of international scientific communication to impact the information

> marketplace, the economic and other forces of natural selection were not allowed to operate. As a result, government fiat has produced a monster that continues to plague libraries, science administrators, and, in the final analysis, the taxpayers who feed the monster. Many research libraries feel forced to buy both Russian and English editions of leading Soviet journals, even though the latter generally appear six months to a year after the original. Those libraries which obtain only the translated version often find they are not used as often as expected. Bibliographically, the situation is a horror.[69]

In effect, in attempting to read Soviet science, the Americans had replicated the most inefficient features of the Soviet science system. Being journals, cover-to-cover fit into libraries' conventional purchasing patterns (unlike translation pools), yet produced bloated budgets and overcrowded libraries without helping anyone locate information, and so added to the glut.[70] The language barrier had been swapped for information overload, on a one-to-one basis. The journals were in English, but no one had time to read them.

Exit Machine

The woes of cover-to-cover translation began to impinge on the other hope for transcending the language barrier: machine translation. Recall that when Léon Dostert organized his program around a corpus of texts in organic chemistry, he selected the *Journal of General Chemistry of the USSR*, so that he could use this ready bilingual corpus to keypunch in terms and work out grammar algorithms. Cover-to-cover would work hand-in-glove with MT, obviating the need to develop translations from scratch. Or so he thought. Dostert quickly learned that the "English-language version was found to be inadequate for machine translation purposes and two persons were assigned the task of preparing a standardized translation which would be free of stylistic idiosyncrasies and as consistent as possible."[71] Precisely because the *Journal* was translated by teams of different people, syntactic and lexical correspondences were not standardized, which was what machines demanded.

Soon, dissidents emerged within the heart of the community, armed with trenchant critiques that threatened to topple the fundamental assumptions of the entire project. By far the most powerful assault came from none other than Yehoshua Bar-Hillel, the Israeli philosopher who had convened the first MT conference at MIT back in 1952. In 1958, Bar-Hillel undertook a tour of all the major Western MT institutes—with especial attention to Georgetown, for "[t]here exists no other group in the United States, or in England for that matter, which has been working on such a broad front"—finding an industry with between 200 and 250 people working full time with an annual outlay of roughly $3 million. Six years earlier, there had been the equivalent of about three full-time researchers with a total budget of $10,000, and the only individual working exclusively on MT had been himself. In 1960 he published a revised version of his working paper about this tour, incorporating his findings from Western accounts of Soviet research, and he came to the conclusion that "fully automatic, high-quality translation (FAHQT)," the stated goal of most research programs, was impossible, "not only in the near future but altogether." Language could not be reduced to algorithmic rules, because humans constantly imported context with serious semantic implications. His chief example, which soon became canonical, was the difference between "The box is in the pen" and "The pen is in the box." We intuitively know that the term "pen" in the first sentence probably is a place with animals, and in the latter might be a writing implement; in both instances, we apply our contextual knowledge of relative size—something a computer could not do. He did not spare scientific language: "Fully automatic, high quality translation is not a reasonable goal, not even for scientific texts."[72] To tell the truth, Bar-Hillel had said something similar as far back as 1953 ("Fully automatic high-accuracy translation seems out of the question in the near future.[. . .] Therefore, either the high accuracy or the complete automatic character of the translation process must be sacrificed"), and even in 1951 had insisted that "high-accuracy, fully automatic MT is not achievable in the foreseeable future."[73] But now, patrons were listening.

Hints of a coming storm emerged in May 1960, when Congress summoned Dostert for testimony in a series of hearings on MT. In September 1959, Dostert had resigned as director of the Institute of Languages and Linguistics so he could devote all of his time to machine translation. In short order, he had an operational system: Georgetown Automatic Translation (GAT). In 1964, Georgetown delivered GAT, designed for the IBM 7090 to translate Russian into English, to the US Atomic Energy Commission and to EURATOM. EURATOM kept using it until 1976, when it was replaced by SYSTRAN, while the Americans maintained GAT until at least 1979. The MT community tended to be rather acerbic about the system; one later analyst complained that "[t]here was no true linguistic theory underlying the GAT design; and, given the state of the art in computer science, there was no underlying computational theory either."[74] GAT's output still required post-editing by a subject specialist, although not necessarily one with knowledge of Russian. Yet its end-users were happy about what they got: 92% of users at Oak Ridge and Ispra considered the results "good" or "acceptable," and 96% said they would recommend MT to a colleague.[75]

Nonetheless, Congress wanted to know what happened to the miracle machine. Dostert marched across town with his team in tow and dazzled the Congressional subcommittee with tales of progress, as well as another demonstration—a spot-translation of a random chemical text—which the system mostly passed (though it took the evaluator four times as long to read the translation as an equivalent English-language text).[76] Dostert had other aces in the hole: two of the Representatives examining him were Georgetown alumni, and one had been his own student.[77] (He always did work best through personal connections.) Dostert secured only a temporary reprieve. Three years later, in 1963, the CIA withdrew all funds from Georgetown's MT program, and that same year Dostert left Georgetown for his other alma mater, Occidental College in Los Angeles.[78] He continued to lobby for MT, however, and in 1963—perhaps implicitly responding to Bar-Hillel's critique—he opined: "We should accept the fact that 'perfect' translation is neither *humanly nor mechanically* achievable.[. . .] What then should we aim for?"[79]

An answer was forthcoming, but it was not what Dostert hoped for. In April 1964, Leland Haworth, director of the NSF, requested the National Academy of Sciences to assemble an Automatic Languages Processing Advisory Committee (ALPAC) "to advise the Department of Defense, the Central Intelligence Agency, and the National Science Foundation on research and development in the general field of mechanical translation of foreign languages." Chaired by John R. Pierce of Bell Laboratories and composed of leading specialists in computer science, linguistics, and even MT (Anthony Oettinger was a member), ALPAC searched through the various funded MT programs looking for progress and efficiency. It concluded that it would be more economical to have specialists invest the short amount of time to come up to speed in Russian. When that failed, commissioning specific translations was cost-effective, and America's stable of translators was more than adequate to the task. "There is no emergency in the field of translation," they insisted. "The problem is not to meet some nonexistent need through nonexistent machine translation."

The prime evidence for the "nonexistence" of MT was, ironically, the very success of the Georgetown-IBM experiment. After providing examples of three translations by different systems of a single Russian passage, all of which were execrable, the report observed that "[t]he reader will find it instructive to compare the samples above with the results obtained on simple, or selected, text 10 years earlier (the Georgetown IBM Experiment, January 7, 1954) in that the earlier samples are more readable than the later ones." The problem with the Georgetown sentences was that they were *too good*: "Early machine translations of simple or selected text, such as those given above, were as deceptively encouraging as 'machine translations' of general scientific text have been uniformly discouraging."[80] Notice the word "deceptively." Dostert's very showmanship had sown the seeds of the collapse.

And collapse it was; seven years after the report's publication in 1966, a survey of the field depicted a wasteland—two years later, the three remaining government-funded MT centers had closed shop.[81] In 1965, the Association of Machine Translation and Computational Linguistics took over the journal *MT*, adding *and Computational Linguistics* to its title; three years later, it removed "Machine Translation" from its own name, and closed the journal down in 1970.[82] The ripples spread across the Iron Curtain as well, as one of the leading researchers of MT in the Soviet Union recalled:

> The effect of the ALPAC report in 1966 was as great in the Soviet Union as in the United States. Many projects were not funded any more; machine translation went into decline. The authorities had seen the ALPAC documents and concluded that if the Americans did not think it worthwhile to support MT, if they did not think there was any hope of MT, then nor should we.[83]

The Soviets and the Americans had goaded each other to invest more in machine translation, and now they would suffer the drought equally. MT as a field would not really recover until the 1980s. Léon Dostert would not live to see it. He died suddenly on 1 September 1971, at a conference in Bucharest, Romania.

Covered

Meanwhile, Earl Coleman was having a very good decade. Consultants Bureau seemed to effortlessly toss off new journals. By 1956, it was easily the biggest producer of cover-to-cover publications in science. Meanwhile, Robert Beyer, the editor of *Soviet Physics JETP*, was overwhelmed with the rapidly expanding Soviet journal. He consulted with AIP director Elmer Hutchisson about subcontracting some of their own journals to Coleman's outfit.[84] Rumors abounded that the translators for Consultants Bureau were incensed at their low pay, but the AIP decided that a limited partnership might be worth doing. After all, reasoned Wallace Waterfall, "the Colemans will undoubtedly do the best possible job for us in order to enhance their own reputations."[85] The AIP farmed out three journals to Coleman, and soon the publisher had 45 full-time employees and translators in the United States, Canada, England, Puerto Rico, and India. In 1958, he became the first Western publisher to offer a royalty to the Soviets for the privilege of translating their science, thus gaining exclusive Western rights (and preempting British publishing mogul Robert Maxwell from horning in on his territory).[86]

The partnership between the professional society (and its state backers) and the private publishing firm was not always smooth. In 1965, Coleman accused Beyer "with vigor and irritation about AIP ruining the translation field—first by paying its translators too much, and second, by getting out too elegant a translation journal."[87] Eventually his feathers returned to their customary unruffled state, and translators continued to be impoverished by Coleman's logic of compensation. In 1966, the AIP abandoned their fledgling efforts at a Chinese cover-to-cover journal, reasoning that Coleman would probably pick it up, and by 1968 Coleman began producing *all* the AIP's journals.[88] By 1970, the renamed Plenum Publishing Corporation continued to tower over the competition. It produced 72 journals, comprising 75,000 pages of text a year—62 of these journals were independent, eight for the AIP, and others for the American Mathematical Society and the American Society of Civil Engineers. The nearest competitor was Faraday Press in New York City, with 29, followed by Scientific Information Consultants in London, with nine. Coleman controlled well over half of the cover-to-cover market.[89]

And that market was poised to take over the world, literally. One has to be careful when estimating the global reach of cover-to-cover translation, for certain features of the industry were peculiarly American. Of the 162 such journals—still only a tiny fraction of the estimated 2,600 to 4,000 Soviet scientific periodicals—published in 1968, some 85% were produced by the Americans. The rest but two were British (the exceptions were a Canadian journal on the Arctic, *Problems of the North*, and the lone non-English journal, the French *Prospection et protection du sous-sol*, on geology).[90] Through the NSF, the American state backed, at least in part, 45 separate ventures in 1960 alone.[91]

But despite the obviously American features of both MT and cover-to-cover translation—the mounds of Cold War money, the insistent focus on Soviet science, the overwhelming emphasis on English—the latter venture proved durable in large part because of factors outside the United States. In 1954, Frances Coleman explained that if the *Journal of General Chemistry of the USSR* had been forced to rely only on domestic markets, it would never have survived. "[B]ut then subscriptions and inquiries began to trickle in from Holland, France, India, Japan, and elsewhere. We realized that these translations would serve a purpose—and have a market—not only in English-speaking countries as we had envisaged, but also in any country where there were chemists who could not read Russian and could read English," she noted. "At the end of our first year more than half of our subscriptions were going to non-English speaking countries."[92] The same year as the Georgetown-IBM experiment, Consultants Bureau sent the journal to seventeen different countries, representing eleven different native languages. The AIP found the same: by 1965 one-third of subscriptions came from outside the United States.[93]

This was a consequential difference from MT. That project, however international, focused entirely on the *production* of texts. Cover-to-cover, as befit its roots in the private sector, was from the beginning worried about generating consumer demand for their product. MT's dependence on state support meant that when ALPAC gave the granting agencies an excuse to pull the plug, there was nothing to fall back on. Léon Dostert's dream of transcending Scientific Babel purely through linguistic means, without attention to the quantity of information or access, foundered. Meanwhile, translation journals spread abroad. If you were a scientist in Pakistan, or Italy, or Brazil, you had to follow both American and Soviet science. Instead of learning two languages, the Americans had made it possible to get by entirely on English, and so it became more and more prevalent as the default language of science—not instead of learning about what the Soviets were doing, but *as a means* of learning what the Soviets were doing.

Interestingly, such an outcome was foreshadowed in the foundational text of American science policy, Vannevar Bush's *Science: The Endless Frontier* (1945). Discussing translations of Russian into English, the text noted that "[s]ince such work would benefit not only science generally in the United States but would very likely promote the use of English in other countries, it seems proper to recommend that the United States Government consider methods by which the cost of such work could be met."[94] In the aftermath of the Second World War, a series of decisions about confronting the challenge posed by Soviet science began to overwhelmingly tip the balance toward a global monopoly of English as a language of science. Nowhere was this more visible than on the Cold War battleground between the Soviets and the Americans, the land whose language used to dominate scientific publications in seemingly every field: Germany.

CHAPTER 10

The Fe Curtain

> Auch zwischen Volks- und Sprachgenossen stehen Schranken, die eine volle Mitteilung und ein volles gegenseitiges Verstehen verhindern, Schranken der Bildung, der Erziehung, der Begabung, der Individualität.*
>
> HERMANN HESSE[1]

It was May 1945, and Germany was broken again. After the Peace of Westphalia ended the religious wars in 1648, the German-speaking principalities that made up the Holy Roman Empire were fragmented to create buffer states throughout the middle of the European continent, and ever since the arrow of history seemed to point toward greater and greater unification. First Prussia swallowed up smaller duchies and kingdoms to grow to a point where it could, by the 1860s, challenge the political focus of the German regions: Vienna, seat of the Habsburg Empire. And then in 1871 most of the German-speaking lands unified into the *Kaisserreich*, a new continental empire to compete with Austria-Hungary, inducing consternation in the French and British. In 1938, Austria was incorporated into a terrifying German Third Reich, spreading a lot more than consternation much farther afield. Now that was all over; Austria was independent and Germany was broken—but no one was yet sure into how many parts.

There were, formally, two possibilities, one or four, but informally—and soon quite forcibly—the answer was definitely two. Technically, Germany had surrendered to the Allies, who governed the occupied country under a council of the four powers: the United States, the Soviet Union, the United Kingdom, and France. Each of those coun-

*"Barriers also stand between national and linguistic peers that prevent full communication and full mutual understanding, barriers of education, of upbringing, of talent, of individuality."

tries also controlled a separate zone (the French zone was carved out of the British one as a gesture toward European comity). Hence, the occupied zones became instantly polyglot, and there was a nationwide boom in German-English, German-French, and German-Russian dictionaries immediately after surrender.[2] For the occupying powers, communicating with each other remained fraught, especially between the Americans and the Soviets. Very few Americans knew Russian, and vice versa; communicating through a third language like German was scarcely more successful.[3]

Yet communication was essential for any kind of postwar settlement. Consider the Soviets, who set up their proxy government, the Soviet Military Administration of Germany (SVAG), on 6 June 1945 under Marshal Georgii Zhukov. Zhukov, both directly and through his deputies Colonels V. D. Sokolovskii and I. A. Serov, commanded 8,000 Soviet troops at their headquarters in Potsdam to the southwest of Berlin, supplemented by 273,000 infantry troops distributed over the entire Soviet Zone, 29,000 air force personnel, 2,700 naval troops, 20,000 special SVAG troops, and 20,000 foot soldiers of the MVD (the security services).[4] These were facts on the ground of the Soviet Zone, and at the center of that ground was the city of Berlin, split into four sectors as a microcosm of defeated Germany. One of the central concerns for the three Western powers, but especially for the Americans, was negotiating access to Berlin, which the Soviets restricted to a single highway and railway line, arguing that Soviet demobilization consumed the remaining transit points.[5] This soon became a perpetual source of conflict, triggering the Soviet closure of access to Berlin on 24 June 1948 and the subsequent Anglo-American airlift to supply the Western zones of the city with food and fuel. The Berlin Blockade was the most evident act that signaled that Germany was to become two nations: the Federal Republic of Germany in the West, and the German Democratic Republic in the East, each under the sway of the United States and the Soviet Union, respectively. Divided Berlin would become the capital of the Cold War, marked in 1961 by the erection of the eponymous Wall as literalization of the "Iron Curtain" posited by Winston Churchill in March 1946.

In the early summer of 1945, when plans for governing Germany in the short term were drawn up, that conflict lay in the future. It will come as no surprise after the previous two chapters that both Americans and Soviets governed their zones through their native tongues. Dealings in American and British offices tended to be conducted almost entirely in

English (except for a few officers who happened to speak German fluently before arriving in-country), not so much because of the ban on fraternization (rescinded, anyway, on 14 July 1945), but more because of social and economic segregation coupled with the intense hostility of some Americans toward the Germans for the Nazi rampage, mirrored by German resentment of occupation.[6]

SVAG also governed mostly monolingually in practice, although there was substantial official discomfort about the fact. SVAG leadership tried repeatedly to force officers to learn German and just as often failed. In March 1946, an order mandating German study was promulgated and rapidly ignored; out of thirty officers signed up for courses in Magdeburg, no more than a third actually came to class. Many of the Soviets who arrived already speaking German labored under a double burden: they were predominantly Jewish, and they often acquired German girlfriends (surely related to their linguistic capacities). As an anti-cosmopolitan campaign unfolded in Stalin's Moscow in mid-1948, fear of Jewish treason and spying led to the recall of many of these officers, purging the one set of officials on the ground who could actually speak to the locals. This was followed by more orders for German study, and more stonewalling.[7]

If Germany's politics were linguistically and politically fragmented, many Germans feared the state of science was even worse: there was nothing to fragment, for science was destroyed. Local German scientists had difficulty assessing the state of affairs because zonal barriers and censorship blocked both travel and mails, and scientific publishing had almost collapsed (exacerbated by a postwar paper shortage).[8] A foreign observer writing in the *Physikalische Blätter*, one of the new periodicals that managed to bloom amidst the rubble, painted a picture of dire need. "In addition there is the most severe lack of all scientific educational and research material," R. C. Evans wrote. "Books are not to be had, the appearance of scientific journals has been stopped, the recurring needs of a laboratory—reagents, apparatus, and everything else, even the simplest material—are almost unattainable, especially if delivery must be obtained from another Zone; the difficulties could not be greater if everything had to be obtained from abroad."* [9] Aside from the

*"Dazu kommt der schärfste Mangel an allem wissenschaftlichen Unterrichts- und Forschungsmaterial. Bücher sind nicht zu haben, wissenschaftliche Zeitschriften haben ihr Erscheinen eingestellt, Reagenzien, Apparate und jedes andere, auch das einfachste Material, der laufende Bedarf des Laboratoriums ist fast unerreichbar,

pockmarked landscape of destruction and massive displacements occasioned by the end of the war, several of the wartime leaders of German science were sitting in Allied detention pending adjudication of responsibility for war crimes. The landscape of German science began to shift; in the West, leafy Göttingen came to displace tense Berlin as the center of physics, for example. Adjusting to the postwar world entailed massive psychological and physical difficulties.[10]

Could German science recover? Part of the answer hinged on whether German as a language of science could survive the shocks of occupation. In 1951, American experts on Soviet bibliography observed that "[t]he influence of German science on Russian research in organic chemistry went down from 59 per cent at the beginning of the industrialization [late 1920s] to 30–36 per cent for the present time. The Russian chemist now uses his own literature at least as much as the German."[11] In the Soviet Union, the collapse of knowledge of the German language occasioned by the advent of the war was so severe that postwar analysts had to discard all statistics about the number of actual speakers and piece together the status quo from guesswork.[12] Meanwhile, the emerging United Nations Organization recognized five official languages—Chinese, English, French, Russian, and Spanish (Arabic was added in 1973)—noticeably leaving German behind. That seemed reasonable, for Germany was defeated and would not function as an architect of the postwar world order. But when the United Nations Educational, Scientific, and Cultural Organization (UNESCO) permitted Italian a limited status and Hindi an official one but explicitly denied any status to German—the language of great educational, scientific, and cultural achievements—it was hard to view this as anything other than punishment by the victorious powers.[13]

While the position of Germany as a country, let alone as a world power, was decidedly gloomy, many German scientists thought they might retain some cultural power for the German language through the reconstruction of German science. That science was rebuilt from the Nazi remnants not once but twice: into a West German science under an increasingly Americanizing (and Anglophone) Western scientific establishment; and into an East German variant that bore numerous stamps of its Soviet patron and that patron's language. The chapter

besonders wenn die Lieferung aus einer anderen Zone erfolgen muß; die Schwierigkeiten könnten nicht größer sein, wenn alles aus dem Ausland bezogen werden müßte."

that follows explores the development of Cold War science outside of the metropoles of the United States and the Soviet Union, chronicling the persistent decline of German as a language of science despite many heroic efforts to salvage it. In the wake of enormous infrastructural and political changes, the long and tumultuous story of scientific German appeared to be coming to a close.

Denazifying the (Mostly Western) Zone

The major mechanism of linguistic transformation throughout this book has been education, and occupied Germany was no different, though there the educational inflections were strongly colored by the unique imperatives of the denazification policy of both the Western Allies and the Soviet Union.[14] Education was a salient instance of the more general postwar reconstruction of science in the image of each superpower. The Americans were deeply invested in building a "Western" science in Europe that was strongly allied with the United States and also predominantly Anglophone. Marshall Plan aid for science, for example, was directly tied to the reorientation of the French infrastructure away from hypercentralization in Paris toward the provinces, and generally toward greater publication in English, while Columbia University physicist I. I. Rabi, a trusted science advisor, lobbied for incorporating science from nascent "West Germany" into collaborations with nations of the North Atlantic Treaty Organization (NATO).[15]

In order to accomplish Rabi's goals, one needed German scientists. The problem, of course, was reconciling the desire to have America-friendly scholars in the universities when those very institutions had been integrated into the Nazi infrastructure, staffed with party members who should at the very least be dismissed from their positions, if not tried for war crimes. (Of course some specialists, most notably rocketry engineers, were "pilfered" by both the Soviets and Americans.) The impact of denazification was massive, leading to twice as many dismissals as had Hitler's 1933 Civil Service Law, for the straightforward reason that there were many more Nazi party members in higher education at the end of the war than there had been Jews in such positions at the dawn of the Third Reich. The University of Heidelberg fired 72 instructors, Frankfurt 33, diminutive Erlangen 30 (representing a full 27% of its teaching staff), and so on down the line.[16] The six remaining major universities in the Soviet Zone—after two universities, Breslau and Königsberg, were ceded to Poland as Wrocław and the Soviet

Union as Kaliningrad, respectively—lost about 75% of their professoriate and roughly 80% of the adjunct teaching staff, a situation that was particularly aggravated in Berlin where many "clean" academics simply decamped for the West. Over 85% of the faculty who were relieved of their jobs never returned to higher education in the Soviet East.[17]

Reopening the universities was urgent, not only to return to a semblance of normality, but also to train cadres who could rebuild the future Germany (or, one should say, German*ies*). The remedies in the Western zones were highly varied owing to the reconstruction of the American, British, and French zones as a *federal* republic, with education a function allocated to the various states (*Länder*, in German) to resolve as each saw fit.[18] The French, for example, recognizing their inability to compete with the Americans politically or economically, emphasized the benignity of their occupation by focusing on "culture," rapidly reconstructing the Kaiser-Wilhelm Institute for Chemistry in Mainz.[19] The rector of Göttingen University after the war estimated that of the sixteen universities and eight Technische Hochschulen (higher technical schools) in the Western zones, only six were relatively unharmed, six could use 50% of their facilities, and the remaining eight were reduced to 25%–30% of their prewar infrastructure. (Münster, for example, was 80% destroyed, Munich 70%, and Würzburg 80%.) The British, in the north, were faced with perhaps the greatest devastation, since the industrial Ruhr area had suffered countless bombing runs; nonetheless, they opened all their universities by the end of 1945, beginning with unscathed Göttingen on 17 September and concluding with Köln on 12 December.[20]

West Berlin represented a unique case; its encirclement by the Soviet Zone promoted a greater degree of autonomy, leading to the most thorough overhaul of the wartime and prewar educational system.[21] The crown jewel of these efforts was the establishment in December 1948, in the midst of the Blockade, of the Freie Universität zu Berlin (the Free University in Berlin) in Dahlem, a tree-lined, somnolent neighborhood that lay right in the center of the American sector and had long been home to an elite scientific tradition. (Fission had been discovered there, for example, in 1938, and Fritz Haber's Kaiser-Wilhelm Institute for Physical Chemistry was nearby.) The Free University was largely a German-organized and German-run affair, and its linguistic emphasis was strongly German. When, after a few years, foreign students trickled in, they were required to take two semesters of German, and almost all instruction took place in that language.[22] Graduate students were admonished that "the dissertation must be written clearly and in good German."*[,23] Visiting scholars in the 1950s, on the other hand, lectured in a range of tongues—a Spanish art historian speaking in Spanish, and many in the humanities lecturing in English. The Natural Sciences Faculty, however, entertained almost exclusively talks in German, preferentially hosting Germanophone Swedes and Americans (many of them émigrés). American policy complemented the mostly grassroots German effort. Instead of deputing American professors to teach in Germany for short periods, preference was given for cycling German students to the United States for study, an ironic inversion of the interwar postdoctoral network that had been ruptured by Nazi protocols.[24]

Scientists active in occupied Germany recognized that the transformations in their country—and especially the emerging split between a communist East and a capitalist West—was beyond their control: "We are able to change nothing about this, and our journal [the *Physikalische Blätter*], that counts the cohesion and connection of German physicists as its noblest tasks, is entirely helpless against this development."†[25] In fall 1946, British authorities allowed the formation at Göttingen of a renewed German Physical Society—die Deutsche Physikalische Gesellschaft in der Britischen Zone. Max von Laue, as one of the "good Germans" who stayed within the Third Reich but did not collaborate with the regime, was made its president, while Otto Hahn was tapped for president of the Max Planck Society, the successor to the Kaiser Wilhelm Society. Both Hahn and von Laue worked within a West German context, and were—to the frustration of the occupying authorities—substantially less interested in punishing past political "mistakes" than in reestablishing a scientific community.[26] Other post–World War I institutions were adapted to this new, constrained Germany. The Notgemeinschaft was eventually transformed into the German Research Society (Deutsche Forschungsgemeinschaft), as a way of providing federal money for research without working through institutions compromised by Nazi affiliations.[27] Finally, new publications, like the *Physikalische Blätter* and the *Zeitschrift für Naturforschung*, leapt into the breach opened up by the delay in approving the denazified continuations of stalwart German journals such as the *Annalen der*

* "Die Dissertation muß klar und in gutem Deutsch geschrieben sein."

† "Wir vermögen nichts daran zu ändern, und unsere Zeitschrift, die den Zusammenhalt und die Verbindung der deutschen Physiker zu ihren vornehmsten Aufgaben zählt, ist gegenüber dieser Entwicklung völlig hilflos."

Physik and *Zeitschrift für Physik*.[28] West German science lost much of its international character. Foreign attendance at meetings of the German Physical Society, for example, began to drop off in 1950.[29] These were now meetings in German and for Germans—well, one half of the Germans.

Stalinizing (Sort of) the East

The year of the Blockade, 1948, not only marked the onset of the rapid differentiation of institutions that would make West Germany characteristically "Western," but also unsurprisingly represented the fulcrum whereupon East Germany pivoted toward Stalinization. The German Democratic Republic was crucial to Soviet designs for consolidation: it was a non-Slavic gateway to the West, a showcase for socialist progress, and an industrial engine that—after suitable reparations were bled out of the Nazi rubble—could power the communist future.[30] It was also the postwar satellite state that had the longest-standing educational infrastructure, and therefore first SVAG and later the Socialist Unity Party (SED)—the communist party that ran East Germany until its collapse—instituted more widespread changes into higher education there than anywhere else in the Warsaw Pact.[31] By way of contrast, the chemical industry retained enormous continuities with its National Socialist predecessor.[32]

The general model was to make East German higher education look like Soviet higher education, which had undergone its own Bolshevik transformation in the 1920s.[33] The consequences in East Germany, after adjusting to the tremendous personnel purges of denazification and simple outmigration, were remarkable indeed. From 1951 to 1955, there was a 463% rise in enrollments in the technical sciences in German higher education. (Lest one consider that simply an adjustment to postoccupation stability, the equivalent for the humanities was only 112%.)[34] The Prussian Academy of Sciences was also entirely refashioned, shifting from an all-German honorific institution to a fully socialist research academy by 1969. In part to counter the visible success of the Free University, the Academy and the University of Berlin—renamed Humboldt University—were endowed with massive resources to promote research and collaboration with an eye toward technological applications and economic growth. Comprising only 131 staff members in 1946, the Academy's payroll reached 12,923 by 1967, and almost doubling again by the moment of collapse in 1989.[35] The East Germans also erected a parallel scientific publication infrastructure to counterbalance the torrent of periodicals and technical books emanating from West Germany: separate professional journals begin to appear in 1951, university journals the following year, and the year after that the ubiquitous Deutscher Verlag der Wissenschaften (German Publisher of the Sciences).[36]

The linguistic hiccups that had hampered SVAG's operation did not magically disappear. Soviet officials neither knew much about German education nor understood the language, so almost all negotiations happened in rudimentary pidgin. When Soviet experts came on lecture tours, even well into the 1950s, they spoke in Russian before uncomprehending audiences; translators had to be rousted at the last minute.[37] On the other hand, those East Germans who did master Russian could attain significant professional advancement.[38] There were, to be sure, some benefits to the stubborn insistence of East Germans on speaking German—it meant they could converse with West German colleagues, blossoming into a vital conduit for the thriving East German specialty of industrial espionage.[39]

Nonetheless, it simply would not do to have the Germans speaking German to the exclusion of everything else. Science, like communism, was international, and surely it would be to the advantage of citizens of the German Democratic Republic if they could access the tremendous contemporary advances in Soviet science. In the case of socialist friends, much like that of the capitalist enemies, the key to learning what the Soviets knew was learning their language. The Soviet leadership was particularly keen to encourage this ambition throughout its sphere of influence. Russian quickly became the first foreign language taught in Eastern Europe. In Hungary, for example, eight years of Russian soon became compulsory and remained so until 1989. The same was true everywhere else, with the exception of Romania, where Russian ceased to be obligatory in 1963; French returned to its traditional place in Romanian education, being preferred to Russian or English by 60% of students.[40]

East Germany was more like Hungary than Romania, and like Hungary the obligation to study Russian did not imply actually *learning* it. As always, some did become enthusiastic about the language, as in this statement drawn from a proposal for a Russian-German dictionary in 1961:

> Russian is one of the leading world-languages. This fact is to be explained through the role of the USSR as a world power and the prestige that it possesses thanks to its achievements and successes in the political, scientific, technical, economic, and cultural arenas as well as in the field of sports. The worldwide interest that is shown to the Russian language grows constantly. Russian is first of all the most important negotiating language of the socialist camp (COMECON, Warsaw Pact, etc.). The Russian language is of an entirely particular significance for the GDR. The necessity of knowing Russian arises for a great part of the population of the GDR objectively from the tight, constantly deepening cooperation and friendship with the Soviet Union.*[41]

Attempts to institute Russian-language pedagogy began with the occupation. SVAG established Slavic Departments in universities across the Soviet Zone, with the explicit aim of producing a team of ready translators from Russian into German (not, generally, vice versa). Translation was also the goal of the Academy of Sciences, whose Institut für Dokumentation began churning out renderings of Russian technical treatises in dry German in 1954.[42] One needs translators only when knowledge of the language is lacking, and Russian did not seem to take (surely at least in part because some identified it as the language of invading occupiers, and understood their own failure to assimilate it as a mark of resistance). Nonetheless, efforts to inculcate the language persisted. In 1958, the Academy of Sciences ran 22 courses with 215 participants; in 1962, it was supporting 37 courses with 370 students.[43] These were all individuals within the Academy hierarchy, and who therefore had passed through higher education. That meant they should have *already* learned the language, since in 1951 it became obligatory in universities, swelling the rosters in Russian classes and generating administrative head-

*"Russisch ist eine der führenden Weltsprachen. Diese Tatsache ist durch die Rolle der UdSSR als Weltmacht und ihr Prestige zu erklären, das sie dank ihrer Leistungen und Erfolge auf politischem, wissenschaftlichem, technischem, wirtschaftlichem und kulturellem Gebiet sowie im Bereich des Sports besitzt. Das weltweite Interesse, das der russischen Sprache entgegenbracht [sic] wird, steigt ständig. Russisch ist vor allem die wichtigste Verhandlungssprache des sozialistischen Lagers (RGW, Warschauer Pakt usw.). Von ganz besonderer Bedeutung ist die russische Sprache für die DDR. Die Notwendigkeit der Kenntnis der russischen Sprache ergibt sich für große Teile der Bevölkerung der DDR objektiv aus der engen, sich ständig vertiefenden Zusammenarbeit und Freundschaft mit der Sowjetunion."

aches.[44] Yet it seems that Russian's impact on the East was rather less significant than English's on the West, in large part because English was also a high-prestige language in the GDR.[45]

In the archives of Humboldt University, the flagship of East German higher education, one can find numerous traces of the struggle to drill perfective verbs and instrumental cases into the heads of reluctant Teutons. In a policy statement of 1957, "foreign language" collapsed into "Russian language" instruction in the space of a breath:

> For modern specialist training the knowledge of at least two world-languages is indispensable.
>
> From the viewpoint of the national interests of the German people and of the further development and flourishing of German science, technology, and culture, knowledge of the Russian language is an absolute necessity for those training specialties at the universities and higher schools. In the same way it is important in the interest of the development of German science, culture, and technology that the scientific, technical, and artistic disciplines assimilate the achievements of other peoples and acquire the knowledge of other world-languages.* [46]

As of 6 June of that year, instruction in Russian as well as another "Weltsprache" (here meaning English, French, Italian, or Spanish) was obligatory, although Russian was substantially emphasized. (One should not assume that the quality of instruction was particularly high. The archive is littered with complaints about the poor level of English teaching, for example.[47]) These languages, however, were not to be learned at the expense of German. As at the Free University in the West, foreign students—in the case of the Humboldt mostly from the Eastern bloc—necessitated making explicit something that had been obvious since the eighteenth century: "German instruction is obligatory for such foreign

*"Für die moderne fachwissenschaftliche Ausbildung ist die Kenntnis von mindestens zwei Weltsprachen unerläßlich.

Vom Gesichtspunkt der nationalen Interessen des deutschen Volkes und der weiteren Entwicklung und Blüte der deutschen Wissenschaft, Technik und Kultur ist die Kenntnis der russischen Sprache für die an den Universitäten und Hochschulen auszubildenden Fachkräfte unbedingte Notwendigkeit. Ebenso ist es im Interesse der Entwicklung der deutschen Wissenschaft, Kultur und Technik wichtig, daß sich die wissenschaftlichen, technischen und künstlerischen Fachkräfte die Errungenschaften anderer Völker aneignen und Kenntnisse in anderen Weltsprachen erwerben."

students and will be taught in the first year of study with six hours per week, in the second year of study with four."*[48]

In 1956 the Humboldt administration commissioned a poll of roughly 150 institutions and departments housed within its walls to see what the most important foreign languages for the various branches of science were, in hopes of allocating its resources accordingly. Even accounting for the biases in data collection and the obvious political skewing of the results, this poll provides a unique snapshot of scientific languages in Berlin—once the epicenter of European science—at the moment the decline of German could no longer be ignored. Of the 100 responses returned, representing 87 disciplines, fully 64 selected English and 42 Russian as "absolutely [necessary]" (*unbedingt*) for mastery of the subject matter of the field. (The remaining numbers were 39 Latin, 17 French, 11 Greek, and 7 Hebrew. One should keep in mind that the German *Wissenschaft* is a more copious term than "science," and that several disciplines selected more than one language.) When asked which languages were also "desirable" (*wünschenswert*), another 23 added English—bringing the number up to the full 87—along with 35 Russian and 55 French. A potpourri of other languages graced this other category, including Italian, Spanish, Danish, Swedish, Norwegian, modern Greek, Polish, and Czech. The results were unequivocal: "*In sum therefore the chief languages appear to be English* 87 x, *Russian* 78 x, *French* 72 x, *Latin* 61 x, *Ancient Greek* 15 x."†[49]

When zeroing in on the Faculty of Mathematical and Natural Sciences, comprised of 17 disciplines, the picture was even more striking. English was required in every single department, Russian was required in eight and optional in the rest, and French and Latin trailed significantly (required in three and two departments, respectively).[50] Adding the fifteen disciplines in the medical faculty raised the figures for Latin (required in 13), but also boosted English (required in eight) at the expense of Russian (only two).[51] The most surprising feature of these numbers is the enthusiasm for Latin, backed even by the chemists.

If German were going to survive as a language of science against the tropical storm of Russian or the hurricane of English, whether in the

*"Der Deutschunterricht ist für solche Auslandsstudenten obligatorisch und wird im ersten Studienjahr mit sechs, im zweiten Studienjahr mit vier Wochenstunden erteilt."

†"*Insgesamt erscheinen also die Hauptsprachen Englisch* 87 x, *Russisch* 78 x, *Französisch* 72 x, *Latein* 61 x, *Altgriechisch* 15 x."

Proletarian East or the Bourgeois West, it was going to need a different strategy than simply business as usual. Unless, that is, the first-mover effect of having once been the dominant language of chemistry could be exploited to maintain a foothold for German among the world's scientists. The solution might be to emphasize not cutting-edge contemporary research, but rather the much less glamorous domain of the stodgy reference work.

Shackled by Abstracts

No chemist has ever read the entire *Chemisches Zentralblatt*, but for well over a century not a single practicing chemist was able to conduct research without it. Founded in 1830, the *Zentralblatt* was the oldest abstract journal in chemistry, offering summaries of what its editors considered to be the most relevant chemical literature. In its very creation, the journal embodied a dominant anxiety of scholars since at least the Renaissance (and likely earlier): there was simply too much to read.[52] Until 1907, when the American Chemical Society assumed control of *Chemical Abstracts*, there was no plausible competitor to the *Zentralblatt* for controlling the torrents of chemical literature, and the American outfit did not become the leading abstract journal until roughly World War II. Thus, the *Zentralblatt* tracks in miniature the rise and then eclipse of German chemistry, and of German as a scientific language.

When the *Zentralblatt*'s first editor, Gustav Theodor Fechner, decided to hang up his spurs in 1834, the journal included roughly 500 abstracts on 950 pages—a hefty tome, true, but also rather wordy abstracts. The journal chewed up a series of editors in the middle years of the nineteenth century, and also shed the references to pharmacy included in its original title. In 1870, the year before the unification of Germany into a powerful nation-state, the format and typesetting were overhauled to account for the journal's continued expansion. Between 1886 and 1887, for example, the contents ballooned from 860 to 1,580 pages, representing abstracts culled from 273 chemistry journals. In 1895, sixty-five years after its inception, the journal seemed too unwieldy, too cowed by the mushrooming researches appearing in ever larger numbers (and numbers of languages) to be continued as a private business venture. The German Chemical Society agreed to bring the journal in-house, and that is where it stayed as the German polity itself underwent shock after shock. In 1929, in an article celebrating the

Zentralblatt's centenary, the language barrier was singled out: "This exchange will only succeed, even with the best intentions of all involved, if all of us in Germany, France, England, America, and in other countries devote more attention and industriousness to foreign languages."*[53] Before World War II, the journal had expanded again by almost 52%. German chemistry may have been largely cut off from the rest of the world during the Third Reich, but foreigners still followed the global literature through the *Zentralblatt*, edited by Maximilian Pflücke in the Hofmann Haus in Berlin. In 1944, a bomb careened into the heart of the building, leaving only rubble. In 1945, the last volume of the *Zentralblatt*, already at the printers, appeared. The journal, like unified Germany, was dead.

Like Germany, it would rise again, curiously schizophrenic. When the fog of the postwar settlement began to dissipate, officials and chemists looked about them and realized that the *Zentralblatt* had gone into abeyance. It had to be revived, declared a top official at the Academy of Sciences: "The reappearance of the *Chemisches Zentralblatt* is necessary if German chemical industry and research are to come up to speed."†[54] But how to do it? The *Zentralblatt* was a production of the German Chemical Society based in Berlin, but no one knew which occupying power controlled it. The Society's headquarters had been located in what was, in 1947, the British Sector of the city, but the editorial offices had been whisked away to the American enclave of Dahlem at war's end. The publisher, Verlag Chemie GmbH, had also once resided in the British Sector but had moved to the American because of war damage. It seemed as though the Americans were going to sponsor the *Zentralblatt*.[55]

The Soviets—or, rather, German chemists in the Soviet Zone—begged to differ. Since the Academy of Sciences was in the Soviet Sector and had on hand a group of former *Zentralblatt* collaborators, it began to put out the 116th volume in 1946, although licensing red tape held it up. In November 1946, the Americans granted Verlag Chemie the authority to publish the journal; the following year the Academy (under authorization from SVAG, issued on 1 July 1947) commissioned

*"Dieser Austausch wird auch beim besten Willen aller Einsichtigen nur gelingen, wenn wir alle in Deutschland, Frankreich, England, Amerika und in den anderen Ländern den Fremdsprachen mehr Aufmerksamkeit und Fleiß zuwenden."

†"Das Wiedererscheinen des Chemischen Zentralblattes ist notwendig, wenn die deutsche chemische Wissenschaft und Forschung in Gang kommen soll."

its in-house Akademie Verlag to do the same. What had once been one unwieldy journal had now become two. "Until the resolution of the matter," wrote Georg Kurt Schauer from Frankfurt am Main, solidly in the American Zone, to the administration of the Akademie Verlag in the East in October 1947, "which the allied command of the occupying powers has reserved for itself, the strange state will continue that one scientific journal will be published by two different publishers, with two different editorial boards and staffs of collaborators, with the same numbering of the volumes in the earlier traditional form."* [56]

There was also a problem with Maximilian Pflücke, editor of the periodical since 1923, who had joined the Nazi Party in 1933. Denazification officials sometimes went easy on individuals who joined for purely opportunistic reasons, and Pflücke might have earned an exemption had he been able to demonstrate the shallowness of his political conviction; unfortunately for him, "he was also actively occupied in a fascist sense during the Nazi regime."† [57] He could hold no public position—and editorship definitely qualified—in the Soviet Zone. But the emergence of the American *Doppelgänger* softened opinions. "[N]ow also upon the existence of a published 'Chemisches Zentralblatt' under an American license," an official noted in 1948, "it seems to us especially important to allow Dr. Pflücke to step in from outside in order to announce that our journal is the old classic 'Chemisches Zentralblatt' under the tried-and-true leadership of 35 years."‡ [58] Nazi or no, Pflücke provided the semblance of continuity which might confute the Americans' claims.

The resolution might very well have resembled the macroscopic outcome: a seemingly permanent division between East and West, with parallel publications mirroring parallel societies. In the event, however, the chemists and scientific publishers opted in 1949 for what had proven unworkable politically. There would be two Germanies, but one

*"Bis zur Entscheidung des Falles, den die alliierte Kommandatur den Besatzungsmächten vorbehalten hat, besteht der seltsame Zustand weiter, dass eine wissenschaftliche Zeitschrift von zwei verschiedenen Verlagen, mit zwei verschiedenen Redaktionen und Mitarbeiterstäben, mit gleicher Bandzahl und Numerierung in der früheren traditionellen Form herausgebracht wird."

†"hat sich auch während des Nazi-Regimes aktiv im faschistischen Sinne betätigt."

‡"jetzt auch auf das Bestehen eines unter amerikanischer Lizenz herauskommenden 'Chemischen Zentralblatts' erscheint es uns als besonders wichtig, Herrn Dr. Pflücke nach aussen hin in Erscheinung treten zu lassen, um so zu betonen, dass unsere Zeitschrift das alte klassische 'Chemische Zentralblatt' unter der 35 Jahre bewährten Leitung ist."

Zentralblatt. Hans Brockmann at Göttingen—an organic chemist who had also joined the Nazi Party in 1933—and Erich Thilo of the Berlin Academy erected a joint *Zentralblatt*, to start appearing as of 1 January 1950. The final compromise established Pflücke as the head of the Eastern office, Eugen Klever as the head of the Western office, and Pflücke as the editor-in-chief, the whole affair organized as a joint East-West production sponsored by the German Chemical Society, the Berlin Academy of Sciences, and the West German Academy of Sciences in Göttingen. (This at a time when West Germany refused to recognize that the German Democratic Republic even existed.) Immediately, Pflücke and his collaborators tackled the enormous backlog that had accumulated, and sixteen supplementary volumes had to be prepared between 1950 and 1954 to clear the docket for business as usual.[59] By 1966, the Berlin Academy proudly lauded the *Chemisches Zentralblatt* as its most important publication.[60] Three years later, it was dead.

The cause of death was both too many languages and one language in particular. Chemists had worried about the mounting linguistic burden for decades, but by the mid-1950s even the stalwart Pflücke wondered whether one could maintain a monolingual abstracts journal in a polyglot world. True, the importance of German was buttressed for a few more years by the fact that chemists worldwide had to consult the *Zentralblatt*, but preserving that German character imposed huge costs. One had to either tame the languages through selective abstracting, or somehow defray the labor costs imposed by the gamut of 36 languages covered by the *Zentralblatt*.[61]

The problem with the one language, English, was obvious. Before World War II, a sizeable chunk of the world's chemical output had been in German. Now that English was swamping something approaching 65% of world publication in chemistry, the German labor force either had to be trained up in this one language, or Anglophone collaborators had to be brought on board—and it was hard enough to maintain the delicate balance of East and West German cooperation in the years before Willy Brandt's *Ostpolitik* (1969–1974) began to normalize relations between the Germanies. The American-run *Chemical Abstracts* faced the inverse of this situation, as the mounting tide of English made their job linguistically *easier* each year. In 1957, the *Abstracts* published 24,600 entries, 41.3% more than the same year's *Zentralblatt*.[62] Two years later, the seventy-year-old Pflücke retired, and Heinrich Bertsch and Wilhelm Klemm attempted to bail out the ship. They succeeded for one more decade. By the 1980s, the infrastructure that had been dedicated to putting out the titanic abstract journal was folded into an input service for *Chemical Abstracts* and a fee-for-service bibliographic resource for industry.[63] If reference works were the best hope to staunch the hemorrhaging of chemical German, the East-West joint venture was not going to get the job done. The West Germans would have to save German on their own.

The Beilstein Gambit

In a wide variety of contexts, West Germany assumed the burdens of prewar Germany. The role of economic juggernaut, the center of intellectual political culture, the guilt and shame of the Holocaust—the Federal Republic of Germany shouldered these as it embarked on the economic miracle, the *Wirtschaftswunder*, of the 1950s. While English began to assume an ever greater role in internal education within West Germany, the government attempted to keep some distance from the Americans driving the development. If English had to be taught, it was going to be British English, and Bonn also continued an assiduous program of promoting the German language abroad that had started with the nineteenth-century *Kaiserreich*.[64] Russian was a significantly smaller concern, although no scientifically active country could completely ignore the language. The West Germans followed the American cover-to-cover journals closely, and also set up some translation ventures and review journals of their own to render Soviet achievements legible on the near side of the Iron Curtain.[65] When West Germans thought about preserving German as a language of science, the enemy was English, and the home of scientific English was now the United States.

The makers of West German science policy undertook two major campaigns to preserve the importance of their country as a scientific metropole and, consequently, their language as an essential mode of communication: one centered on personnel, and one on publication. Today, the problem with personnel would be called "brain drain," but no one labeled it that in the 1950s. In truth, it was not a new phenomenon, but rather a continuation of the bleeding that had begun with the economic crises of the 1920s and the 1933 Civil Service Law. German scientists were leaving Central Europe and heading to sunnier climes, principally the United States; afterward, they tended to speak and publish in English. The German Research Council estimated that between 1950 and 1967, about 1,400 scientists were lost from West Germany through outmigration.[66] Internal estimates by the Max Planck Society—the leading research establishment in the country—were even graver: between 1957 and 1964, it counted 973 natural scientists out of a total of 3,400 scientists and engineers who had emigrated to the United States alone.[67] It seemed, however, that by 1968 the flow out had been balanced by economic émigrés returning to assume jobs within the Federal Republic.[68] This was the result of an active plan to recruit émigrés whenever high-level posts became available, securing the best and brightest for German science.[69]

That, however, would only preserve German-speaking science for German speakers. How might one persuade foreign scientists to learn the language? The obvious answer, to those at the Max Planck Society, was to follow what had convinced past generations of foreign scientists to learn German: provide them with a quality product they could access only in the language. The most explicit instantiation of this strategy had its roots in Imperial Russia in the 1860s and 1870s, born of the very characters we first met in the stormy priority dispute over Dmitrii Mendeleev's periodic system.

It is the tale of a man who became a book, a very large and important reference work that became indispensible for practicing organic chemists for about a century. The man was Friedrich Konrad Beilstein, and we first encountered him as an editor of the *Zeitschrift für Chemie*, the one charged with translating Mendeleev's Russian-language abstract into German for foreign consumption, and who botched the affair by entrusting it to a graduate student. Beilstein was born in St. Petersburg in February 1838 to a family of German-speakers who had migrated eastward to try their fortunes in the growing Imperial capital. Raised bilingually, he acquired several more languages during his later scientific training in the German states, eventually securing a post at the University of Göttingen before he was summoned back to St. Petersburg's Technological Institute in 1866, an offer he accepted to succor his family after his father's sudden death. Beilstein thrived in St. Petersburg: he retired with honors from the Technological Institute and was elected to the Imperial Academy of Sciences in 1886. Yet it was a difficult environment for him—he felt isolated because of his German name, habits, and language in an increasingly nationalist environment.[70]

He turned that isolation to good purpose, converting the organic chemistry textbook he had been working on into an index of all carbon-containing molecules that had yet been discovered, complete with detailed properties and accurate citations to the relevant scholarly literature. The first volume of his *Handbuch der organischen Chemie* (*Handbook of Organic Chemistry*) appeared in 1881, a two-volume behemoth consisting of 2,200 pages and detailing roughly 15,000 organic compounds.[71] It instantly brought him accolades from across Europe. (Nearer to home, however, the nationalist gibes would not stop. As he wrote to Heinrich Göppert in 1881: "Even *the* fact that my large just-published *Handbuch der organischen Chemie* appeared in the *German* language (buyers would be lacking for a Russian work) brought me the censure of the patriots."[*72]) Immediately, Beilstein turned to work on a second edition, which appeared in three volumes between 1886 and 1889. The third edition of 1893–1899, spanning 6,800 pages and an additional 50,000 compounds, consisted of four tomes, and Beilstein had had enough. There were too many new organic molecules being discovered, the literature was unimaginably vast, and he was growing old.

So this project, begun as a lone venture by a Germanophone scientist living in the Russian capital, was catapulted to the heart of Berlin. In 1896 Beilstein began to make arrangements for the German Chemical Society to undertake the publication of future editions, and he deputed Paul Jacobson, then 36 years old and soon to be appointed as the general secretary of the Society, as editor.[73] Not everyone was happy with the arrangement; Jakob Volhard, for example, argued that "[i]n my opinion, one would better leave both the *Beilstein* and the *Centralblatt* to private industry. But Beilstein was so hot and heavy for this plan that the further editions would be edited by the Chemical Society that there was no setting out of reasons against it[. . .]."[†74] Beilstein died in 1906, pleased that his magnum opus would live on. In 1914 the Society in turn entrusted the printing of the fourth edition, due to contain all molecules discovered before 1 January 1910, to the Springer publishing firm. That same year, of course, the Great War erupted.

Amazingly, the war had very little impact on the progress of the *Beilstein*, as the book came to be universally known. By the middle of 1916, the entire file of material for the fourth edition was assembled in 123

*"Selbst *die* Thatsache, daß mein so eben erscheinendes großes Handbuch der organischen Chemie in *deutscher* Sprache erscheint (für ein russisches Werk würde es an Abnehmern fehlen), hat mir den Tadel der Patrioten zugezogen."

†"Sowohl den Beilstein als auch das Centralblatt hätte man meiner Meinung nach besser der Privatindustrie überlassen. Aber Beilstein war so Feuer und Flamme für diesen Plan, daß die weiteren Ausgaben von der chemischen Gesellschaft herausgegeben werden, es war kein Aufkommen mit Gründen dagegen[. . .]."

fireproof filing cabinets in Hofmann Haus (the same location as the *Chemisches Zentralblatt*), with a photograph of the entire manuscript for backup. The first volume was sent to the printers that November, and it was completed by 1918 despite a paper shortage.[75] Though the Entente fumed about the dominance of German science, *Beilstein* was specifically cited in April 1918 as a reason why "a reading knowledge of German, with French if possible," was required for a chemist.[76] An attempt to translate *Beilstein* into French in the interwar period self-destructed.[77] As German chemists became pariahs in the wake of the Boycott, Jacobson hoped that "[m]aybe it is granted to the 4th edition[. . .] to assume for itself [the task] and thus help peoples to come closer to each other on a common path in the pursuit of scientific progress!"*[78] He died in 1923, three years before the boycott was rescinded. Jacobson's long-time collaborator Bernhard Prager was now joined by Friedrich Richter, who oversaw the appearance of the final (27th!) volume of the fourth edition in 1937. The stormiest period of *Beilstein*'s existence was just about to begin.

Prager was summarily dismissed in 1933 on political grounds; he died the following year. Six Jewish collaborators were also sacked, and Richter feared he could not keep to his publication deadline and begged the board to postpone further dismissals until 1936. That year and the following, four central employees and five additional workers of Jewish extraction were fired. Losing 30% of its total staff (28 in 1933 and 31 in 1937, including the editor) was crippling.[79] Richter kept his skeleton crew working throughout the war, although they abandoned Hofmann Haus in 1943 because of air raids—presciently, given the devastation the following year that caught the *Zentralblatt* unawares—and carried their library to Zobten, near Breslau in Silesia. The staff, burdened by their massive library, then retreated before the encroaching Red Army and settled in Tharandt until January 1945, when Soviet incursions forced them back to Berlin. The Americans happily welcomed them in July and set them to work.

Beilstein moved into the former offices of the Kaiser-Wilhelm Institute for Biology. By 1946, the staff had been reduced to a paltry seven people. In 1951 the rechristened "Beilstein Institute for the Literature of Organic Chemistry" had built up to a healthy complement of 39,

*"Vielleicht ist es der 4. Auflage beschieden[. . .], auf sich zu ziehen und daran mitzuhelfen, daß die Völker sich einander wieder in der Verfolgung des wissenschaftlichen Fortschritts auf gemeinsamen Wegen nähern!"

newly settled far away from the Soviets and East Germans in Frankfurt-Höchst, and moving six years later into the newly constructed Carl-Bosch-Haus in Frankfurt am Main itself.[80] All of this cost a good deal of money, and resources were hard to find in postwar West Germany. As the official history of the International Union of Pure and Applied Chemistry (IUPAC) noted, "there was a widely held view that these publications"—*Beilstein* and its sister for inorganic chemistry, the *Gmelin*—"were regarded by some nations as spoils of war; thus it was vital that the Union should take an active part in ensuring that the whole chemical community could benefit from information gathered by their editorial staff."[81] The financial burden was massive, but the elites who crafted science were convinced—just as they had been after World War I, but with perhaps a greater degree of urgency—that *Beilstein*'s "role as a German bearer of culture abroad today is of especial significance for us Germans[. . .]."* [82]

Beilstein could bring redemption, and who better to offer it than Otto Hahn, recent Nobel laureate for the discovery of fission and a "good German" who had weathered the Nazi onslaught with minimal compromises to his good name. As the new president of the Max Planck Society, Hahn was willing to adopt *Beilstein* if he could thereby stave off the collapse of German as a scientific language. He wrote to the Ministry of Economy in 1952 for 83,000 Deutschmarks (about $177,000 in 2014 dollars—a princely sum in the circumstances) to be disbursed to the Beilstein Institute. "Precisely the fact that a standard work of chemistry appears in the German language is of especial value for the return to recognition of German in the scientific field," he explained. "I have often sadly had to notice that precisely the retreat of the German language at international congresses has in the end damaging effects for the economy and for the image of Germany in general. It seems to me therefore especially desirable that the Beilstein Institute, the leader of which enjoys the greatest recognition and esteem, also receives further support from the part of your ministry."† [83]

---

* "Für uns Deutsche ist seine Rolle als deutscher Kulturträger im Ausland in der heutigen Zeit von besonderer Bedeutung[. . .]."

† "Gerade die Tatsache, dass ein Standardwerk der Chemie in deutscher Sprache erscheint, ist für die Wiedergeltung der deutschen Sprache auf dem wissenschaftlichen Gebiet von besonderem Wert. Ich habe oft leider feststellen müssen, dass gerade der Rückgang der deutschen Sprache auf internationalen Kongressen sich letztenendes auch schädlich für die Wirtschaft und für das Ansehen Deutschlands überhaupt auswirkt. Es scheint mir deshalb besonders erwünscht, dass das Beilstein-

Hahn understood that working chemists needed to consult *Beilstein*'s hefty volumes constantly, and to do so they needed at least a modicum of German. A whole raft of handbooks were produced—some in German, some in English, many bilingual—to teach the uninitiated enough "*Beilstein* German" to make headway.[84] Springer distributed a slim 2,000-word dictionary free of charge.[85] In the end, one English guide pointed out, "even students whose ability to cope with the German introduction is very meager will be found to have little difficulty with the technical vocabulary of the main part of the work."[86] As long as *Beilstein* was indispensible, so was the German language.

But all was not quiet on the European scientific-publishing front. In 1950 Elsevier—Springer's Dutch arch-competitor—announced that it was contemplating putting out an encyclopedia in organic chemistry, duplicating much of *Beilstein*, in English. (Ironically, the project was itself an outcome of the exile of German chemists; fired and displaced employees of *Beilstein* would form the core of its work force.) Hahn convened a meeting in Frankfurt to discuss the implications and insisted repeatedly that *Beilstein* must remain in German in order to retain the support of his Society. "Besides," he continued, "it would be good if the 'American boys' at least still had in *Beilstein* the opportunity to practice German. Such a work would be *good* propaganda for Germany."* [87] Dr. R. Fraser of UNESCO, who attended the meeting, also insisted that "the Beilstein in any event will be published in the German language. Not just for the reason that thus the 'American youth' will learn German, but because Beilstein has always been a German undertaking and it belongs to the German language."† [88] Richter, the Third Reich's *Beilstein* editor and now also West Germany's, lamented that "interest in and knowledge of the German language abroad has greatly fallen off." Would *Beilstein* save scientific German? "In the end," he continued, "decisive for the sales of the Beilstein Handbuch is however the

Institut, dessen Leiter international grösste Anerkennung und Wertschätzung geniesst, auch von seiten Ihres Ministeriums eine weitere Unterstützung erfährt."

*"Ausserdem waere es gut, wenn die 'American boys' wengistens noch im Beilstein Gelegenheit haetten, sich in der deutschen Sprache zu ueben. Ein solches Werk waere eine *gute* Propaganda fuer Deutschland."

†"der Beilstein auf alle Faelle in deutscher Sprache gedruckt wird. Nicht gerade aus dem Grunde, damit die 'amerikanischen Jungens' auch Deutsch lernen, sondern weil Beilstein immer ein deutsches Unternehmen gewesen ist und es zur deutschen Sprache gehoert."

high quality that will retain for the Handbuch its uniqueness and indispensability."* [89] *Beilstein* would remain German.

Until 1981, when it was decided to render all future volumes of *Beilstein* in English.[90] By then, the Beilstein Institute had expanded to a staff of 160, 110 of whom possessed a PhD in chemistry, and their work was supplemented by roughly 350 outside contributors (mostly West Germans) who had a higher degree in chemistry. These individuals understood the shape of the chemical literature, saw the almost miniscule contribution that appeared in German, and bowed to what seemed inevitable. On 1 April 1978, Reiner Luckenbach succeeded H. G. Boit as editor of *Beilstein*, and he moved what had been a personal project of a nineteenth-century subject of the Tsar into an avatar of the digital age.[91] *Beilstein* was plagued by delays and exorbitant prices as long as it stayed a serial monograph. Shifting, albeit slowly, to an electronic search engine resolved a host of orthographic and especially language-barrier difficulties and made the high price tag—by the 1990s, it cost more than $30,000—worth the investment.[92] In 1998, a chemist who had been using *Beilstein* for decades applauded the transformation; while piecing together the components of a reaction had once been a tiresome slog through hardbound volumes, "now those activities take a few seconds, because the database is computerized and the information is essentially all in English."[93]

Götterdämmerung

One would not recognize in today's *Beilstein* any of the traces of this long and complex trek through the history of scientific German. After the end of the Cold War, with the breaching of the Berlin Wall in November 1989 and the reunification of Germany the following year, the Beilstein Institute remained in Frankfurt am Main but the entire enterprise "remade itself into a commercial venture, and it is run as a business, in a most businesslike manner," to quote one commentator. "Virtually nothing but the name and high quality are the same after this massive reorganization effort."[94] While it had once been funded by Springer and the German state, the efforts of the Institute were now en-

* "das Interesse an der deutschen Sprache und ihre Kenntnis im Ausland sehr zurückgegangen seien. Letzten Endes entscheidend für den Absatz des Beilstein-Handbuchs sei aber eine hohe Qualität, die dem Handbuch seine Einmaligkeit und Unentbehrlichkeit wahren werde."

tirely controlled by a private company, Information Handling Systems, complete with a new bureaucratic structure. The name of the informational service changed accordingly. With full computerization, the system was dubbed "CrossFire Beilstein," and in 2009 its content was subsumed into Elsevier's "Reaxsys," its German origins subsumed within a trademark neologism of the age of globalization.

The stories in this chapter have been episodic, tacking between institutions, publishing ventures, abstract journals, and countries, but each occupied the same territorial space, often only a kilometer or two apart in the center of the city that defined for much of the Northern Hemisphere the meaning of "Cold War." Told through the eyes of scientists who (for the most part) sincerely believed that they worked above ideology and outside of narrow geopolitical interests, the narrative differs from conventional stories of the Cold War. There aren't many spies and there is surprisingly little overt grandstanding, but nonetheless the choices made by Otto Hahn with *Beilstein*, the East German academy with the denazification of Pflücke, the universities East and West as they struggled to staff their courses and simultaneously adapt to a new, post–Third Reich Germany (and, eventually, Germanies), relate a story of Europe for the modern age. Scientific German provides a less dramatic take, granted, than John Le Carré or Ian Fleming might have, but perhaps that is because the ordinary life of scientists attempting to reconcile with the past, communicate with their present peers, and plan for the future represents the lived reality of the Cold War for the vast majority.

Not only was the state of scientific German hard to characterize by the end of the 1970s, so was German itself. Always a pluricentric language—think of the distinctions between German in Berlin and in Munich, in Dresden and in Köln, not to mention Austria (Salzburg vs. Vienna), Switzerland, Lichtenstein, Pennsylvania German, and so on—the surprising stability of the Cold War prompted discussion of the division of German itself. A vigorous sociolinguistic debate grappled with the question of whether East and West German were becoming two distinct dialects or even languages. Given the ideological context of the times, it is unsurprising that characteristic patterns emerged, with East Germans highlighting variations not only in lexicon but also in syntax to argue for the development of a distinctive socialist culture, and West Germans attempting to minimize these as unimportant variations in the face of a common linguistic bond.[95] By the 1970s, this debate had become like the Cold War: static and without resolution.

The fuzzy national status of German—in the postwar period, it was an official language in six countries and enjoyed subordinate (minority or regional) status in Belgium, Italy, and Namibia besides—was to some extent an advantage. German was a capitalist language, as represented by the Federal Republic. It was a socialist language, as evidenced by the Democratic Republic. It was politically neutral, thanks to Switzerland. It enjoyed, therefore, a marked capacity to serve as a passport between different worlds, facilitating a resurgence of the language as a vehicle for international trade.[96] For many, however, hopes for the rehabilitation of the language to its former international dominance rested with science, for this was an area (unlike politics or economics) where German dominance was not resented in the contemporary world, and in which the achievements of the past retained value. The rare optimistic article proclaiming a renaissance in German to lie right around the corner would always cite science—the Max Planck Society, the excellent universities, the prestigious journals and publishers—as the vehicle for future growth.[97] More realistic sociolinguists, however, recognized that the state of German in the sciences was locked into a zero-sum relationship with English. As Ulrich Ammon, the foremost scholar of the present-day status of German as a scientific language, noted in 1990: "the ground lost by German has been gained virtually exclusively by English."[98] There is no other place to bring our story to conclusion than the language in which it has been written.

CHAPTER 11

Anglophonia

The language in which we are speaking is his before it is mine. How different are the words *home, Christ, ale, master*, on his lips and on mine! I cannot speak or write these words without unrest of spirit. His language, so familiar and so foreign, will always be for me an acquired speech. I have not made or accepted its words. My voice holds them at bay. My soul frets in the shadow of his language.

JAMES JOYCE[1]

On New Year's Day, 2012, science reached the end of its Latin. As of that date, the International Code of Botanical Nomenclature, the official record of plant species, declared as no longer obligatory the long-established practice of requiring not only that the Linnaean binomial classification, but also the description of candidates for new species (how many stamens, the shape of the leaves, and so forth) be in Latin. You could still submit descriptions in Latin if you wished—perhaps to keep up skills from primary school, or to continue a pleasant association with the classicists across campus—but from this date onward English would also be acceptable. Descriptions in Latin became a requirement in 1906, in response to a request by Spanish botanists to allow their language as legitimate for botanical diagnoses alongside French, English, German, and Italian. The reaction was predictable: to avoid an incipient Babel of too many languages, the international organization insisted upon the language of the Romans, perceived as neutral. The custom was reaffirmed in Article 37 of the International Code, published in 1961. And now, in 2012, Latin was perceived as unwieldy and backward, and the new language of neutrality was one of the very tongues the Spanish delegates had protested: English.[2] This outcome is probably no surprise to you. The only question is why it took so long.

Today, English is not only the dominant form of international scientific publication and oral communication at conferences and in multinational laboratories—it is almost always the *only* language of such communication. There are many ways to illustrate this, from grabbing your nearest scientist and simply inquiring to perusing the shelves of scientific journals in any technical library, but the quickest way of surveying the extent of the transformation is with numbers. As in the graph presented in the introduction, the evidence of the past half century is unequivocal. If one counts the cover-to-cover translations of Soviet journals as "English" articles—and one really should, since this was how most Soviet science was consumed abroad—then already by 1969 fully 81% of the physics literature appeared in English. More conservatively, *Chemical Abstracts* recorded in 1980 that 64.7% of the articles it abstracted appeared originally in English, 17.8% in Russian, and 5.2% in Japanese, followed by smaller numbers in German and French (with Polish next in line, at 1.1%), a dramatic transformation from the triumvirate that had opened the twentieth century. Between 1980 and 1996 German dropped from 2.5% to 1.2% across all the natural sciences and Russian equivalently moved from 10.8% to 2.1%; English, on the other hand, had jumped from 74.6% to 90.7%.[3] That data, however, does not fully take into account the consequences of the collapse of the Soviet Union or the globalization of China and India. It is hard to measure the total output now, but in elite journals across the natural sciences, no matter the country of origin, well over 98% of publication—a sum that has, recall, been steadily increasing over time—is in English.[4] There is an absolute flood of natural knowledge being produced in a language once confined to the southern part of one particular island in the North Sea.

It is not just a question of *how much* English, but also *what kind* of English we are talking about. English, like any other language, shows enormous (and constant) diversification and divergence, differentiated by geography, social class, race, and other factors. These distinctions range not only from the obvious markers of accent or word choice ("flat" vs. "apartment"), but to dramatic rearrangements of syntax. That is, the tension we have explored between opting for "identity" (expressing yourself in the idiom most comfortable to you) and "communication" (attempting to reach the broadest audience possible) remains an issue for English speakers even when talking to others who speak ostensibly "the same" language.[5] If you believe that all "English" speakers are mutually intelligible, you need to get out more.

Nonetheless, there is a "standard" English that facilitates communication around the world, although it is not regulated by any official state body as in the case of French or Modern Hebrew, and the English spoken and written by scientists is an even more rigorously standardized and specialized variant.[6] The peculiar features of international scientific English, the particular history of its emergence, and the impact of its growth upon the other dominant languages of science, are the subjects of this chapter, concluding the history of scientific languages we have traced through the centuries. It seems that the oscillation between communication and identity seems to have settled, for the present, very definitively upon the "communication" side of the spectrum. English's rise has received its greatest push not from native speakers, but from *non*-native Anglophones (the majority of scientists and engineers in the world) using the language to reach the broadest audience. This has happened largely because English has come to be seen—rather surprisingly, given its history—as a "neutral" international mode of communication, whereas using French or Russian or Japanese is interpreted as a gesture directed at domestic audiences. This perception of neutrality has been the engine enabling English's omnipresence in international science.

How Widespread Will English Become?

The development of English—from the arrival of Angles and Saxons to the British Isles, the intermixing of Scandinavian influences due to Viking invasions, the Norman conquest of 1066 and the grafting of French forms onto the dominant Anglo-Saxon, the constant presence of Latin (from Roman centurions to medieval monks), down to the flourishing of Geoffrey Chaucer's Middle English and William Shakespeare's Modern English—has been extensively studied, and this is not the place to rehearse well-worn milestones.[7] Although one might presume that the position of English in the world today is most heavily indebted to imperial expansion, it is rather the case, as linguist Robert Kaplan has observed, that "the spread of English is a relatively modern phenomenon;[. . .] most of the spread has occurred since the end of World War II."[8] The rise of English not only in the sciences but in other areas actually postdates the high-water mark of the British Empire in the eighteenth and nineteenth centuries.

For most of the period when the speaker base of English was centered in Britain, the prospects for its diffusion were not particularly rosy. In 1582, Richard Mulcaster, often considered the founder of English lexicography, sadly noted that "[t]he English tongue is of small reach, stretching no further than this island of ours, nay not there over all."[9] A century later, after Britain's first wave of overseas colonization, there were only an estimated eight million speakers of the language worldwide.[10] Foreign correspondence, both mercantile and diplomatic, took place in the dominant vehicular languages of the early modern age: Latin, French, and Dutch. As late as 1714, when the posthumous edition of French linguist Giovanni Veneroni's dictionary of the chief languages of Europe was published, English was not considered important enough to include beside French, German, Italian, and Latin.[11] Later, provinciality was displaced by an abiding certainty that French-speakers and German-speakers (let alone the rest of the planet) would not accede calmly to the enormous advantage the global spread of English would give to British and American national interests.[12]

Geopolitics was one strike against English's dominance; another, perceived by many to be much more serious, was the sheer difficulty of the language. English was too hard to be global. In 1886, Scottish phonetics pioneer Alexander Melville Bell—whose much more famous son and namesake is credited with the invention of the telephone—published an ambitious pamphlet entitled *World-English*. Writing in the shadow of ever-growing Volapük, he was mostly optimistic: "No language could be invented for International use that would surpass English, in grammatical simplicity, and in general fitness to become the tongue of the World. The only drawback to extension of English has been its difficult and unsystematic spelling."[13] Orthographic reform could remove the final roadblock—although the sheer bizarreness of his proposed new letters to represent specific sounds (which I would love to reproduce but am prevented by typographic constraints) may give the modern reader pause. (It is striking, in fact, how rarely spelling comes up as an obstacle in contemporary discussions of scientific English, probably largely because the lexicon is so circumscribed for each subdiscipline.)

The much more common diagnosis of English's difficulty stressed the sheer variety of words covering similar notions, and the proliferation of grammatical exceptions that often obscured what boosters saw as the tongue's essential simplicity. The most prominent critique in this direction was C. K. Ogden and I. A. Richards's Basic English. According to Richards, among the leading literary critics of his generation, the idea occurred to Ogden while the two were writing their book on semantics, *The Meaning of Meaning*.[14] While exploring definitions of various abstract terms, Ogden was struck by "the fact that whatever you are defining, certain words keep coming back into your definitions. Define them, and with them you could define anything."[15] The solution to a universal language might be to preserve the simple grammar of English—no gender, limited agreement, fixed word order—and cap the vocabulary.

Basic English, Ogden and Richards would insist, was nothing more than English with fewer words, 850 to be exact: 600 names of things, 150 names of qualities, and 100 "operations," a catch-all category that lumped verbs together with prepositions.[16] Basic English, Ogden proclaimed, "is an English in which 850 words do all the work of 20,000, and has been formed by taking out everything which is not necessary to the sense. *Disembark*, for example, is broken up into *get off a ship*. I *am able* takes the place of I *can*; *shape* is covered by the more general word *form*; and *difficult* by the use of *hard*."[17] Richards—the St. Paul for Ogden's language—held that 850 was the perfect number: "It would be easy to cut Basic English down to 500 words, but then it would depart from Standard Usage and at the same time the strain of making the limited language cover the needs of its users would increase prohibitively."[18] Perhaps, but is "umbrella" essential? Is "dance" superfluous? Ogden and Richards were inflexible about the core vocabulary, but they admitted that specialist activities—importantly including science—demanded supplemental vocabularies, which could be added on to the basic word list or simply defined upon their first use in terms of the original 850.[19]

According to its advocates, Basic English solved every difficulty that beset English. First, it was "not greatly different from ordinary standard English."[20] This meant that, unlike a pidgin or simplified language, there was nothing to unlearn in moving from Basic to Standard English—the former was a proper subset of the latter; nothing that was grammatical in the first would be unintelligible to native speakers. Second, by quickly enabling students to maneuver with the language, it would lessen feelings of "intellectual, technological, or other domination" by English speakers.[21] That said, in 1943 the British War Cabinet began active promotion of Basic, which they hoped would prevent the disintegration of the language into pidgins and dialects as the British Empire continued to occupy disparate regions of the world. Basic English, under the active personal promotion of Richards, even had a distinguished career in Republican China before the Japanese invasion at the start of World War II.[22]

Basic English did not lack for contemporary critics. One of the most forceful arguments came from Lancelot Hogben—British biologist, statistician, and science popularizer with a knack for languages. "With due recognition of [Ogden's] unique achievement," he wrote, "it is therefore important to state charitably at the outset why Basic was bogus."[23] The answer was what he called "mnemonic load." While Ogden and Richards claimed they could minimize learning difficulties by reducing vocabulary, was replacing "belittle" with "make light" and "manifest" with "come to be" really a savings? For a word, one must now memorize a compound phrase. The real work lay in the metaphors behind the words, a point essentially conceded by Richards, who noted that "it must not for a moment be supposed that Basic leaves it to the learner to invent and experiment with these metaphors at random. The greatest part of the labour of producing Basic did in fact go to the thorough inventory of these metaphors."[24] More sniping followed: Basic sounded wooden; it functioned as a pidgin; it was helpful for reading but not writing or speaking; it did nothing for pronunciation; and it merely delayed the inevitable need to learn English.[25] If English was inhibited because it was too hard, too verbose, too difficult to spell—then Basic English would not help.

It was not obvious even in the wake of World War II that English would take over the way it has. In a history of "scientific English" penned in 1947, a curmudgeonly author anticipated the continuation of the triumvirate: "Thus it is that every scholar today is trilingual, perhaps lamely so but still struggling valiantly toward that end. Three instead of one linguae francae for science are a burden."[26] And even those who recognized that English was on the rise—and a cursory examination of abstract journals would tell any scientist that, as of September 1949, 57% of all scientific articles were published in English—there were still fears that "Russian, Chinese, or Urdu" would eventually supplant this dominance.[27] With the benefit of hindsight, we know this did not happen. Not by a long shot.

When Did English Come to Seem Inevitable?

It is broadly assumed that the greatest hostility to the omnipresence of English is based in Paris, a reasonable inference based on the visibility of excoriations of English by politicians and intellectuals in the Fifth Republic. Yet even as far back as 1982, the dominant attitude in French scientific periodicals toward the growth of English publishing was resignation. "Despite the fact that French is still the language of scientific work in West Africa, in the countries of the Maghreb, in Quebec, and in certain francophone European countries," noted a special commission of the *Comptes rendus*, the journal of the storied Académie des Sciences, "English is today the international language of science; it could become its sole language very soon."*[28] The commission wondered then—and some French commentators have continued to ponder today—whether there was even any point in maintaining French-language scientific journals, dressing up as serious research what might be better understood as "popularization."[29] When did such views become reasonable to their proponents? When and why, that is, did *non*-native speakers of English begin to see the position of English in the sciences as a fait accompli?

One dominant factor is the sheer size of the scientific vocabulary in English. There are more words in English dedicated to the various sciences than for any other function, as a casual glimpse on almost any page of a reasonable dictionary will make abundantly plain.[30] (There are also more scientific words in English that have at least partly Ancient Greek roots than there are words in Ancient Greek.[31]) The size of the vocabulary not only indicates that it is possible to conduct research in any science in English; with each word that is developed for English alone, it becomes harder to repeat the Russians' accomplishment from chapter 3 and engineer a scientific language in, say, Tagalog or Swahili or Malay. One would need to develop a standard term for every scientific notion, publicize it, and get it into use. The cost of this, as well as developing the full complement of publishing houses, is overwhelming, even increasingly for languages like German, which have a healthy scientific vocabulary of their own.[32]

For most scientists, the sharpest evidence of Anglification has been in the contents of scientific journals, the main outlet by which findings about the natural world are disseminated. The pattern has become so routine as to be almost cliché: first, a periodical publishes only in a particular ethnic language (French, German, Italian); then, it permits publication in that language and also a foreign tongue, always including English but sometimes also others; finally, the journal excludes all other languages but English and becomes purely Anglophone, regardless of whether it is published in Milan, or Marseilles, or Mainz, or Mexico City. As one (English-speaking) chemist put it: "[O]nce an editorial committee decides to allow the use of English in the pages of its jour-

*"Bien que la français soit encore la langue de travail des scientifiques en Afrique occidentale, dans les pays du Maghreb, au Québec et dans certains pays européens francophones, l'anglais est dès aujourd'hui la langue internationale de la science; elle pourrait devenir très prochainement sa langue unique."

nal, it finds that it has invited a cuckoo into its nest that pushes the native fledglings aside."[33] Foreign publishers, adjusting to copyediting and production in English, have incurred higher costs in hiring editors with the obligatory native or near-native English skills.[34] Often, but not always, the name of the journal changes as well, the bland English moniker hiding any trace of the national origin of the periodical. A scattershot survey will suffice: *Die Heidelberger Beiträge zur Mineralogie und Petrographie*, founded in 1947, became *Contributions to Mineralogy and Petrology* in 1966; *Mineralogische Mittheilungen*, founded in 1871, became *Mineralogy and Petrology* in 1987; *Zeitschrift für Tierpsychologie*, which published 100% of its articles in German in 1950 (although English, French, and Italian were acceptable), began to shift to English already in 1955, and changed its name to *Ethology* in January 1986; the storied *Annales de l'Insitut Pasteur* became *Research in Immunology* in 1989; the Mexican *Archivos de Investigación Médica* transitioned gradually in the 1980s into the *Archives of Medical Research*; the *Archiv für Kreislaufforschung* is now *Basic Research in Cardiology*; the *Zeitschrift für Kinderheilkunde* is now the *European Journal of Pediatrics*; *Gastroenterologia* became *Digestion*; and the official organ of the Japanese Society of Plant Physiologists is called *Plant and Cell Physiology*.[35]

Simply relating a list of titles does not, however, give a sense of how this transformation was experienced by the non-Anglophone contributors to these journals. An instructive case in point is *Psychologische Forschung*, founded in 1921 by Kurt Koffka and several other giants of German psychology, which rechristened itself *Psychological Research* in 1974, adding the subtitle "An International Journal of Perception, Learning, and Communication." The transition to a fully English journal had been in the cards for a while. In 1971–1972, the journal published 24 articles, 13 of which were by German-speaking authors, yet 18 of the articles appeared in English and only six in German. Only eight of the English articles displayed German abstracts. After the language change, German abstracts atrophied. Examining the "Instructions for Authors" published in the journal's paratext gives some indication of why. In *Psychologische Forschung*, potential contributors were informed, in German, that "Contributions will be accepted in German, English and French. It is requested that *manuscripts* be composed *in English as far as possible*"* (emphasis in original); but in *Psychological Research*

*"Es werden Beiträge in deutscher, englischer und französischer Sprache angenommen. Es wird gebeten, die *Manuskripte möglichst in Englisch* abzufassen."

the English-language equivalent declared: "*Papers should be preferably written in English*" and also that "[e]ach paper should be preceded by a *summary* of the main points. . . . Papers in French and German should also have the title and summary in English." Werner Traxel, an irate Germanophone psychologist, wrote to the publisher in April 1975 and asked whether, as the instructions implied, the journal would still accept German articles, and if so whether they would facilitate a translation into English. An editor responded that English articles were strongly encouraged, and that if Traxel felt uncomfortable in the language "[p]erhaps you have an English-speaking colleague who can be helpful to you in the translation. Insofar as this is not the case, in exceptional cases there is the possibility of sending the manuscript to an editor of the journal who lives in Germany . . . to proofread the English."* [36] A linguistic tradition in psychology was at an end, and no resources were provided to guide stragglers into the new standardized scientific communication.

Standards demand conformity. Just as there had earlier been manual upon manual to teach Anglophones how to read chemical German or technical Russian, new handbooks instructed scientists in "scientific English," all published, *natürlich*, in English.[37] That is, if you wanted, like most researchers in Helsinki, to compose your articles in English instead of having them translated (expensively) from Finnish, you had to be fluent enough in English to fully comprehend the guide that would help you accomplish your goal.[38] The format of scientific articles had, over the past two centuries, become increasingly regularized until it reached its homogeneous postwar rubric of Introduction, Methods, Results, and Discussion (IMRAD); this much was obvious.[39] Less apparent but no less real was that the *English* too had compressed to very limited variation. Scientific English, even more so than scientific French and German, was characterized by a uniformity of style: "relatively short, syntactically simple sentences containing complex noun phrases with multiple modification, verbs in the passive voice, noun strings, technical abbreviations, quantitative expressions and equations, and citational traces."[40] The standard English of scientific prose, distinct from the "standard English" of Hollywood and the financial

*"Vielleicht haben Sie einen englischsprachigen Kollegen, der Ihnen bei der Übersetzung behilflich sein kann. Sofern dies nicht der Fall ist, besteht in Ausnahmefällen die Möglichkeit, das Manuskript an einen in Deutschland lebenden Herausgeber der Zeitschrift . . . zur Überprüfung des Englischen zu senden."

press, amounted to a new dialect for nonnative speakers (and for native speakers too, as anyone who has attempted to write scientific prose has learned).

Alongside the hegemony of English in written science, its prominence in spoken science as, in most instances, the only language of international scientific conferences, is just as striking, and substantially more burdensome. Translation is expensive, and so only manageable at selected large meetings; most scientific gatherings are of smaller scale and take place without the benefit of professional interpreters.[41] This difficulty is exacerbated by the fact that many native speakers of English, unaware of the height of the language barrier or its radical asymmetry, often make little to no accommodation to the linguistic capacities of their audience. Although most international scientists consistently self-identify English as their best foreign language, multiple studies indicate that nonnative speakers are "*handicapé par la langue*" in oral communication.[42]

Science policy makers in traditionally strong scientific countries have been making significant adjustments to this emergent Anglophone world for decades. Japanese researchers functioned since the mid-nineteenth century with the knowledge that they would need some vehicular language besides Japanese if they wanted to be understood abroad, a need also reflected in a tradition of publication in European languages as well as Japanese (especially for graphs and figures). Now English is the chief language deployed. RIKEN, Japan's premier research institute, reported the publication of just under 2,000 research reports in English, but only 174 in Japanese in 2005, and even domestic scientific gatherings are using English.[43]

Anglophonia is starting ever earlier, saturating education at lower levels. The Jacobs University in Bremen, Germany, offers all of its instruction in English. In a newspaper interview, an undergraduate biology student who studied *E. coli* was excited by this development. "I find it convenient that there is a single leading scientific language," he told reporters. "Only when I tell my grandparents about my studies does it sometimes become complicated. Then I have to translate twice—first from scientific language into lay language, and then again into German."* In fact, "English was one of the chief reasons for me to begin my studies at Jacobs University. Because I definitely want to go into re-

*"Ich finde es angenehm, dass es eine einzige führende Wissenschaftssprache gibt. [. . .] Nur wenn ich meinen Großeltern von meinem Studium erzähle, wird es manch-

search, and as we all know everything there runs in English."* [44] In about half of the international bachelors programs for the German Academic Exchange Service (DAAD), education is exclusively in English, while 460 of 640 International Masters Programs in Germany use English as the exclusive language, up from 250 in 2007. Advanced science students are obligated to use English-language textbooks regardless of the language of instruction.[45] Already in the 1980s, eight German universities permitted scholars to submit dissertations in English, and this is now essentially universal in the natural sciences and increasingly common in the social sciences and humanities.[46] As a consequence, as a manifesto of academics declared in 2005, "the use of the English language conveys the impression that in Germany one can no longer formulate and express ideas as before. Students and scientists would prefer to study, research, and teach in the Anglo-American original than in such a country."† [47]

As a final illustration of the ubiquity of English, consider the stories of an admittedly biased population: Nobel laureates in Chemistry in the twenty years since the collapse of the Soviet Union. Between 1992 and 2011 there were 45 laureates, and of course this limited sample is profoundly unrepresentative. In fact, they were awarded the prize *because* their work was deemed exceptional, and many of them led unusual careers. It is important to remember that these scientists were awarded the Nobel for work performed many years, sometimes decades, earlier, often during the height of the Cold War. The English in their backgrounds illustrates how long ago Anglification became a dominant feature of the landscape of chemistry. There is no reason to expect other sciences to be significantly different; in the case of physics, the effect is likely even more pronounced.

What does this sample tell us? Of the 45 laureates, 19 (42%) were not native speakers of English, an indication of the enormous resources poured into science by the United Kingdom but especially the United

mal kompliziert. Dann muss ich zwei Mal übersetzen—erst von der Wissenschafts- in die Laiensprache und dann noch ins Deutsche."

*"Für mich war das Englische einer der Hauptgründe, an der Jacobs University mein Studium zu beginnen. Denn ich will unbedingt in die Forschung, und da läuft nun mal alles auf Englisch."

†"vermittelt der Gebrauch der englischen Sprache den Eindruck, man könnte in Deutschland neue Ideen nicht mehr als erste formulieren und aussprechen. Ein solches Land wird für Studenten und Wissenschaftler studieren, forschen und lehren daher lieber gleich beim angloamerikanischen Original."

States. (The most pronounced minority is women: only one in the entire set.) Only three of these chemists, according to their official autobiographies, passed their entire career without ever having studied or worked in an Anglophone context. All of them knew English in order to keep up with the literature, but this small number indicates that by far the most common way of securing a command of the language was to spend time in an environment entirely surrounded by it. World War II marked most of these laureates, many of whom were refugees, the children of refugees, or otherwise affected by the conflict. With that, the commonalities end; each chemist had an idiosyncratic path to science.

A surprising number brought up language in their autobiographies. Although most of these were originally written in English, there is an undertone of other languages studied with hopes of breaking into science. For example, Mario J. Molina, born in Mexico City in 1943 and laureate in 1995, "was sent to a boarding school in Switzerland when I was 11 years old, on the assumption that German was an important language for a prospective chemist to learn."[48] German was studied by several, and yet almost none who were not native speakers of the language published in it. On the other hand, English is everywhere. As one of the three who had no direct Anglophone exposure, Jens Skou—born in Denmark in 1918, laureate in 1997, and a resistance fighter against Hitler's occupation of his homeland during the war—noted that his 1954 dissertation was published in Danish "and written up in 6 papers published in English" immediately afterward.[49] Publishing in English seems to have been most crucial (and challenging) for the Japanese laureates. Koichi Tanaka, born in 1959 and co-recipient of the 2002 prize, at first studied German in university, although he lamented his poor grades in the subject. The major transition in his career came in September 1987, at the Second Japan-China Joint Symposium on Mass Spectrometry in Takarazuka, Japan, when "we announced our results in English for the first time." That is, even though this was a meeting primarily for Chinese and Japanese, English was the crucial language. "There is a double significance here," he continued, "in that not only were the research results written in English, I actually presented the results in English for the first time. Although my English was far from good, my meaning was well enough understood by Professor [Robert] Cotter [of Johns Hopkins University] for him to make the results known around the world."[50] A similar story, characterized by diligent study of the language, was expressed by El-Ichi Negishi, born in

Japanese-occupied China in 1935 and laureate in 2010, whose English was further strengthened by obligatory classes associated with his Fulbright award at the University of Pennsylvania.[51] Perhaps the most poignant, however, were the recollections of Ahmed H. Zewail, born in 1946 in Damanhur, Egypt, who was the sole recipient of the Nobel in 1999. He studied in the United States and now works there, but his arrival was rough:

> I had the feeling of being thrown into an ocean. The ocean was full of knowledge, culture, and opportunities, and the choice was clear: I could either learn to swim or sink. The culture was foreign, the language was difficult, but my hopes were high. I did not speak or write English fluently, and I did not know much about western culture in general, or American culture in particular.[52]

The Nobel population, though not typical, is indicative of some major trends. As the Nobel Prizes have been awarded since 1901 by the Swedish Academy of Sciences, the history of the science prizes can tell us a lot about how Swedish science—that is, science in a small, wealthy, geopolitically peripheral nation—fits into the global context. Until the end of World War II, the overwhelming tendency was to award prizes to German scientists. On the one hand, this was a reflection of the tremendous ferment in German science in that period. On the other, it is also an indication of educational patterns: Swedish scientists were often educated in Germany and German was their most comfortable vehicular language. They therefore read German publications and nominations with greater ease. The same characteristics of both quality and concomitant linguistic familiarity can be read in the dominance of Anglophone publications and scholars after the Second World War. In the Nobel population, the point of transition seems to be 1920; scientists born after that date lived in an Anglophone world.[53] When this pattern is broken, there is usually an interesting story to be found. For example, Soviet chemist Nikolai Semenov won the Chemistry Nobel in 1956 largely as the result of persistent lobbying by Lars Gunnar Sillén, professor of inorganic chemistry at the Royal Swedish Institute of Technology in Stockholm, who happened to know Russian and was committed to improving Swedish-Soviet relations.[54] Without an inside advocate possessing an unusual linguistic profile, the message is quite clear: if you aim for the Prize, aim in English.

Why Did This Happen?

As with any historical change on such a broad scale, involving many thousands of scientists spanning the entire globe across more than a century, there were many causes of this compression to a single language. First, the triumvirate had to be displaced before English could break away from French and German. The initial destabilizing impulse came with the rise of nationalist ambitions from large scientific cohorts such as the Russophone one, which challenged the tight strictures around three dominant languages, but this was overshadowed by the refusal of the largest and richest Anglophone population, that in the United States of America, to continue learning foreign languages. These factors were supplemented by geopolitical developments by which American science was lifted with American power, and the English language alongside both.

English itself was not responsible—that is, English does not possess specific qualities that make it particularly well suited for scientific research. Most linguists today would shudder at the notion that any language is intrinsically suited for, say, chemistry, not least because languages themselves are subject to constant modifications and interactions. Yet repeatedly one finds claims that this must be what lies behind English's victory. For example, Max Talmey, the prime advocate of Ido in interwar America and a native speaker of German, considered English "far richer, far more expressive than any other language. Far more often than with any other tongue one meets, in a comparison pertaining to expressiveness, with concepts each expressible in English by a single word and only by a circumlocution in any other language," while even a French scholar considered English "more malleable, more plastic than French as far as being a vehicular language."* [55] Other paeans to its simplicity and "masculinity" can also be set aside.[56]

More powerful than any intrinsic linguistic advantage to English was a definite *political* backlash against German in the wake of National Socialist brutality. The reaction against German in the United States happened earlier, with the nihilistic enthusiasm of anti-Teutonic sentiment uncorked by the Great War; to a lesser degree, other nations dismantled their German educational structures in the late 1940s. Even Germanophilic Sweden, neutral in the global conflagration, replaced German with English as the first foreign language taught to children, beginning

*"plus malléable, plus plastique que le français en tant que langue véhiculaire."

in their fifth year of school, in 1947. Two years later, students destined for higher education could begin to study a second foreign language, typically German, but those headed for practical training only learned English. Similar patterns, ramified across the globe, wrought enormous damage on knowledge of German.[57] Even Indonesia had banned German instruction in 1940, and when it was reintroduced to the schools in 1945 it was met with remarkable lack of enthusiasm on the part of both students and teachers.[58]

While German as a global language of scholarship suffered because of the actions of Germanophone political leaders, to a limited degree English benefited from active promotion by Anglophone governments, especially the United States. On 11 June 1965, American President Lyndon B. Johnson declared that the promotion of English was now "a major policy," and the Peace Corps, the US Agency for International Development, and other organizations encouraged study of the language.[59] Likewise, the American (and British and Canadian) leadership of the North Atlantic Treaty Organization (NATO), two out of five Anglophone seats on the United Nations Security Council, and American sponsorship of the International Monetary Fund (IMF) surely did not hurt the status of English as an international language.[60] Although, as of 1 January 1975, English was the sole official language in only 21 countries, it was recognized alongside a local language in 11 more and grew increasingly popular in international organizations, becoming by 2004 the official language of 85% of the 12,500 such organizations worldwide. (French came second at 49%.)[61] Yet the Anglification of the sciences *preceded* many of these policy measures, and much of the enthusiasm for the language stems not from top-down political promotion, but from the ground up.

These trends notwithstanding, there was a good deal of pessimism in the 1960s, especially in the United Kingdom, about the future of English as a language of science. Such gloominess seems like lunacy in the face of that curve of ever-increasing English abstracts that we saw in the introduction, but a closer look reveals a plateau in the 1960s combined with an uptick in Japanese and Russian—not just foreign tongues, but written in impenetrable scripts to boot! As one study from 1962, bemoaning the slippage of English, put it: "It seems wise to assume that in the long run the number of significant contributions to scientific knowledge by different countries will be roughly proportional to their populations, and that except where populations are very small contributions will normally be published in native languages."[62] In addition, there were certainly plenty of predictions, even into the late 1970s, that decolonization would produce a new international Babel that would have made the inferno of the nineteenth-century age of nationalism seem like a brush fire.[63] Given the global reach of the British Empire, hostility to English was particularly noticeable.[64]

One of the most curious features of the rise of English is the fact that political resistance from rapidly decolonizing nations did not provide effective resistence either in diplomacy or in scholarship. Decolonization did little to staunch the spread of Anglophonia, often because of the unavoidable necessity of some vehicular or auxiliary language. For example, at the 1955 Bandung Conference, where nonaligned nations proclaimed their autonomy both from former colonial masters and the growing US-Soviet Cold War, participants settled, after intense and often heated discussion, on one official language: English.[65] Meanwhile, developing nations in what was once called the "Third World" have repeatedly opted for English. Ethiopia, for example, never colonized by a European power, added English as an official language, signaling a much broader trend. Students from most decolonizing countries, especially in the sciences, often selected their destination of foreign study precisely to gain fluency in English. The United States was particularly desirable, as students flowed across the network established by Cold War foreign policy. Seven thousand foreign students studied in the United States in 1943, growing to 26,000 in 1949 and 140,000 in 1971, an expensive enterprise funded in part by the United States government but increasingly by grants from home countries.[66]

The crucial shift was the transition from a triumvirate that valued, at least in a limited way, the expression of identity within science, to an overwhelming emphasis on communication and thus a single vehicular language. The very same arguments that had been pooh-poohed when voiced by Esperantists and Idists at the dawn of the twentieth century came to be unquestioned axioms by century's end. Most sociologists and applied linguists who have examined the hegemony of scientific English have pointed to English's ubiquity as the almost accidental outcome of computerized reference tools and the inexorable and omnipresent gravitational pull generated by the wealth and scientific prominence of the United States.[67] Early computerized databases privileged English; it was estimated in 1986 that fully 85% of the information available in worldwide networks was already in English.[68] Database followed database, and the advent of hyperinfluential metrics such as the Science Citation Index and "impact factors" only increased the first-mover advantage that had accrued to American indices. Publishing in English placed the lowest barriers toward making one's work "detectable" to researchers.[69]

The jig was therefore already up when the second most popular—but grossly diminished—scientific language faced a terminal crisis on Christmas Day, 1991: the dissolution of the Soviet Union. Russian had been losing status in the sweepstakes of scientific languages since the early 1970s upon the dramatic international success of American cover-to-cover translation efforts. It had until then been boosted as a scientific language by precisely the same kind of graduate-study and postdoctoral flows from the Third World that were so significant for English, but the numbers—7,600 students from Latin America at the height in 1985—were a drop in the bucket compared with those headed to the United States, and the bucket had developed a serious hole in the bottom: Eastern Europe, which began to calve off the Soviet Bloc with rapidity as that decade reached its end.[70] Russian was stripped of its special status everywhere but Romania—which had abandoned obligatory Russian decades earlier—prompting a deluge of students into German and English. (German was a surprise beneficiary of the decline of Russian from the Baltic to the Balkans.)[71] Although post-Soviet researchers expressed a reluctance to publish in English for several years after the extent of the damage was made clear, the realities of the new vehicular language set in. In 1991, the Soviet Academy of Sciences set up the Nauka/Interperiodica International Publishing House in cooperation with an American firm. Their mission: to make English versions of 88 academic journals, translated by experts and edited by Americans. Cover-to-cover had ceased to be a stopgap, catch-up scramble by American scientists to handle Russian work, and was now the official outlet of Russian scientists trying to be heard overseas.[72] The mutable evolving story of scientific languages seemed to have reached equilibrium, or stasis.

Is This a Good Thing?

The answer, of course, depends on your view of science and your attitude toward English. Aside from those native Anglophones who breathe a sigh of relief at no longer having to struggle through manuals of scientific French, the most salient argument in favor of the developments described in this chapter has been that English is "neutral." English is a rather funny language in many respects. It has proven enormously pliant over the centuries, absorbing words, idioms, even syntax from dozens of different languages. It has no centralized academy to regulate usage. It does not even have a single dominant country, for surely the burly United States is at least strongly counterbalanced by the United Kingdom, not to mention India, which possibly has more speakers of the language than both combined. As Sabine Skudlik, one of the leading scholars of English as a scientific language, has noted: "This is the really new thing, the essential marker of modern Anglophony in science: that it not only bridges the differences of languages, but neutralizes all separating differences, whether of a linguistic or a more generally cultural kind."* [73]

English—whose dominance as a language of science we have seen to be intimately linked to geopolitics, personal preferences, economic pressures, and a host of contingent twists and turns—is understood as neutral ground, even by critics of the virtual disappearance of German, French, Russian, and Japanese.[74] How is it possible, even reasonable, to come to this conclusion? Perhaps it was not so much that English was seen as neutral and therefore appropriate for scientific interchange, but rather that the association with science, long famed for objectivity and impartiality, endowed Anglophony with neutrality, American hegemony trailing behind an Erlenmeyer flask. Any aura of neutrality has been enabled by native speakers of other languages—especially so-called "minor languages," like Dutch or Danish—who prefer it to German or French. (As is easy to observe, English benefits greatly from being "not French" or "not Russian.") In fact, these days, publishing science *not* in English is seen as marked, and is almost always done only by a native speaker of the language in question; if you see physics published in Russian, odds are that a Russian is the author.[75] Yet evidence that English is not neutral is remarkably easy to find. The most obvious asymmetry is that a certain segment of the community learns the language effortlessly as children; the rest—the majority—struggle through years of education. Their goal is not just to be able to muddle through an English article, dictionary in hand, to extract a general sense, but to acquit themselves orally under the intense pressure of hostile interrogation at a conference. Scientists are typically not gentle in their probing

*"Das ist das eigentlich neue, das wesentliche Kennzeichen der modernen Anglophonie in der Wissenschaft: daß sie nicht nur Sprachunterschiede überbrückt, sondern alle trennenden Verschiedenheiten, ob sprachlicher oder allgemein kultureller Art, neutralisiert."

of their colleagues, and a failure of fluency can be a fatal handicap for one's theories, or one's career.

"Struggle" is the correct word for many scientists' encounters with learning English. French and German are both closely related to English—the former contributing to the shaping of Middle English after the Norman Conquest, and the latter by virtue of kinship within the Germanic language family. Yet native-speaking scientists of these two languages express significant frustration with even the reduced structures of scientific English. Germans seem to Americans to have an amazing command of the English language, but that is partially a product of selection bias: the conversations you remember are the ones that you actually manage to have. In a 1995 survey of scientists at the University of Duisburg in northwestern Germany, 25% reported trouble reading English science, 38% had problems with speech, and 57% were challenged by writing.[76] This among a set of professionals who have the broadest exposure to the language and in a country where English study is obligatory. Werner Traxel, who protested the linguistic overhaul of *Psychological Research*, attributed the flaw to the English language itself: "Above all however English is not only an extremely flexible and nuance-rich language, but also one that, for the most part, cannot be described with fixed rules (in contrast to the Romance languages). Thus it gives us the impression that it is relatively imprecise, and indeed constructions that go against the logic of the language appear not infrequently in the specialist terminology in English."*[77] But just as English is not uniquely suited to science, it cannot be uniquely *ill*-suited either. A response to Traxel noted an interesting ambiguity about provincialism and internationality. On the one hand, knowing only one language can be seen as provincial, and insisting on diversity can enhance international exchange; on the other, if that one language is English, resisting it might be a knee-jerk provincial response. In short, Traxel had just better get over it.[78]

Not that he had a choice: English had become a seemingly permanent fixture of the intellectual landscape. The efficiency gains seem to

* "Vor allem aber ist das Englische nicht nur eine überaus flexible und nuancenreiche, sondern auch eine weitgehend nicht in feste Regeln faßbare Sprache (im Unterschied etwa zu romanischen Sprachen). Daher erscheint es uns auch leicht als relativ unpräzise, und in der Tat sind sprachlogische Fehlbildungen in der englischen Fachterminologie nicht selten."

have been tremendous, since essentially all elite natural science now appeared in one language without the tedious process of translating. Of course, the current state of affairs seems more efficient only to the native speakers of English; the gains have come at the cost of *everyone else* learning fluent English. Weighing the costs and benefits is a tricky affair, but it appears that the bump under the rug has only been moved around rather than smoothed out. The flip-side to all this English learning, adding insult to injury for many foreign scientists, is that most English-native scientists have given up all pretense of learning foreign languages.[79] Beginning in the 1960s, foreign-language requirements for graduate study in various sciences began to be eliminated—initially dropping from two to one, and then by the 1980s from one to zero.[80] This change confronts us with a chicken-egg dilemma in terms of causation. The absence of a language requirement obviously meant that ever fewer students would be equipped to consult foreign scientific literature, tilting the Anglophones ever more strongly into English; on the other hand, the requirements were eliminated in large part because they were no longer seen as necessary. Whichever way you understand it, there was no arguing with the consequences. "Those who speak English may get the impression of being—more or less—at home everywhere," wrote one rare Anglophone observer who noticed the asymmetry. "This helps to be quicker, more mobile and more efficient, which corresponds to modern ideals of life and work. To superficial observers the whole world seems to be steeped in English. It is an impression which may breed irritation."[81]

Just so. Lingering behind objections to scientific Anglophonia lies a nagging sense of the unfairness of it all. German scientists, to take a prominent example, have to make the difficult choice between identity and communication, between supporting journals and educational institutions in their native language or disseminating cutting-edge research to the broadest-possible readership. Anglophones don't; there is no dilemma, because identity and communication are the same.[82] The inequities extend beyond psychological comfort, because native speakers of English, by virtue of not having to spend time learning languages, have more time to study science, research, and publish. As a result, native speakers of English are overrepresented in the scholarly literature. Even though there are more nonnative-Anglophone scientists than vice versa, one study has found that only about 20% of the global quantity of English scientific works are produced by those individuals.[83] One political theorist has even suggested that the unfairness might be moderated by a progressive taxation scheme, whereby Anglophone scientists might pay slightly higher page-costs for their publications, which in turn would be used to subsidize the copyediting of nonnative submissions.[84]

It seems evident that Anglophone hegemony in the sciences would be disadvantageous for some *scientists*, but surely this is simply the luck of the draw. Americans used to be disadvantaged by the German dominance in chemistry, for example. A more abstract but potentially more serious question remains: is the current system bad for *science*? Or for *English*? The questions are related, but the arguments they raise are slightly different, so we will take them in turn.

First, is English bad for science—not because it is English, but because it is a single language? Does science benefit when it is multilingual? The contrary position—that it is simpler to have one vehicular language than to have three, let alone dozens—although ignored when Esperantists proferred it, now seems to hold sway. There are plenty of examples of facts delayed in transit, as when it took the rest of the world several years to catch up to what the Japanese were finding out about the plant hormone gibberellin, simply because the publications were trapped in *kanji* and *katakana*.[85] So maybe everyone wins when communication expands.

Or do they? The earliest losers in the lottery of scientific languages are younger students. Imagine a child in sub-Saharan Africa who is being taught chemistry. In what language is the class? If in a Bantu language, who translated the word for "oxygen"? Such a concept has been around for long enough that it might have filtered down to local languages around the world. But how about more contemporary concepts, like ozone depletion, or the Planck length, or object-centric debugging? Educational research to date indicates that children understand scientific concepts better when presented in their native language, but that requires textbooks and lesson plans in all the world's languages.[86] Those don't exist. The further one advances in science, the greater the scarcity of non-English pedagogical materials. If you want to study topological theory or stereochemistry in college, your English needs to be up to snuff. How many students are lost not because of weak scientific skills, but weak linguistic ones?

In the less mathematical sciences, even professional scientists—those who have already cleared the hurdles of advanced education and who presumably are more than passingly familiar with English texts—sometimes suggest that something has been lost with monolingualism.

All science develops through making connections between seemingly unrelated phenomena, and much of this work begins through linguistic metaphors. "If everyday speech is no longer the source of the specialized languages, the linguistic images will be lacking which are necessary to make something novel vividly understandable," noted one frustrated German scientist. "Since every language affords a different point of view onto reality and offers individual patterns of argumentation, this leads to a spiritual impoverishment if teaching and research are hemmed into English."*[87] This resembles the Whorfian hypothesis—that languages carve up nature, and we all live in different worlds shot through with our native languages—but it is hardly so ambitious. Rather, the claim is that insights come more quickly in words that are more familiar. It is, simply, a plea for identity. One might also anticipate deleterious consequences for public policy. It is challenging enough to persuade politicians to act on scientific, technological, or medical evidence given the paucity of public officials with scientific training and the difficulty of understanding the nuances of the data. Add to this a language barrier, and the situation rapidly worsens.[88] These are problems only for the non-Anglophones, but there are burdens on the other side as well, as native speakers of English are imposed upon to translate or correct their peers' papers, and locked out of private foreign-language conversations between lab-mates and at conferences.

Does the English language itself suffer when, as is currently the case for perhaps the first time in history, nonnative speakers of a living language start to greatly outnumber native speakers? If you wanted to isolate an effect, science would be a good place to look, because it has been Anglophone longer and more completely than any other domain of cultural endeavor. The "English" that is used in scientific communication—particularly in written form, but also quite often in oral interchange—is simplified, reduced, stereotyped to highlight communication and minimize stylistic nuance. German sociologist Wolf Lepenies has called this dialect "Englisch II," which another commentator worries has become nothing more than "a practical, reduced communications code."†[89]

*"Wenn die Quelle für die Fachsprachen nicht mehr die Alltagssprache ist, werden die Sprachbilder fehlen, die nötig sind, um Neues anschaulich begreiflich zu machen. Da jede Sprache einen anderen Blickwinkel auf die Wirklichkeit zulässt und individuelle Argumentationsmuster bietet, läuft es auf eine geistige Verarmung hinaus, wenn Lehre und Forschung auf das Englische eingeengt werden."

†"ein praktischer, reduzierter Kommunkationscode"

Imagine one ironic outcome: To the extent Scientific English resembles Basic English, and Basic English was dreamed up in part to minimize the "pidginization" of English in colonial contexts, Scientific English might itself become the pidgin. "Under certain circumstances English as a scientific language in non-English-speaking countries would degenerate into a cookie-cutter-language," linguist Sabine Skudlik observes, "in cases where constant feedback from mother-tongue speakers is not to be expected. This development would be desirable for nobody."*[90]

Almost certainly true, if the effect is in fact happening. The reader may have noticed that for the last several pages an odd thing has occurred to this manifestly historical book: we seem to have lost the past and moved instead to scientists' and linguists' rampant speculations about the future, ill-disguised as a conversation about the present. There seems no way to talk about Anglophonia in science without willy-nilly drifting into ruminations over where this all might lead. Before fully indulging that impulse, it is important to not lose the central lesson of the journey so far: English has attained its current position owing to a series of historical transformations that it also in turn shaped, exploiting a perception of neutrality that it gained through being distinctly non-neutral in either its British or American guise. There is a circularity to studying language and history together, scrambling our notions of time even in the buttoned-down domain of science. The history of scientific languages ends here, until it no longer does.

*"Unter Umständen würde das Englische als Wissenschaftssprache in den nicht-englischsprachigen Ländern zu einer Schablonensprache verkümmern, falls nicht eine ständige Rückmeldung von Muttersprachlern zu erwarten ist. Diese Entwicklung wäre für niemanden wünschenswert."

CONCLUSION

Babel Beyond

Consider a quotation and a story. The quotation comes from Edward Sapir, one of America's leading linguists before the Second World War. In 1921, in the midst of the wholesale destruction of the teaching of the German language in the United States and an international boycott of scientists from the Central Powers, he wrote:

> A scientific truth is impersonal, in its essence it can be untinctured by the particular linguistic medium in which it finds expression. It can readily deliver its message in Chinese as in English. Nevertheless it must have some expression, and that expression must needs be a linguistic one. Indeed the apprehension of the scientific truth is itself a linguistic process, for thought is nothing but language denuded of its outward garb. The proper medium of scientific expression is therefore a generalized language that may be defined as a symbolic algebra of which all known languages are translations. One can adequately translate scientific literature because the original scientific expression is itself a translation.[1]

The historical record shows that the actual state of affairs is and has been—to say the least—more complicated. While it might in principle be the case that the same scientific truths hold no matter which language they are expressed in, as a matter of daily experience the choice of a *specific* language has had enormous bearing on the capacity of scientific messages to be "readily delivered." The friction of translation between Russian and German powered the priority dispute over the periodic system of chemical elements, for example, and replicating Lavoisier's French nomenclature proved problematic both in Swedish and in Russian, not to mention Ido and Esperanto. Sapir also put forward a second claim, about a metalanguage of scientific truth, which brings us to the story.

In 1957, H. Beam Piper published "Omnilingual" as one of a collection of tales from *Astounding Science Fiction*. In view of the Cold War crisis of scientific language imposed by the deluge of scientific Russian pouring into the United States—which prompted new courses in recently established Slavic departments, witnessed the origins of Machine Translation, and finally settled into a steady, and gigantic, cover-to-cover translation industry—Piper's imaginary voyage to Mars assumes deeper resonances.

We do not see the journey in "Omnilingual"; the story begins with a crack team of scientists already combing the surface. They encounter the remnants of an advanced civilization, long extinct. The group included Martha Dane, an archaeologist with a particular fixation: learning to read the language of the Martians. Finding markings on the walls that she took to be writing, she developed a systematic transliteration into the Latin alphabet, transcribing Martian via a syllabary consisting of vowel-consonant pairs. Like the astounding "cracking" of Linear B by Michael Ventris in the early 1950s—just a few years before Piper put pen to paper—Dane hoped to figure out the referents for these signs. Ventris was aided by some inspired guessing about place names and then the surprising discovery, against expectation, that Linear B was simply a syllabic rendering of Greek.[2] Dane had no such luck. She needed something like a code-book, an analog to the Rosetta Stone, whose inscriptions in Egyptian hieroglyphs, Demotic Egyptian, and Greek enabled Jean-François Champollion to become, in 1824, the first human in over a millennium to actually read ancient Egyptian script. Sadly, for Martha Dane, "There is no Rosetta Stone, not anywhere on Mars.[. . .] We'll find one. There must be something, somewhere, that will give us the meaning of a few words, and we'll use them to pry meaning out of more words, and so on."[3] But since there could be no possible bilingual text between Martian and *any* Earth language, her colleagues dismissed her dream as a fantasy.

The team entered a massive building, which they speculated must have been something like a university. The group split up to explore different wings, and Dane wandered through rooms, transliterating Martian along the way, until she turned a corner and stood struck still by an inscription. "There was something familiar about the table on the left wall," the narrator tells us. "She tried to remember what she had been taught in school about physics, and what she had picked up by accident afterward. The second column was a continuation of the first: there were forty-six items in each, each item numbered consecutively—"

Dane counted the number of cells, and reached 92. That seemed interesting. What consisted of 92 items? The number of naturally occurring chemical elements, capped by uranium! She started at the top: "Hydrogen was Number One, she knew; One, *Sarfaldsorn*. Helium was Two; that was *Tirfaldsorn*."[4] From here on, nothing could stop Dane; she had found her Rosetta Stone, and from there could move to other scientific "bilinguals"—"astronomical tables, tables in physics and mechanics, for instance—in which words and numbers were equivalent." Most of her teammates were instantly convinced. Yet the leader Selim, a scholar of ancient Hittite, expressed some skepticism about the Martian Rosetta: "How do you know that their table of elements was anything like ours?" The three natural scientists on the team stared at him in disbelief. One, Mort Tranter, responded: "That isn't just the Martian table of elements; that's *the* table of elements. It's the only one there is."[5] (Dmitrii Mendeleev and Lothar Meyer might have begged to differ.) Finally, the colleague who had given Dane the hardest time, Hubert Penrose, granted her the highest of praise: "This is better than a bilingual, Martha. Physical science expresses universal facts; necessarily it is a universal language."[6]

We have come a long way from worries about whether one could even *do* science in Latin rather than Greek, or whether science would be destroyed by a Babel of languages. Rather than language serving as a barrier to block transmission of science (as the Esperantists and Idists argued at the dawn of the twentieth century) or averting Babel through the imposition of a unifying natural language (English, say, at the moment that I am writing this, or Latin several centuries earlier), we see Sapir and Piper, writing on either side of the linguistic chasm of World War II, expressing a common assumption: Science isn't just written *in* language, it is itself a language. The continually evolving and dynamic history of languages and the science conducted within them indicates that this proposition is most likely false, or at least deeply ambivalent. Nonetheless, the idea that mathematics or the facts of the physical and biological sciences alone might prove a "universal language" is omnipresent; the idea itself emerges from the rise and fall of Scientific Babels across the centuries. To further explore this notion, we must shift our gaze away from the past and examine how contemporaries think about language and science in our future.

The most common question along these lines is whether something can displace English from its current dominance in the natural sciences (and perhaps soon in the social sciences and humanities). All anyone can do is guess about the way this situation might evolve, because there is no historical precedent for today's Anglophonia.[7] There are essentially three ways of thinking through the possibilities. The first is that the status quo will continue into the future.[8] This is entirely possible, although of course languages change over time, and when someone says "English" will continue to be the dominant language of the sciences, this allows for both the possibility that the current reduced dialect of "scientific English" will persevere and the alternative that scientific English will mirror inevitable changes in global English. In the annals of science fiction, it takes an apocalypse on the order of global thermonuclear war, a genetically engineered plague, an alien invasion, or some combination of the three to displace English, breaking it up into mutually unintelligible daughter languages.[9]

A second view is that scientific English will be replaced by a scientific dialect of another language, so that science would remain monoglot, just in a different tongue. This was the aspiration of the Esperantists, and all the other visionaries who hoped that a constructed auxiliary would eliminate the Babel generated by the ethnic languages. Those who anticipate this possibility have one candidate in mind: Chinese![10] (They apparently refer to Mandarin Chinese in its Beijing variant. Chinese is no less fluid and multivariant than English.[11]) The major argument for Chinese being the single language of future science is based on population and geopolitical power, yet there are two problems with these inferences: one empirical, and the other theoretical. The empirical problem is that, despite the rapidly increasing number of Chinese scientists and engineers, they are actually a major component in the contemporary growth of *English*, because most of their publications appear in that language, not Chinese.[12] The theoretical problem is more to the point: why on earth should we expect that science will be monolingual in the future? It certainly was not the case in the past. Even Latin, recall, was not the sole vehicular language in Europe except for the high Renaissance. As for reasoning based on population—if that were sufficient, then surely Spanish would have occupied a place as one of the major languages of science. That this has not been the case is telling.

A widespread, but controversial, way of thinking about linguistic diversity is to make an analogy to ecology.[13] A brief reflection brings up dozens of examples, ranging from language growth and competition, to endangered languages and language death. But languages are not precisely analogous to biological species. Languages do not "die" or go "extinct"; the native-speaking people using the language do, sometimes violently.[14] The language, if it is documented, can still be used. (Witness, again, our old friend Latin.) To those who think of languages as functioning in a global ecology, however, the transition to a single dominant language for science is a linguistic Green Revolution equivalent to the eradication of traditional agricultural systems, imposing monoculture for the sake of efficiency but potentially imperiling precisely the intellectual diversity (shades of Whorf here!) that can generate new scientific ideas.[15] This is an alluring argument, but it is impossible to evaluate without a sense of how science might have looked had English not become the single global scientific language over the past half century. Barring a counterfactual crystal ball, we can simply observe that an awful lot of science is currently being done, and scientists do not seem overly concerned about a dearth of new ideas. Behind this worry about monocultures is an abiding worry about monolingualism, one which adheres to the assumption that such a state is perhaps inevitable into the distant future. It is, in short, a lament.

It seems just as, if not more, likely, projecting the past into the future, that we will have several languages of science, not one—that if English were to lose its dominance, it might follow the pattern of Latin and break up into several vehicular languages, while still retaining significant currency. (That would be essential in order to access past secondary literature: as demonstrated by the attempts to preserve German in the Cold War, or the retention of Latin long after the onset of the Protestant Reformation.) One could imagine a future of Chinese, English, and Spanish or Portuguese. Would it look so different from our past? There would surely be hand-wringing about the lost position of English, new schemes for artificial languages to blend the dominant tongues, and a lot of effort expended in language learning and translation.

All of this assumes, of course, that the history we have seen in this book is irreversible. Yet there remain those (although fewer every year) who hope that we might be able to re-Babelize science just a smidgen, just as far back as restoring the triumvirate of English, French, and German (and skipping, *bien sûr* and *natürlich*, impossible languages like Russian). To the extent that words are met with deeds in this regard, we see an inverse of the Cold War pattern; at that time, the French by and large acquiesced to the eclipse of their language as a vehicular tongue for science, while the Germans attempted to staunch the damage they perceived as caused by Hitler's regime. Today, there is some backlash against English as the sole scientific language within German-speaking Europe—the topic surfaces periodically in newspapers, especially as it concerns science education in secondary schools—but the state has put only limited resources behind promoting German abroad, and German-speaking scientists continue to publish in English.[16]

In France you are more likely to come across metaphors of cultural genocide: "It would be a national drama of incalculable consequences to remove from the French language its character as a scientific language."*[17] Despite the obvious fact that most Francophone scientists today are publishing in English, a discourse of resistance (often harkening back to the public mythology of the French Resistance against the Nazis) crops up fairly regularly.[18] While French truly is the only language besides English to have a global reach and a distinguished, centuries-long tradition in the natural sciences, nonetheless a French author trades communication for identity when she publishes in her native language—with certain notable exceptions.[19] Mathematics is a field where publication in French is still quite common. Laurent Lafforgue—winner of the 2002 Fields Medal—notes that French math is so strong that people will still read French to get at it; in fact, "it is to the degree that the French mathematical school remains attached to French that it conserves its originality and its force. *A contrario*, France's weaknesses in certain scientific disciplines could be ascribed to linguistic dereliction."†[20] The richness of metaphor and quickness of thought in one's native language enable creative work; identity should not be sacrificed without a fight. Yet dialing the hands of the clock back to the mid-nineteenth century seems extremely improbable.

The alternative is less likely to be full-blown multilingual publication than computer-mediated Machine Translation among several different tongues. In January 2012, former President of Harvard University, Lawrence Summers, famously dismissed "the substantial investment necessary to speak a foreign tongue" as not "universally worthwhile" given "English's emergence as a global language, along with the rapid progress in machine translation and the fragmentation of languages spoken worldwide."[21] Bypassing the non sequitur of how global fragmentation of languages would aid communication, Summers's point about MT seems to many a reasonable solution to the tensions explored

*"Ce serait un drame national aux conséquences incalculables que d'enlever à la langue française son caractère de langue scientifique."

†"c'est dans la mesure où l'école mathématique française reste attachée au français qu'elle conserve son originalité et sa force. *A contrario*, les faiblesses de la France dans certains disciplines scientifiques pourraient être liées au délaissement linguistique."

in this book: you can keep your identity by using your native tongue, and let computers take care of the communication. When we last left MT, it had collapsed into disgrace following the disappointed censure of the ALPAC report of 1966. Obviously, a lot has changed in the digital world since then. In particular, the single greatest roadblock for Léon Dostert's brand of MT—the scarcity of memory—has vanished. Memory has become dirt cheap. The speed of computation, expressed in the frenetically doubling euphoria of "Moore's Law"—which enthusiasts for an MT-utopia believe will continue indefinitely—has enabled completely different statistical approaches to computerized translation of natural languages, such as Google Translate, which relies not on an algorithmic decision-tree but on brute-force statistical comparison.[22] To casual observers, it looks like the language barrier is a thing of the past; computers are no longer "English-only," and even though computer languages and the language of computer science are dominated by English, the monolingual stranglehold on this area appears to be weakening.[23] Yet it seems less than certain that problems of Scientific Babel, and its current solution in English, can or will be transcended through these means. The significant challenges of access to computing technology in the poorer regions of the world probably matter less for the admittedly elite community of scientists, but the substantial infrastructure of education and publication already extant in English does likely entail that continuing to learn English will be more economical than translating everything multiple times among several thousand languages. There is a yet deeper difficulty with Summers's vision, which concerns how statistical MT actually works. Push a little harder on Google Translate, and one thing is evident: it is utterly dependent on human translation to provide the bilingual texts for statistical comparison. Hidden beneath our current MT, in other words, is more cover-to-cover. *Plus ça change.*

Set aside the future, and let's return to the present on our way back to the past. Both Sapir and Piper insisted upon the idea that science itself—whether expressed in terms of mathematics or purely in the sense impressions beloved of Logical Positivists in interwar Vienna—can serve as a kind of language to enable communication. Scientists are currently (and have been for fifty years) operationalizing this postulate into the foundation of one of the most breathtakingly visionary of contemporary scientific ventures: the Search for Extra-Terrestrial Intelligence (SETI). For the purposes of the subfield, "intelligence" is essentially synonymous with "ability to communicate," in part for epistemological reasons: we search for life in the cosmos by monitoring various frequency bands in all directions that are deemed to be the most likely carrier waves for deliberate interstellar communication; that implies that making contact is identical to receiving a message. This is, in short, a judgment about language.[24]

A fundamental postulate of SETI is that the intellectual problems that we have in composing a message for the heavens and understanding an incoming one are symmetric. That is, if we have difficulty assembling a text that can be understood *as a message in language* by intelligent beings that share neither our genetic capacities for tongues or any of the historical flotsam and jetsam of our present-languages, then so will the aliens. The quest quickly reduces to finding a metalanguage beyond our contingent languages and then monitoring the skies for any messages that might be broadcast in such a metalanguage. Already in 1921, Edward Sapir suggested that science might be that medium. Or, as expressed by a leading SETI practitioner in 2010: "By common consent, mathematics, being culturally neutral and forming the basis of the universal laws of nature, would be the lingua franca of interstellar discourse."[25]

This scientific-mathematical linguistic assumption—that is, that science and mathematics *are* a language—brings us to one fitting place to close this history of the languages in which modern science has been done. SETI is science being pursued today across the globe for the purposes of transcending an even more ineradicable language barrier than that which confronted Wilhelm Ostwald or Lise Meitner or Antoine Lavoisier in the past. There are, of course, thousands of objections that one might level at this enterprise, including the obvious fact that we have not yet figured out how to communicate terribly effectively with relatively intelligent animals occupying our own planet, not to mention other humans who share your scientific mindset and disciplinary training but happen to have been born in, say, Prague instead of San Francisco.[26] But rather than closing this book with a recitation of problems and rebuttals, let us return to the history of science.

In 1960, a Dutch mathematician named Hans Freudenthal published (in English, of course) the last constructed language we will take up in this volume, dubbed "Lincos" (for "Lingua Cosmica," a nod to Latin). If SETI scientists were looking for a likely interstellar signal, Freudenthal proffered the text—or at least the language in which such a text could be written. This was a language expressed through symbols and devoid of all of the features of either "natural" or "artificial" ones (Freudenthal's terms) besides semantics. This was a language about conveying *meaning*; everything else was superfluous. Yet even Freudenthal would not go so far as to think of mathematics as a language in itself: "It is true that mathematical language as written in textbooks still parasitizes on natural languages. The text surrounding the mathematical formulas is usually written in an idiom that bears the characteristics of the vernacular, to which it belongs in the ordinary sense."[27] But he needed to get beyond vernacular, to convey "in principle the whole bulk of our knowledge," not just selected proofs. Lincos would be considered understood by the recipient if he (Freudenthal's choice of pronoun) could "operate on it," manipulate it to generate other phrases in it—a decent enough definition of scientific language, come to think of it.[28] Lincos, in being communicated, taught itself through itself, building on "facts which may be supposed to be known to the receiver."[29]

What might those be? Carl Sagan and Iosif Shklovskii—guiding spirits of American and Soviet SETI, respectively—praised Freudenthal's efforts with Lincos and speculated about how precisely we might begin our Lincos messages. Pictures might be best, assuming that vision was a reasonably likely evolutionary trait no matter where you were in the universe, and there was one picture that seemed particularly apt. "For example, Mendeleyev's periodic system of the elements could be pictured, accompanied by the corresponding words in Lincos," they wrote. "The number and distribution of electrons, of course, would indicate the nature of the atom. Then, a graph of the number of protons in the nucleus versus the number of neutrons could be transmitted. By this time, the cosmic discourse is well along into atomic and nuclear physics."[30]

Would Mendeleev and Meyer, who wrangled about which words could and should properly characterize the periodic system in Russian and German, be flattered or flummoxed that the system they fought over was now understood to be beyond language, beyond Earth? Is the idea so strange? H. Beam Piper imagined Martha Dane communicating with dead aliens by using lists of elements. We can indeed picture such an eventuality in a time far in the future, or on a world millions of miles away. Yet for the present, as in the past, we remain bound to the constraints of history, to the shackles of the words in human languages: untranslatable yet intelligible, frustrating yet infinitely beguiling.

ACKNOWLEDGMENTS

This book has been in progress—in some form or another, often without my being conscious that I was obsessed with the history of scientific languages—for over fifteen years, and the list of debts accumulated in that span is proportionately large. I am unlikely to be comprehensive in my expression of gratitude here, but I will try.

First, I would like to extend my thanks to the archivists, librarians, and historians who made the materials for this history available: Joe Anderson and Greg Good at the Niels Bohr Library at the Center for the History of Physics; Marc Rothenberg at the National Science Foundation; Lynn Conway at Georgetown University Archives; Vera Enke at the Berlin-Brandenburg Akademie der Wissenschaften; Bernd Hoffmann at the Archives of the Max-Planck-Gesellschaft; Winifred Schultze at the Archives of Humboldt University in Berlin; Irene Jentzch, Gerd Walter, and Birgit Rehse at the Archives of the Freie Universität in Berlin; and the very helpful staff at the Esperantomuseum at the Austrian National Library in Vienna, at the National Archives and Records Administration in College Park, at the Library of Congress, at the Massachusetts Institute of Technology Special Collections, and at Princeton University's indispensible Article Express and Interlibrary Loan offices. I am especially grateful to Igor S. Dmitriev at the Archive-Museum of D. I. Mendeleev and the archivists in St. Petersburg, Russia, where this project began during the past century.

The research and writing of this project were supported by the generosity of Princeton University as well as two external funding agencies (the National Endowment for Humanities, grant 72879 in 2010, and the John Simon Guggenheim Memorial Foundation in 2011), which provided me with the time away from academic obligations to complete the writing.

Parts of four chapters of this book were published as scholarly articles, and I acknowledge the journals who printed them and their granting of the right to republish some of that material here, in revised form: "Translating Textbooks: Russian, German, and the Language of Chemistry," *Isis* 103 (2012): 88–98; "The Table and the Word: Translation, Priority, and the Periodic System of Chemical Elements," *Ab Imperio*, no. 3 (2013): 53–82; and "The Dostoevsky Machine in Georgetown: Scientific Translation in the Cold War," *Annals of Science* 72 (forthcoming 2015).

This book benefitted enormously from conversations with dozens of colleagues and friends, many of whom were generous enough with their time to read part or all of the manuscript. Among their number are Mitchell Ash, Melinda Baldwin, Michael Barany, Deborah Coen, Angela Creager, Lorraine Daston, Peter Galison, Yael Geller, Katja Guenther, Evan Hepler-Smith, Matthew Jones, Simeon Koole, Robert MacGregor, Patrick McCray, Projit Mukharji, Carla Nappi, Matthew Stanley, Jennifer Tomlinson, Daniel Trambaiolo, and Keith Wailoo. Ulrich Ammon, David Bellos, David Kaiser, Jan Surman, Marc Volovici, and Nasser Zakariya provided detailed comments on the entirety of the manuscript, and I accommodated their very astute comments as much as I was able. I am particularly grateful to Joshua Katz's guidance throughout this project; my capacity to say anything reasonable about language owes much to conversations and co-teaching with him throughout the conception and writing of this book.

Over the past few years, I have presented elements of this argument at talks at several institutions, and gained a good deal from the critical feedback at those events. I would like to thank the participants at the seminars and colloquia at Vanderbilt University (especially Paul Kramer, Ole Molvig, and Leor Halevi), the Penn Humanities Forum, Dartmouth University, New York University (especially Ken Alder and Myles Jackson), the University of Vienna, the Princeton Society of Fellows, the Princeton History Department Works-in-Progress series, the University of Wisconsin-Madison's History of Science Department and the Mellon T3 workshop, Harvard University, Indiana University, the Chemical Heritage Foundation, the University of California-Berkeley, the University of British Columbia, the Zentrum für Literatur- und Kulturforschung in Berlin, and the European University in St. Petersburg.

The final stages of writing coincided with my term as the inaugural Director of the Fung Global Fellows Program at the Princeton Institute for International and Regional Studies, a year of collaborative discussions and seminars that centered around the topic of "Languages and

Authority." I owe an enormous debt to Beate Witzler, who made this environment of intensive study of languages possible, and to the first cohort of Fung Global Fellows, who sharpened many of my thoughts and arguments: Adam Clulow, Helder De Schutter, David E. Kiwuwa, Pritipuspa Mishra, Brigitte Rath, and Ying Ying Tan.

Carsten Reinhardt and Arika Okrent refereed this manuscript for University of Chicago Press; I have always learned much from their writings, and am even happier to have had the benefit of their critical counsel on an earlier version of this text. My editor at University of Chicago Press, Karen Merikangas Darling, has been especially supportive of this project throughout its gestation. Levi Stahl and Mary Corrado in turn helped tame the manuscript and transform it into the present book. Meg Jacobs brought this project to the attention of the Wylie Agency, who in turn introduced it to Andrew Franklin at Profile Books. I thank all of them.

As always, my greatest debt is to Erika Milam, who has patiently listened to more half-baked ideas than anyone should ever be exposed to, and always had the discernment to point me in the right direction.

Finally, I would like to dedicate this book to all those who have taught me languages. It is a long list, only incompletely presented here. I hadn't always appreciated the dedication and persistence that you demonstrated in showing me the richness of speech, but I assure you that I do now. Sadly, I have been unable to reconstruct all your names, but at the very least I would like to explicitly thank those I could: Yelena Baraz, Natalia Chirkov, Maria Garcia, Nora Hampl, David Keily, James Lavine, and Sharon Muster. For those who are missing, I apologize; I assure you my gratitude is undiminished. I cannot, of course, omit my first and most important language teachers: Gila and Rafael Gordin. All of you have given me the greatest of gifts, and these words are but poor compensation for that.

ARCHIVES

The notes make reference to several archival collections, sometimes in abbreviated form. The full citations to all archives are presented here, with their accompanying abbreviation, when necessary.

Austria

Esperantomuseum of the Österreichische Nationalbibliothek, Herrengasse 9, 1010 Vienna.

Germany

[AMPG] Archives of the Max-Planck-Gesellschaft, Boltzmannstraße 14, 14195 Berlin.

[BBAW] Archives of the Berlin-Brandenburgische Akademie der Wissenschaften, Jägerstraße 22, 10117 Berlin.

[FUA] University Archives of the Freie Universität zu Berlin, Malteserstraße 74–100, Bldg. L, 12249 Berlin.

[HBA] Hofbibliothek und Stiftsbibliothek Aschaffenburg, Hugo-Dingler Stiftung, Schloßplatz 4, 63769 Aschaffenburg.

[HUA] Archives of the Humboldt-Universität zu Berlin, Eichborndamm 113, 13403 Berlin.

Russia

[ADIM] Archive-Museum of D. I. Mendeleev, St. Petersburg State University, Mendeleevskaia liniia 2, St. Petersburg 199034.

[PFARAN] St.-Peterburg Branch of the Archive of the Russian Academy of Sciences, Universitetskaia nab. 1, St. Petersburg 199034.

[TsGIASPb] Central State Historical Archive of St. Petersburg, Pskovskaia ulitsa 18, St. Petersburg 190121.

[*ZhRFKhO*] *Zhurnal Russkogo Fiziko-Khimicheskogo Obshchestva* [Journal of the Russian Physico-Chemical Society].

United States of America

[AEDA] Albert Einstein Duplicate Archive, Princeton University Library, Rare

Books and Special Collections, Princeton University, Princeton, New Jersey 08544.

[AIP] Niels Bohr Library and Archives, American Institute of Physics, 1 Physics Ellipse Drive, College Park, Maryland 20740.

[APSL] American Philosophical Society Library, 105 S. 5th Street, Philadelphia, Pennsylvania 19106.

[GUA] Georgetown University Archives, Lauinger Library, Georgetown University, 3700 O Street NW, Washington, DC 20057.

[-SLL] School of Languages and Linguistics

[-MTP] Machine Translation Papers

[LOC] Library of Congress Manuscript Division, 101 Independence Avenue SE, Room LM 101, James Madison Memorial Building, Washington, DC 20540.

[MIT] Massachusetts Institute of Technology Archives and Special Collections, Building 14N-118, 77 Massachusetts Avenue, Cambridge, MA 02139.

[NARA] National Archives and Records Administration, 8601 Adelphi Road, College Park, Maryland 20740.

NOTES

Introduction

1. Jean Le Rond D'Alembert, *Discours préliminaire de l'Encyclopédie*, ed. Michel Malherbe (Paris: J. Vrin, 2000), 137.

2. The invisibility of translation (and translators) is the central point in Lawrence Venuti, *The Translator's Invisibility: A History of Translation*, 2d ed. (London: Routledge, 2002 [1995]). See also David Bellos, *Is That a Fish in Your Ear? Translation and the Meaning of Everything* (New York: Faber and Faber, 2011). Although most works in translation studies focus on humanistic or literary translation, there is no reason to believe that the general issues do not also apply in the sciences. See I. J. Citroen, "The Myth of the Two Professions: Literary and Non-Literary Translation," *Babel* 11 (1965): 181–188.

3. On medicine, see John Maher, "The Development of English as an International Language of Medicine," *Applied Linguistics* 7 (1986): 206–218.

4. On the particular difficulties of "*Wissenschaft*" in German, see Denise Phillips, *Acolytes of Nature: Defining Natural Science in Germany, 1770–1850* (Chicago: University of Chicago Press, 2012), 4.

5. On the equivalent trends in the social sciences, see Abram De Swaan, "English in the Social Sciences," in Ulrich Ammon, ed., *The Dominance of English as a Language of Science: Effects on Other Languages and Language Communities* (Berlin: Mouton de Gruyter, 2001), 71–83; Ulrich Ammon, "Kaum noch ein Prozent Weltanteil in den Naturwissenschaften: Über Deutsch als Wissenschaftssprache," *Forschung und Lehre* (June 2010): 318–320, on 319.

6. I build gratefully on earlier scholarship on these questions. Most of it, with the exception of Scott L. Montgomery, *Science in Translation: Movements of Knowledge through Cultures and Time* (Chicago: University of Chicago Press, 2000), has approached the issue of language barriers and language spread in the sciences from a more sociological viewpoint, not a historical one. A partial list of the most important works includes Ammon, *The Dominance of English as a Language of Science*; J. A. Large, *The Foreign-Language Barrier: Problems in Scientific Communication* (London: André Deutsch, 1983); Conseil de la langue française, Gouvernement du Québec, *Le français et les langues scientifiques de demain: Actes du colloque tenu à l'Université du Québec à Montréal du 19 au 21 mars 1996* (Quebec, Canada: Gouvernment du Québec, 1996); Sabine Skudlik, *Sprachen in den Wissenschaften: Deutsch und Englisch in der internationalen Kommunikation* (Tübingen: Gunter Narr, 1990); and Hubert Fon-

din, "La langue de la publication scientifique: la préponderance de l'anglais et la recherche," *Documentation et bibliothéques* (June 1979): 59–69.

7. This problem has been the central concern of the entire theoretical field of hermeneutics. For an especially clear introduction, see Naoki Sakai, *Translation and Subjectivity: On 'Japan' and Cultural Nationalism* (Minneapolis: University of Minnesota Press, 1997), 1–17.

8. Clarence Augustus Manning, "Language and International Affairs," *Sewanee Review* 32, no. 3 (July 1924): 295–311, on 296.

9. A corollary of this point is that all contemporary scientists except Anglophones have to be bilingual, and even native Anglophones have to use a highly specialized and stereotyped scientific English, which makes them diglossic. On the notion of "diglossia"—typically used to refer to related dialects/languages stratified by class (Swiss-German/German, Haitian-Creole/French)—see the classic essay by Charles A. Ferguson, "Diglossia," *Word* 15 (1959): 325–340. On the general prevalence of multilingual rather than monoglot behavior, see John McWhorter, *The Power of Babel: A Natural History of Language* (New York: Perennial, 2001), 63; and Craig Calhoun, *Nationalism* (Minneapolis: University of Minnesota Press, 1997), 19, 41.

10. Derek J. de Solla Price, *Science since Babylon*, enlarged ed. (New Haven: Yale University Press, 1975 [1961]).

11. Minoru Tsunoda, "Les langues internationales dans les publications scientifiques et techniques," *Sophia Linguistica* (1983): 70–79.

12. Rainer Enrique Hamel, "The Dominance of English in the International Scientific Periodical Literature and the Future of Language Use in Science," *AILA Review* 20 (2007): 53–71, on 63; and J. Garrido, "Scientific and Technical Publications in the Lesser Known Languages," *Science East to West* 5, no. 14 (April 1964): 1–6, on 2.

13. For some studies which use language in this more metaphorical sense, see Matthias Dörries, "Language as a Tool in the Sciences," in Dörries, ed., *Experimenting in Tongues: Studies in Science and Language* (Stanford: Stanford University Press, 2002): 1–20; Peter Galison, "Trading Zone: Coordinating Action and Belief," in Mario Biagioli, ed., *The Science Studies Reader* (New York: Routledge, 1999): 137–160; Theodore H. Savory, *The Language of Science: Its Growth, Character, and Usage* (London: André Deutsch, 1953); and Maurice P. Crosland, *Historical Studies in the Language of Chemistry* (Cambridge, MA: Harvard University Press, 1962).

14. I also manage Spanish, Modern Hebrew, and some Czech, but those are, for reasons this book addresses, not dominant scientific languages.

15. On the essential distinction between written and spoken competence, see Skudlik, *Sprachen in den Wissenschaften*, 25; and Herbert Newhard Shenton, *Cosmopolitan Conversation: The Language Problems of International Conferences* (New York: Columbia University Press, 1933).

16. Ralph A. Lewin and David K. Jordan, "The Predominance of English and the Potential Use of Esperanto for Abstracts of Scientific Articles," in M. Kageyama, K. Nakamura, T. Oshima, and T. Uchida, eds., *Science and Scientists: Essays by Biochemists, Biologists, and Chemists* (Tokyo: Japan Scientific Societies Press, 1981): 435–441, on 438. The results are very robust. See, for example, Ulrich Ammon, "Linguistic Inequality and Its Effects on Participation in Scientific Discourse and on Global Knowledge Accumulation—With a Closer Look at the Problems of the Second-Rank Lan-

guage Communities," *Applied Linguistics Review* 3, no. 2 (2012): 333–355; Large, *The Foreign-Language Barrier*, 32; Graham K. L. Chan, "The Foreign Language Barrier in Science and Technology," *International Library Review* 8 (1976): 317–325, on 321–322; and C. M. Louttit, "The Use of Foreign Languages by Psychologists, Chemists, and Physicists," *American Journal of Psychology* 70, no. 2 (June 1957): 314–316.

17. Richard B. Baldauf, Jr. and Björn H. Jernudd, "Language of Publications as a Variable in Scientific Communication," *Australian Review of Applied Linguistics* 6 (1983): 97–108, on 97.

18. Savory, *The Language of Science*, 113; a similar statement occurs on p. 107.

19. See, for example, Hans Niels Jahnke and Michael Otte, "On 'Science as a Language,'" in Jahnke and Otte, eds., *Epistemological and Social Problems of the Sciences in the Early Nineteenth Century* (Dordrecht: D. Reidel, 1981): 75–89.

20. Sundar Sarukkai, *Translating the World: Science and Language* (Lanham, MD: University Press of America, 2002), 7; Montgomery, *Science in Translation*, 254.

21. S. Chandrasekhar, *Newton's* Principia *for the Common Reader* (Oxford: Clarendon Press, 1995), 36.

22. Isaac Newton, *The* Principia*: Mathematical Principles of Natural Philosophy*, tr. I. Bernard Cohen and Anne Whitman (Berkeley: University of California Press, 1999), 700. Emphasis in original.

23. Isaac Newton, *Philosophiae naturalis principia mathematica*, 2d ed. (Cambridge, 1713), 571.

24. On multilingualism among mathematicians in this period, see Jeremy J. Gray, "Languages for Mathematics and the Language of Mathematics in a World of Nations," in Karen Hunger Parshall and Adrian C. Rice, eds., *Mathematics Unbound: The Evolution of an International Mathematical Research Community, 1800–1945* (Providence: American Mathematical Society and London Mathematical Society, 2002): 201–228.

25. Henri Poincaré, *Sechs Vorträge über ausgewählte Gegenstände aus der reinen Mathematik und mathematischen Physik, auf Einladung der Wolfskehl-Kommission der Königlichen Gesellschaft der Wissenschaften gehalten zu Göttingen vom 22.-28. April 1909* (Leipzig: B. G. Teubner, 1910), 51. I thank Michael Barany for bringing this citation to my attention.

26. I call this view "Whorfian" rather than the "Sapir-Whorf Hypothesis," because it is not at all clear that Whorf's teacher, the distinguished linguist Edward Sapir, actually held it. On the history of the hypothesis, see John E. Joseph, "The Immediate Sources of the 'Sapir-Whorf Hypothesis,'" *Historiographia Linguistica* 23, no. 3 (1996): 365–404.

27. Benjamin Lee Whorf, "Science and Linguistics," *Technology Review* 42 (1940), reproduced in John B. Carroll, ed., *Language, Thought, and Reality: Selected Writings of Benjamin Lee Whorf* (Cambridge, MA: MIT Press, 1956), 214.

28. For a summary of the data, see Paul Kay and Willett Kempton, "What Is the Sapir-Whorf Hypothesis?," *American Anthropologist* 86 (1984): 65–79. Guy Deutscher, in *Through the Language Glass: Why the World Looks Different in Other Languages* (New York: Metropolitan Books, 2010), argues for some Whorfian effects. The arguments pro and con are usefully and impartially parsed in G. E. R. Lloyd, *Cognitive Variations: Reflections on the Unity and Diversity of the Human Mind* (Oxford: Clarendon, 2007).

29. See, for example, Jessica Riskin, "Rival Idioms for a Revolutionized Science

and a Republican Citizenry," *Isis* 89 (1998): 203–232; Lissa Roberts, "Condillac, Lavoisier, and the Instrumentalization of Science," *Eighteenth Century* 33 (1992): 252–271; idem, "A Word and the World: The Significance of Naming the Calorimeter," *Isis* 82, no. 2 (June 1991): 198–222; Trevor H. Levere, "Lavoisier: Language, Instruments, and the Chemical Revolution," in Levere and William R. Shea, eds., *Nature, Experiment, and the Sciences* (Dordrecht: Kluwer Academic, 1990): 207–223; and Marco Beretta, *The Enlightenment of Matter: The Definition of Chemistry from Agricola to Lavoisier* (Canton, MA: Science History Publications, 1993). For a general survey of the varied historiography of this episode, see John G. McEvoy, *The Historiography of the Chemical Revolution: Patterns of Interpretation in the History of Science* (London: Pickering & Chatto, 2010).

30. Antoine Lavoisier, "Mémoire sur la nécessité de réformer et de perfectionner la nomenclature de la chimie," in *Œuvres de Lavoisier* (Paris: 1864–1893): V:354–364, on 356.

31. See, among others, Marc Fumaroli, *When the World Spoke French*, tr. Richard Howard (New York: New York Review Books, 2011 [2001]).

32. David C. Gordon, *The French Language and National Identity (1930–1975)* (The Hague: Mouton, 1978), 22–27.

33. Fumaroli, *When the World Spoke French*, xviii.

34. Gordon, *The French Language and National Identity*, 35; R. E. Keller, *The German Language* ([Atlantic Highlands], NJ: Humanities Press, 1978), 486.

35. Comte de Rivarol, *L'Universalité de la langue française* (Paris: Arléa, 1991 [1784]), 27.

36. On Schwab, see Edwin H. Zeydel, "A Criticism of the German Language and Literature by a German of the Eighteenth Century," *Modern Language Notes* 38, no. 4 (April 1923): 193–201; idem, "Johann Christoph Schwab on the Relative Merits of the European Languages," *Philological Quarterly* 3 (1924): 285–301; and Freeman G. Henry, "From the First to the Fifth Republic: Antoine de Rivarol, Johann Christoph Schwab, and the Latest 'Lingua Franca,'" *French Review* 77, no. 2 (December 2003): 312–323.

37. Rivarol, *L'Universalité de la langue française*, 72–73. Emphasis in original.

38. J. C. Schwab, *Le Grand Concours: "Dissertation sur les causes de l'universalité de la langue française et la durée vraisemblable de son empire,"* tr. Denis Robelot, ed. Freeman G. Henry (Amsterdam: Rodopi, 2005), 142.

39. David A. Bell, "Lingua Populi, Lingua Dei: Language, Religion, and the Origins of French Revolutionary Nationalism," *American Historical Review* 100, no. 5 (December 1995): 1403–1437; idem, *The Cult of the Nation in France: Inventing Nationalism, 1680–1800* (Cambridge, MA: Harvard University Press, 2001), chapter 6; Patrice L.-R. Higonnet, "The Politics of Linguistic Terrorism and Grammatical Hegemony during the French Revolution," *Social History* 5, no. 1 (January 1980): 41–69; Peter Flaherty, "*Langue nationale/langue naturelle*: The Politics of Linguistic Uniformity during the French Revolution," *Historical Reflections/Réflexions historiques* 14, no. 2 (Summer 1987): 311–328; Martyn Lyons, "Politics and Patois: The Linguistic Policy of the French Revolution," *Australian Journal of French Studies* 18 (1981): 264–281; and Jean-Yves Lartichaux, "Linguistic Politics during the French Revolution," *Diogenes* 25 (1977): 65–84.

40. Robert E. Schofield, *The Enlightenment of Joseph Priestley: A Study of His Life and Work from 1733 to 1773* (University Park: Pennsylvania State University Press, 1997), 79–80, 232.

41. Maurice Crosland, *In the Shadow of Lavoisier: The* Annales de Chimie *and the Establishment of a New Science* (Oxford: Alden Press, 1994), 88.

42. Arthur Donovan, *Antoine Lavoisier: Science, Administration and Revolution* (Cambridge: Cambridge University Press, 1993), 30.

43. J. B. Gough, "Lavoisier's Early Career in Science: An Examination of Some New Evidence," *British Journal for the History of Science* 4, no. 1 (June 1968): 52–57; McEvoy, *The Historiography of the Chemical Revolution*, 98. See also J. B. Gough, "Lavoisier and the Fulfillment of the Stahlian Revolution," *Osiris* 4 (1988): 15–33.

44. Donovan, *Antoine Lavoisier*, 95; Henry Guerlac, *Lavoisier—The Crucial Year: The Background and Origin of His First Experiments on Combustion in 1772* (Ithaca: Cornell University Press, 1961), 13, 15–16, 28, 52, 65.

45. Robert E. Schofield, *The Enlightened Joseph Priestley: A Study of His Life and Work from 1773 to 1804* (University Park: Pennsylvania State University Press, 2004), 105.

46. Jan Golinski, "The Chemical Revolution and the Politics of Language," *Eighteenth Century* 33, no. 3 (1992): 238–251, on 241; Jean-Pierre Poirier, *Lavoisier: Chemist, Biologist, Economist*, tr. Rebecca Balinski (Philadelphia: University of Pennsylvania Press, 1996 [1993]), 180. Madame Lavoisier also translated Italian commentaries on chemistry for her husband, as discussed in Marco Beretta, "Italian Translations of the *Méthode de Nomenclature Chimique* and the *Traité Élémentaire de Chimie*: The Case of Vincenzo Dandolo," in Bernadette Bensaude-Vincent and Ferdinando Abbri, eds., *Lavoisier in European Context: Negotiating a New Language for Chemistry* (Canton, MA: Science History Publications, 1995): 225–247, on 228.

47. Translator's preface to Richard Kirwan, *Essai sur le phlogistique, et sur la constitution des acides, traduit de l'anglois de M. Kirwan; avec des notes de MM. de Morveau, Lavoisier, de la Place, Monge, Berthollet, & de Fourcroy* (Paris: Rue et Hôtel Serpente, 1788), vii.

48. Donovan, *Antoine Lavoisier*, 175.

49. Bensaude-Vincent and Abbri, *Lavoisier in European Context*; Crosland, *Historical Studies in the Language of Chemistry*, 191, 208–209; Beretta, *The Enlightenment of Matter*, 302–303, 319.

50. Albert Léon Guérard, *A Short History of the International Language Movement* (London: T. Fisher Unwin, 1922), 88.

Chapter One

1. Lucretius, *De rerum natura*, ed. W. H. D. Rouse and Martin Ferguson Smith (Cambridge, MA: Harvard University Press, 1992 [1924]), I:136–139.

2. This litmus-test quality is nicely observed in Leslie Dunton-Downer, *The English Is Coming! How One Language Is Sweeping the World* (New York: Touchstone, 2010), 200. For those who see English as functioning very much like Latin, see Roger Balian, "Le physicien français et ses langages de communication," in Conseil de la langue française, Gouvernement du Québec, *Le français et les langues scientifiques de demain: Actes du colloque tenu à l'Université du Québec à Montréal du 19 au 21 mars*

1996 (Quebec, Canada: Gouvernment du Québec, 1996): 43–53, on 43; Clarence Augustus Manning, "Language and International Affairs," *Sewanee Review* 32, no. 3 (July 1924): 295–311, on 309; and James Clackson and Geoffrey Horrocks, *The Blackwell History of the Latin Language* (Malden, MA: Blackwell, 2007), 77. For scholars who dispute the analogy, see Sabine Skudlik, *Sprachen in den Wissenschaften: Deutsch und Englisch in der internationalen Kommunikation* (Tübingen: Gunter Narr, 1990), 9–10; and Hanno Helbling, "Aspekte des Verhältnisses von Wissenschaft und Sprache," in Hartwig Kalverkämper and Harald Weinrich, eds., *Deutsch als Wissenschaftssprache: 25. Konstanzer Literaturgespräch des Buchhandels, 1985* (Tübingen: Gunter Narr, 1986): 151–153, on 152.

3. On the terminological issues associated with "universal languages" and "lingua franca," among other categories, see Ulrich Ammon, "International Languages," in R. E. Asher, ed., *The Encyclopedia of Language and Linguistics*, vol. 4 (Oxford: Pergamon Press, 1994): 1725–1730; Conrad M. B. Brann, "Lingua Minor, Franca & Nationalis," in Ulrich Ammon, ed., *Status and Function of Languages and Language Varieties* (Berlin: Walter de Gruyter, 1989): 372–385; Henry Kahane and Renée Kahane, "*Lingua Franca*: The Story of a Term," *Romance Philology* 30 (August 1976): 25–41; and Nicholas Ostler, *The Last Lingua Franca: English until the Return of Babel* (New York: Walker, 2010). On English as Latin rather than a lingua franca, see Hans Joachim Meyer, "Global English—a New Lingua Franca or a New Imperial Culture?," in Andreas Gardt and Bernd Hüppauf, eds., *Globalization and the Future of German* (Berlin: Mouton de Gruyter, 2004): 65–84, on 72–73.

4. The history of Latin is very well documented, thanks to a continuous tradition of writing dating from antiquity and the assiduousness of generations of scholars. For a more formal study, see Clackson and Horrocks, *The Blackwell History of the Latin Language*; and L. R. Palmer, *The Latin Language* (London: Faber and Faber, [1954]). For more accessible accounts, see Nicholas Ostler, *Ad Infinitum: A Biography of Latin* (New York: Walker, 2007); Joseph B. Solodow, *Latin Alive: The Survival of Latin in English and the Romance Languages* (Cambridge: Cambridge University Press, 2010); and Tore Janson, *A Natural History of Latin* (New York: Oxford University Press, 2004). I draw extensively from all of these.

5. On this question, see J. N. Adams, *The Regional Diversification of Latin 200 BC–AD 600* (Cambridge: Cambridge University Press, 2007).

6. Clackson and Horrocks, *The Blackwell History of the Latin Language*, 79.

7. J. N. Adams, *Bilingualism and the Latin Language* (Cambridge: Cambridge University Press, 2003); Clackson and Horrocks, *The Blackwell History of the Latin Language*, 189; Erich Auerbach, *Literary Language and Its Public in Late Latin Antiquity and in the Middle Ages*, tr. Ralph Manheim (Princeton: Princeton University Press, 1965 [1958]), 248–249.

8. Geoffrey Horrocks, *Greek: A History of the Language and Its Speakers*, 2d. ed. (Malden, MA: Wiley-Blackwell, 2010 [1997]), 110; William V. Harris, *Ancient Literacy* (Cambridge, MA: Harvard University Press, 1989), 175; L. D. Reynolds and N. G. Wilson, *Scribes and Scholars: A Guide to the Transmission of Greek and Latin Literature*, 3d. ed. (Oxford: Clarendon Press, 1991 [1968]), 55.

9. I choose not to provide the diacritic over the "e," but readers should be aware that many accounts do, and that the word is pronounced with two syllables.

10. Bruno Rochette, *Le latin dans le monde grec: Recherches sur la diffusion de la*

langue et des lettres latines dans les provinces hellénophones de l'Empire romain (Brussels: Latomus, 1997), especially 70 and 139.

11. On Latin and early Christianity, see Christine Mohrmann, *Latin vulgaire, Latin des chrétiens* (Paris: Librairie C. Klincksieck, 1952).

12. This quotation is from J. M. Millas-Vallicrosa, "Translations of Oriental Scientific Works (to the End of the Thirteenth Century)," tr. Daphne Woodward, in Guy S. Métraux and François Crouzet, eds., *Evolution of Science: Readings from the History of Mankind* (New York: New American Library, 1963): 128–167, on 128. On the "poverty topos" frequently invoked by Latins vis-à-vis Greek, see Joseph Farrell, *Latin Language and Latin Culture from Ancient to Modern Times* (Cambridge: Cambridge University Press, 2001), 28; and Ostler, *Ad Infinitum*, chapter 5.

13. On the Romans' natural knowledge—whether written in Latin or Greek—see Daryn Lehoux, *What Did the Romans Know? An Inquiry into Science and Worldmaking* (Chicago: University of Chicago Press, 2012). I am indebted for the point about popularization to David C. Lindberg, *The Beginnings of Western Science: The European Scientific Tradition in Philosophical, Religious, and Institutional Context, Prehistory to A.D. 1450*, 2d. ed. (Chicago: University of Chicago Press, 2007 [1992]), 135; and Scott L. Montgomery, *Science in Translation: Movements of Knowledge through Cultures and Time* (Chicago: University of Chicago Press, 2000).

14. Cicero, *Tusculan Disputations*, ed. J. E. King (Cambridge, MA: Harvard University Press, 1950 [1927]), I.i.1.

15. Cicero, *Academica*, I.ii.4, in Cicero, *De natura deorum. Academica*, ed. H. Rackham (Cambridge, MA: Harvard University Press, 1951 [1933]), 414.

16. Cicero, *Academica*, I.iii.10, in Cicero, *De natura deorum. Academica*, 420.

17. Cicero, *Academica*, I.vii.25, in Cicero, *De natura deorum. Academica*, 434.

18. D. R. Langslow, *Medical Latin in the Roman Empire* (New York: Oxford University Press, 2000), esp. 35–36 for the contrast with Cicero; Rebecca Flemming, "Galen's Imperial Order of Knowledge," in Jason König and Tim Whitmarsh, eds., *Ordering Knowledge in the Roman Empire* (Cambridge: Cambridge University Press, 2007): 241–277, on 269.

19. Horrocks, *Greek*, 197, 207; Henry Kahane and Renée Kahane, "Decline and Survival of Western Prestige Languages," *Language* 55 (March 1979): 183–198, on 183–186.

20. Gilbert Dagron, "Formes et fonctions de pluralisme linguistique à Byzance (IXe-XIIe siècle)," *Travaux et mémoires* 12 (1994): 219–240.

21. Lindberg, *The Beginnings of Western Science*, 159.

22. On the rareness of knowledge of Greek in the West, see F. A. C. Mantello and A. G. Rigg, eds., *Medieval Latin: An Introduction and Bibliographical Guide* (Washington, DC: Catholic University of America Press, 1996), 718; Bernhard Bischoff, "The Study of Foreign Languages in the Middle Ages," *Speculum* 36, no. 2 (April 1961): 209–224, on 215; Reynolds and Wilson, *Scribes and Scholars*, 118–119; A. C. Dionisotti, "On the Greek Studies of Robert Grosseteste," in Dionisotti, Anthony Grafton, and Jill Kraye, eds., *The Uses of Greek and Latin: Historical Essays* (London: The Warburg Institute, 1988): 19–39; and idem, "Greek Grammars and Dictionaries in Carolingian Europe," in Michael W. Herren, ed., *The Sacred Nectar of the Greeks: The Study of Greek in the West in the Early Middle Ages* (London: King's College London Medieval Studies, 1988): 1–56.

23. Pierre Riché, "Le grec dans les centres de culture d'Occident," in Herren, ed., *The Sacred Nectar of the Greeks* (1988): 143–168; Marie-Thérèse D'Alverny, "Translations and Translators," in Robert L. Benson and Giles Constable, eds., *Renaissance and Renewal in the Twelfth Century* (Cambridge, MA: Harvard University Press, 1982): 421–462, on 427.

24. Lindberg, *The Beginnings of Western Science*, 147–148, 197.

25. Bischoff, "The Study of Foreign Languages in the Middle Ages," 209; Auerbach, *Literary Language and Its Public in Late Latin Antiquity*, 119–120, 269; Charles Homer Haskins, *The Renaissance of the Twelfth Century* (Cambridge, MA: Harvard University Press, 1955 [1927]), 127.

26. Dimitri Gutas, *Greek Thought, Arabic Culture: The Graeco-Arabic Translation Movement in Baghdad and Early 'Abbāsid Society (2nd–4th/8th–10th Centuries)* (New York: Routledge, 1998), 2.

27. Montgomery, *Science in Translation*, 106.

28. Reynolds and Wilson, *Scribes and Scholars*, 109; William Chester Jordan, *Europe in the High Middle Ages* (New York: Viking, 2001), 116.

29. Lindberg, *The Beginnings of Western Science*, 224; Mantello and Rigg, *Medieval Latin*, 506.

30. On the importance of science in this first generation of translations, see D'Alverny, "Translations and Translators," 451; Mantello and Rigg, *Medieval Latin*, 343.

31. On the Toledo translations, see Mantello and Rigg, *Medieval Latin*, 724–725; F. Gabrieli, "The Transmission of Learning and Literary Influences to Western Europe," in P. M. Holt, Ann K. S. Lambton, and Bernard Lewis, eds., *The Cambridge History of Islam*, v. 2 (Cambridge: Cambridge University Press, 1970): 851–889; Millas-Vallicros, "Translations of Oriental Scientific Works"; George F. Hourani, "The Medieval Translations from Arabic to Latin Made in Spain," *Muslim World* 62 (1972): 97–114.

32. John Murdoch, "Euclid: Transmission of the Elements," *Complete Dictionary of Scientific Biography* (Detroit: Charles Scribner's Sons, 2008), IV: 437–459; William R. Newman, *Promethean Ambitions: Alchemy and the Quest to Perfect Nature* (Chicago: University of Chicago Press, 2004), 43–44. On quality, see R. W. Southern, *The Making of the Middle Ages* (New Haven: Yale University Press, 1953), 65.

33. Haskins, *The Renaissance of the Twelfth Century*, 301. Emphasis in original.

34. Ostler, *Ad Infinitum*, 217; Reynolds and Wilson, *Scribes and Scholars*, 121.

35. Benedict Anderson, *Imagined Communities: Reflections on the Origin and Spread of Nationalism*, rev. ed. (London: Verso, 1991 [1983]), 38. This claim is accurate, but Anderson goes on to declare: "Then and now the bulk of mankind is monoglot." According to most definitions of linguistic competence, this is demonstrably untrue.

36. Peter Burke, *Languages and Communities in Early Modern Europe* (Cambridge: Cambridge University Press, 2004), 44–45.

37. See, especially, Michael Baxandall, *Giotto and the Orators: Humanist Observers in Italy and the Discovery of Pictorial Composition, 1350–1450* (Oxford: Clarendon Press, 1971), 9, 46. For more on this point, see Paul Botley, *Latin Translation in the Renaissance: The Theory and Practice of Leonardo Bruni, Giannozzo Manetti and Desiderius Erasmus* (Cambridge: Cambridge University Press, 2004), 152; Christopher S. Celenza, *The Lost Italian Renaissance: Humanists, Historians, and Latin's Legacy* (Baltimore: Johns Hopkins University Press, 2004), 144, 146.

38. Mantello and Rigg, *Medieval Latin*, 76; Jozef IJsewijn, *Companion to Neo-Latin Studies, Part I: History and Diffusion of Neo-Latin Literature*, 2d ed. (Leuven: Leuven University Press, 1990), 22.

39. Johan Huizinga, *Erasmus and the Age of Reformation* (New York: Harper & Brothers, 1957), 43.

40. Anthony Grafton, *Defenders of the Text: The Traditions of Scholarship in an Age of Science, 1450–1800* (Cambridge, MA: Harvard University Press, 1991), 166–167; Southern, *The Making of the Middle Ages*, 16.

41. On Sanskrit, see especially Sheldon Pollock, *The Language of the Gods in the World of Men: Sanskrit, Culture, and Power in Premodern India* (Berkeley: University of California Press, 2006). On the combination of European vehicular languages and regional vernacular translations of modern science in South Asia, see Michael S. Dodson, "Translating Science, Translating Empire: The Power of Language in Colonial North India," *Comparative Studies in Society and History* 47, no. 4 (October 2005): 809–835; and Gyan Prakash, *Another Reason: Science and the Imagination of Modern India* (Princeton: Princeton University Press, 1999), 62–63.

42. Frits Staal, "The Sanskrit of Science," *Journal of Indian Philosophy* 23 (1995): 73–127; Sheldon Pollock, "The Languages of Science in Early-Modern India," in Karin Preisendanz, ed., *Expanding and Merging Horizons: Contributions to South Asian and Cross-Cultural Studies in Commemoration of Wilhelm Halbfass* (Vienna: Verlag der Österreichischen Akademie der Wissenschaften, 2007): 203–220.

43. Hermann Jacobi, "Über den nominalen Stil des wissenschaftlichen Sanskrits," *Indogermanische Forschungen* 14 (1903): 236–251.

44. Otto Jespersen, *The Philosophy of Grammar* (Chicago: University of Chicago Press, 1992 [1924]), 139.

45. The classic and comprehensive history of traditional Chinese science is, of course, the work begun by Joseph Needham and continued by generations of scholars: Joseph Needham with Wang Ling, *Science and Civilisation in China* (Cambridge: Cambridge University Press, 1954–). On Chinese encounters with Catholic Jesuits and English Protestants from the sixteenth to the nineteenth century, and how the translations shaped this traditional learning, see Benjamin A. Elman, *On Their Own Terms: Science in China, 1550–1900* (Cambridge, MA: Harvard University Press, 2005).

46. For a general introduction to Chinese and common misconceptions about how it works, see John DeFrancis, *The Chinese Language: Fact and Fantasy* (Honolulu: University of Hawaii Press, 1984). On the extent to which written Chinese can be thought of as a lingua franca, see the helpful essay by Victor H. Mair, "Buddhism and the Rise of the Written Vernacular in East Asia: The Making of National Languages," *Journal of Asian Studies* 53, no. 3 (August 1994): 707–751. The general information in this paragraph about the circulation and limits of Classical Chinese, as well as the analogy with Latin, are drawn from Peter Kornicki, "The Latin of East Asia?," Lecture 1 of the 2008 Sandars Lectures in Bibliography, Cambridge University, 10 March 2008, available at http://www.lib.cam.ac.uk/sandars/kornicki1.pdf, accessed 10 August 2013.

47. For an introduction to the Chinese script, see Oliver Moore, *Chinese* (Berkeley: University of California Press, 2000). For how early modern Europeans understood Chinese writing, see the detailed discussion in Bruce Rusk, "Old Scripts, New

Actors: European Encounters with Chinese Writing, 1550–1700," *East Asian Science, Technology, and Medicine* 26 (2007): 68–116.

48. Reproduced in Marin Mersenne, *Correspondance*, ed. Cornelis de Waard, II (1628–1630) (Paris: Presses Universitaires de France, 1945), 324–328, quotation on 328.

49. For both surveys and detailed studies, see James Knowlson, *Universal Language Schemes in England and France, 1600–1800* (Toronto: University of Toronto Press, 1975); M. M. Slaughter, *Universal Languages and Scientific Taxonomy in the Seventeenth Century* (Cambridge: Cambridge University Press, 1982).

50. On the dissatisfaction with Latin, see Knowlson, *Universal Language Schemes in England and France*, 8; J. A. Large, *The Foreign-Language Barrier: Problems in Scientific Communication* (London: André Deutsch, 1983), 138; Hans Aarsleff, *From Locke to Saussure: Essays on the Study of Language and Intellectual History* (London: Athlone, 1982), 260; and Peter Dear, *Mersenne and the Learning of the Schools* (Ithaca: Cornell University Press, 1988), 170, 229.

51. On the connection of Chinese to the enthusiasm for philosophical languages, see Jonathan Cohen, "On the Project of a Universal Character," *Mind* 63, no. 249 (January 1954): 49–63, on 51; Barbara J. Shapiro, *John Wilkins, 1614–1672: An Intellectual Biography* (Berkeley: University of California Press, 1969), 47; and Knowlson, *Universal Language Schemes in England and France*, 25.

52. Knowlson, *Universal Language Schemes in England and France*, 108–109.

53. Shapiro, *John Wilkins*, 46–47.

54. See especially Clark Emery, "John Wilkins' Universal Language," *Isis* 38, no. 3–4 (February 1948): 174–185; Benjamin DeMott, "The Sources and Development of John Wilkins' Philosophical Language," *Journal of English and Germanic Philology* 57 (1958): 1–13; and the delightful account in Arika Okrent, *In the Land of Invented Languages: Esperanto Rock Stars, Klingon Poets, Loglan Lovers, and the Mad Dreamers Who Tried to Build a Perfect Language* (New York: Spiegel & Grau, 2009), chapter 1.

55. John Wilkins, *An Essay Towards a Real Character, and a Philosophical Language* (London: John Martin, 1668), 10. Emphasis in original.

56. Slaughter, *Universal Languages and Scientific Taxonomy in the Seventeenth Century*, 176; Knowlson, *Universal Language Schemes in England and France*, 140.

57. Parry Moon and Domina Eberle Spencer, "Languages for Science," *Journal of the Franklin Institute* 246, no. 1 (July 1948): 1–12, on 5.

58. See, for example, Robert Boyle's objections to Latin, as described in Richard Jones, "Science and Language in England of the Mid-Seventeenth Century," *Journal of English and Germanic Philology* 31, no. 3 (1932): 315–331, on 319.

59. Galileo was translated into Latin by his friend C. Bernegger and then published in Holland by Elsevier. Jozef IJsewijn and Dirk Sacré, *Companion to Neo-Latin Studies, Part II: Literary, Linguistic, Philological and Editorial Questions*, 2d. ed. (Leuven: Leuven University Press, 1998), 494. Just because Galileo did not want to talk to the rest of Europe does not mean the rest of Europe did not want to hear him.

60. Peter Burke, "Translations into Latin in Early Modern Europe," in Burke and R. Po-Chia Hsia, eds., *Cultural Translation in Early Modern Europe* (Cambridge: Cambridge University Press, 2007): 65–80, on 73–74; Isabelle Pantin, "The Role of Translations in European Scientific Exchanges in the Sixteenth and Seventeenth Centuries," in ibid.: 163–179, on 166, 172.

61. On the Reformation, see Françoise Waquet, *Latin, or the Empire of a Sign: From the Sixteenth to the Twentieth Centuries*, tr. John Howe (London: Verso, 2001 [1998]), 21; Peter Burke, *The Art of Conversation* (Ithaca: Cornell University Press, 1993), 35–39; and IJsewijn, *Companion to Neo-Latin Studies, Part I*, 48. On female readers, see Burke, *The Art of Conversation*, 64.

62. IJsewijn and Sacré, *Companion to Neo-Latin Studies, Part II*, 324.

63. On translation into Latin, see Burke, *The Art of Conversation*, 41–42; idem, "Cultures of Translation in Early Modern Europe," in Burke and Hsia, eds., *Cultural Translation in Early Modern Europe* (2007): 7–38, on 15, 20; Waquet, *Latin, or the Empire of a Sign*, 85; Augustinius Hubertus Laeven, *The "Acta Eruditorum" under the Editorship of Otto Mencke (1644–1707): The History of an International Learned Journal between 1682 and 1707*, tr. Lynne Richards (Amsterdam: APA-Holland University Press, 1990), 51; and W. Leonard Grant, "European Vernacular Works in Latin Translation," *Studies in the Renaissance* 1 (1954): 120–156.

64. Anthony Grafton, *Bring Out Your Dead: The Past as Revelation* (Cambridge, MA: Harvard University Press, 2001), 170–171.

65. Howard Stone, "The French Language in Renaissance Medicine," *Bibliothèque d'Humanisme et Renaissance* 15, no. 3 (1953): 315–346. See also Waquet, *Latin, or the Empire of a Sign*, 81–82, on the transition from Latin to French in the sixteenth century.

66. Jean-Baptiste Du Hamel, *Regiae Scientarum Academiae historia* (Paris: Etienne Michallet, 1698), unpaginated preface. I am grateful to Anita Guerrini for bringing this passage to my attention.

67. The information in this paragraph is drawn from J. R. Partington, *A History of Chemistry*, vol. 3 (London: Macmillan, 1962), 179–180; and Hugo Olsson, "Torbern Bergman, 1735–1784," in Göte Carlid and Johann Nordström, eds., *Torbern Bergman's Foreign Correspondence. Volume One: Letters from Foreigners to Torbern Bergman* (Stockholm: Almqvist & Wiksell, 1965): xi–xviii. For Bergman's own account, see his relatively sparse autobiographical essay, composed shortly before his death: Torbern Bergman, "Självbiografi," in *Äldre svenska biografier*, v. 3–4 (Uppsala: Almqvist & Wiksell, 1916): 83–103.

68. Marco Beretta, *The Enlightenment of Matter: The Definition of Chemistry from Agricola to Lavoisier* (Canton, MA: Science History Publications, 1993), 93, 317; Evan M. Melhado, *Jacob Berzelius: The Emergence of His Chemical System* (Stockholm: Almqvist & Wiksell International, 1981), 63. On eighteenth-century Swedish science in general, see Andreas Önnerfors, "Translation Discourses of the Enlightenment: Transcultural Language Skills and Cross-References in Swedish and German Eighteenth-Century Learned Journals," in Stefanie Stockhorst, ed., *Cultural Transfer through Translation: The Circulation of Enlightened Thought in Europe by Means of Translation* (Amsterdam: Rodopi, 2010): 209–229.

69. Marco Beretta, "T. O. Bergman and the Definition of Chemistry," *Lychnos* (1988): 37–67, on 40.

70. Lisbet Koerner, *Linnaeus: Nature and Nation* (Cambridge, MA: Harvard University Press, 20), chapter 1 (esp. p. 28 on his ignorance of French).

71. Johann Gottlieb Georgi to Bergman, 9 August 1768, reproduced in Carlid and Nordström, *Torbern Bergman's Foreign Correspondence*, 67.

72. Richard Kirwan to Bergman, 20 January 1783, reproduced in Carlid and Nordström, *Torbern Bergman's Foreign Correspondence*, 182. This is Kirwan's fifth letter.

73. Franz Xaver Schwediauer to Bergman, 3 July 1780, reproduced in Carlid and Nordström, *Torbern Bergman's Foreign Correspondence*, 329–330.

74. Fausto de Elhuyar to Bergman, 15 January 1784, reproduced in Carlid and Nordström, *Torbern Bergman's Foreign Correspondence*, 58.

75. Guyton de Morveau to Bergman, 10 October 1781, reproduced in Carlid and Nordström, *Torbern Bergman's Foreign Correspondence*, 119; and Editors' introduction, in ibid., xxxv, xxxviii.

76. Franz Xaver Schwediauer to Bergman, 14 February 1874, reproduced in Carlid and Nordström, *Torbern Bergman's Foreign Correspondence*, 381–382. See also the similar complaint in Ignaz von Born to Bergman, 10 August 1777, reproduced in ibid., 6.

77. Johan Ditlev Breckling Brandt to Bergman, 22 November 1770, reproduced in Carlid and Nordström, *Torbern Bergman's Foreign Correspondence*, 11.

78. For Sweden, see Margareta Benner and Emin Tengström, *On the Interpretation of Learned Neo-Latin: An Explorative Study Based on Some Texts from Sweden (1611–1716)* (Göteborg: Acta Universitatis Gothoburgensis, 1977). On Latin outside the universities, see Isabelle Pantin, "Latin et langues vernaculaires dans la littérature scientifique européenne au début de l'époque moderne (1550–1635)," in Roger Chartier and Pietro Corsi, eds., *Sciences et langues en Europe* (Paris: European Communities, 2000 [1994]): 41–56. For an excellent general picture of Latin's endurance, see Ann Blair, "La persistance du latin comme langue de science à la fin de la Renaissance," in ibid.: 19–39.

79. Alix Cooper, *Inventing the Indigenous: Local Knowledge and Natural History in Early Modern Europe* (Cambridge: Cambridge University Press, 2007), 78–79.

80. On libraries, see Jonathan I. Israel, *Radical Enlightenment: Philosophy and the Making of Modernity 1650–1750* (New York: Oxford University Press, 2001), 137; on Voltaire, see Burke, *The Art of Conversation*, 53.

81. Janson, *A Natural History of Latin*, 159.

82. See the explanation in Beretta, "T. O. Bergman and the Definition of Chemistry," 53–54; and Beretta, *The Enlightenment of Matter*, 102, 148, 155.

83. Torbern Bergman, *Meditationes de systemate fossilium naturali* (Florence: Typis Josephi Tofani, 1784), 123–124. This is the first Italian edition; the work was originally published in *Nova acta Regiae Societatis Scientiarum Upsalensis* 4 (1784): 63–128.

84. Maurice P. Crosland, *Historical Studies in the Language of Chemistry* (Cambridge, MA: Harvard University Press, 1962), 135–136, 164; Beretta, "T. O. Bergman and the Definition of Chemistry," 55; Beretta, *The Enlightenment of Matter*, 139–140, 318.

85. Trevor Williams, "Scientific Literature: Its Influence on Discovery and Progress," *Interdisciplinary Science Reviews* 2, no. 2 (1977): 165–172, on 165.

86. IJsewijn and Sacré, *Companion to Neo-Latin Studies, Part II*, 258–259.

Chapter Two

1. Mendeleev, *Neftianaia promyshlennost' v severo-amerikanskom shtate Pensil'vanii i na Kavkaze* (St. Petersburg: Obshchestvennaia pol'za, 1877), reproduced in D. I. Mendeleev, *Sochineniia*, v. 10: *Neft'* (Moscow: Izd. AN SSSR, 1949), 153.

2. On the creation of the periodic system, see Michael D. Gordin, *A Well-Ordered*

Thing: Dmitrii I. Mendeleev and the Shadow of the Periodic Table (New York: Basic Books, 2004), chapter 2, and references therein.

3. D. I. Mendeleev, *Novye materialy po istorii otkrytiia periodicheskogo zakona*, ed. N. A. Figurovskii (Moscow: Izd. AN SSSR, 1950), image 2.

4. For an introduction to these three conflicts, see, respectively, Domenico Bertoloni Meli, *Equivalence and Priority: Newton versus Leibniz* (Oxford: Clarendon Press, 1993); Thomas S. Kuhn, "Energy Conservation as an Example of Simultaneous Discovery," in Kuhn, *The Essential Tension: Selected Studies in Scientific Tradition and Change* (Chicago: University of Chicago Press, 1977): 66–104; and Janet Browne, *Charles Darwin: The Power of Place* (Princeton: Princeton University Press, 2002), chapter 1.

5. See Michael D. Gordin, "The Textbook Case of a Priority Dispute: D. I. Mendeleev, Lothar Meyer, and the Periodic System," in Jessica Riskin and Mario Biagioli, eds., *Nature Engaged: Science in Practice from the Renaissance to the Present* (New York: Palgrave Macmillan, 2012): 59–82.

6. J. W. van Spronsen, *The Periodic System of Chemical Elements: A History of the First Hundred Years* (Amsterdam: Elsevier, 1969), 1, and 142–143.

7. Minutes of the Russian Chemical Society meeting of 6 March 1869 (O.S.), *ZhRFKhO* 1 (1869), 35. In the nineteenth century, the Russian calendar lagged twelve days behind the new-style Gregorian calendar standard in Western Europe.

8. Mendeleev's findings were also reported in German in the flagship journal of the fledging German Chemical Society, but in much briefer form: V. von Richter, "[Correspondence from St. Petersburg]," *Berichte der Deutschen Chemischen Gesellschaft zu Berlin* 2 (1869): 552–554.

9. On Beilstein's relationship with Meyer, see his letter to Jakob Volhard lamenting the death of his friend, on 30 May/11 June 1895, reproduced in Elena Roussanova, *Friedrich Konrad Beilstein, Chemiker zweier Nationen: Sein Leben und Werk sowie einige Aspekte der deutsch-russischen Wissenschaftsbeziehungen in der zweiten Hälfte des 19. Jahrhunderts im Spiegel seines brieflichen Nachlasses*, vol. 2 (Hamburg: Norderstedt, 2007), 429.

10. Lothar Meyer, *Die modernen Theorien der Chemie und ihre Bedeutung für die chemische Statik* (Breslau: Maruschke & Berendt, 1864), 136.

11. Karl Seubert, "Zur Geschichte des periodischen Systems," *Zeitschrift für Anorganische Chemie* 9 (1895): 334–338.

12. D. Mendelejeff, "Ueber die Beziehungen der Eigenschaften zu den Atomgewichten der Elemente," *Zeitschrift für Chemie*, N.S. 5 (1869): 405–406, on 405.

13. Lothar Meyer, "Die Natur der chemischen Elemente als Function ihrer Atomgewichte," *Annalen der Chemie und Pharmacie*, Supp. VII (1870): 354–364, on 355–356, 358.

14. Mendeleev, "Sootnoshenie svoistv s atomnym vesom elementov," *ZhRFKhO* 1 (1869): 60–79, on 76, reproduced in Mendeleev, *Periodicheskii zakon. Klassiki nauki*, ed. B. M. Kedrov (Moscow: Izd. AN SSSR, 1958), 30. Emphasis in original.

15. Mendeleev himself would highlight the damage of this translation error in a German article in 1873: D. Mendelejeff, "Zur Frage über das System der Elemente," *Berichte der Deutschen Chemischen Gesellschaft* 4 (1871): 348–352, on 351. This issue has only rarely and all-too-briefly been noted in the massive scholarship on the history of the periodic system, and its implications have never been fully explored. See V. A. Krotikov, "Dve oshibki v pervykh publikatsiiakh o periodicheskom zakone

D. I. Mendeleevym," *Voprosy istorii estestvoznaniia i tekhniki,* no. 4 (29) (1969): 129–131; and Van Spronsen, *The Periodic System of Chemical Elements,* 127.

16. Quoted in K. Bening, *D. I. Mendeleev i L. Meier* (Kazan: Tsentral'naia tip., 1911), ii.

17. Mendeleev, "Ob atomnom ob"eme prostykh tel" (1870), in Mendeleev, *Periodicheskii zakon. Klassiki nauki,* 48–49; and Mendeleev, "O meste tseriia v sisteme elementov" (1870), in ibid., 59.

18. Mendeleev to Erlenmeyer, [August 1871?], in Otto Krätz, "Zwei Briefe Dmitri Iwanowitsch Mendelejeffs an Emil Erlenmeyer," *Physis* 12 (1970): 347–352, on 351.

19. Mendelejeff, "Die periodische Gesetzmässigkeit der chemischen Elemente," reproduced in Mendeleev, *Nauchnyi arkhiv, t. 1. Periodicheskii zakon,* ed. B. M. Kedrov (Moscow: Izd. AN SSSR, 1953), on 361. Emphasis in original.

20. Lothar Meyer, *Die modernen Theorien der Chemie und ihre Bedeutung für die chemische Statik,* 3d. ed. (Breslau: Maruschke & Berendt, 1876), 291n. On the second edition, see idem, *Die modernen Theorien der Chemie und ihre Bedeutung für die chemische Statik,* 2d. ed. (Breslau: Maruschke & Berendt, 1872), 298.

21. Adolphe Wurtz to Mendeleev, 27 July 1877, ADIM I-V-23-1-27.

22. Adolphe Wurtz, *La théorie atomique* (Paris: Librairie Germer Ballière et Cie., 1879), 112. On Meyer, see pages 118 and 122.

23. Adolf Wurtz to German Chemical Society, 29 December 1879, as printed in minutes of the meeting of 11 January 1880, *Berichte der Deutschen Chemischen Gesellschaft* 13 (1880); 6–7, on 7.

24. Adolphe Wurtz to German Chemical Society, 1 March 1880, as printed in *Berichte der Deutschen Chemischen Gesellschaft* 13 (1880): 453–454.

25. Lothar Meyer to the Vorstand of the German Chemical Society, 25 January 1880, as printed in *Berichte der Deutschen Chemischen Gesellschaft* 13 (1880): 220–221, on 221.

26. Lothar Meyer, "Zur Geschichte der periodischen Atomistik [I]," *Berichte der Deutschen Chemischen Gesellschaft* 13 (1880): 259–265, on 261 and 259 ("judge").

27. Mendeleev, "Spisok moikh sochinenii," reproduced in S. A. Shchukarev and S. N. Valk, eds., *Arkhiv D. I. Mendeleeva, t. 1: Avtobiograficheskie materialy, sbornik dokumentov* (Leningrad: Izd. Leningradskogo gosudarstvennogo universiteta imeni A. A. Zhdanova, 1951), 67.

28. D. Mendelejeff, "Zur Geschichte des periodischen Gesetzes," *Berichte der Deutschen Chemischen Gesellschaft* 13 (1880): 1796–1804, on 1799n1, 1800n3, 1797, and 1801.

29. Lothar Meyer, "Zur Geschichte der periodischen Atomistik [II]," *Berichte der Deutschen Chemischen Gesellschaft* 13 (1880): 2043–2044, on 2043. Emphasis in original.

30. Iu. I. Solov'ev, *Istoriia khimii v Rossii: Nauchnye tsentry i osnovnye pravleniia issledovaniia* (Moscow: Nauka, 1985), 79–81.

31. See Nathan M. Brooks, "Russian Chemistry in the 1850s: A Failed Attempt at Institutionalization," *Annals of Science* 52 (1995): 577–589.

32. *Khimicheskii zhurnal N. Sokolova i A. Engel'gardta* 1 (1859), front cover.

33. N. Sokolov and A. Engel'gardt, "Ot redaktsii," *Khimicheskii zhurnal N. Sokolova i A. Engel'gardta* 1 (January 1859): i–xvi, on ix.

34. August Hofmann, "O dvuatomnykh i triatomnykh ammiakakh," *Khimicheskii*

zhurnal N. Sokolova i A. Engel'gardta 3 (1860): 55–74; Edward Frankland, Auguste Cahours, and George Buckton, "O metalloorganicheskikh soedineniiakh," *Khimicheskii zhurnal N. Sokolova i A. Engel'gardta* 3 (1860): 109–129.

35. For example, A. M. Butlerov, "O nekotorykh produktakh deistviia al'kogoliata natriia na iodoform," *Khimicheskii zhurnal N. Sokolova i A. Engel'gardta* 3 (1860): 340–351.

36. N. N. Sokolov to Mendeleev, 28 January 1860 (O.S.), St. Petersburg, ADIM I-V-44-1-12.

37. Quoted in Richard Anschütz, *August Kekulé*, 2 v. (Berlin: Verlag Chemie, 1929), I:130. See also the opening editorial in the first issue: A. Kekulé, G. Lewinstein, F. Eisenlohr, and M. Cantor, "[Editorial Announcement]," *Kritische Zeitschrift für Chemie, Physik und Mathematik* 1 (1858): 3–7.

38. Otto Krätz, ed., *Beilstein-Erlenmeyer: Briefe zur Geschichte der chemischen Dokumentation und des chemischen Zeitschriftenwesens* (Munich: Werner Fritsch, 1972), 11.

39. On Erlenmeyer's sense of humor, see Richard Meyer, "Emil Erlenmeyer," *Chemiker-Zeitung* 23, no. 19 (13 February 1909): 161–162, on 161. Intrusive editorializing was not atypical in several of the most prominent scientific journals of the day. On the norms and practices of midcentury chemical publishing, see J. P. Phillips, "Liebig and Kolbe, Critical Editors," *Chymia* 2 (1966): 89–97. For contemporary complaints, see the letter from Kekulé to Erlenmeyer, 8 November 1871, quoted in Anschütz, *August Kekulé*, I:407; and Beilstein to Butlerov, 24 November 1866 (O.S.), reproduced in G. W. Bykow and L. M. Bekassowa, "Beiträge zur Geschichte der Chemie der 60-er Jahre des XIX. Jahrhunderts: II. F. Beilsteins Briefe an A. M. Butlerow," *Physis* 8 (1966): 267–285, on 281.

40. On the Russian contributions to the *Zeitschrift*, and a general history, see G. V. Bykov and Z. I. Sheptunova, "Nemetskii 'Zhurnal khimii' (1858–1871) i russkie khimiki (K istorii khimicheskoi periodiki)," *Trudy Instituta istorii estestvoznaniia i tekhniki* 30 (1960): 97–110. On Russian students in Heidelberg, see Gesa Bock, "Studenten des russischen Reichs an der Universität Heidelberg (1862/63–1914)" (Diplomarbeit, Institüt für Übersetzen und Dolmetschen, Universität Heidelberg, 1991); Willy Birkenmaier, *Das russische Heidelberg: Zur Geschichte der deutsch-russischen Beziehungen im 19. Jahrhundert* (Heidelberg: Wunderhorn, 1995); and Annette Nolte, *D. I. Mendeleev in Heidelberg, Russica Palatina* 22 (1993). On Borodin, see Michael D. Gordin, "The Weekday Chemist: The Training of Aleksandr Borodin," in Jed Z. Buchwald, ed., *A Master of Science History: Essays in Honor of Charles Coulston Gillispie, Archimedes* 30 (Berlin: Springer, 2012): 137–164.

41. Otto Krätz, "Emil Erlenmeyer, 1825–1909," *Chemie in unserer Zeit* 6 (1972): 52–58, on 55; M. Conrad, "Emil Erlenmeyer," *Berichte der Deutschen Chemischen Gesellschaft* 43 (1910): 3645–3664, on 3647.

42. Beilstein to Butlerov, 24 November 1866 (O.S.), reproduced in Bykow and Bekassowa, "Beiträge zur Geschichte," 281. On postal and ordering issues, see Beilstein to Butlerov, 29/17 January 1865, reproduced in ibid., 270; and V. V. Markovnikov to A. M. Butlerov, 10 December [1867, O.S.], reproduced in G. V. Bykov, ed., *Pis'ma russkikh khimikov k A. M. Butlerovu, Nauchnoe Nasledstvo*, v. 4 (Moscow: Izd. AN SSSR, 1961), 248. See also the discussion in Alan J. Rocke, *The Quiet Revolution: Hermann Kolbe and the Science of Organic Chemistry* (Berkeley: University of California Press, 1993), 255, 258.

43. Erlenmeyer to Butlerov, 25 March 1864, reproduced in G. W. Bykow and L. M. Bekassowa, "Beiträge zur Geschichte der Chemie der 60-er Jahre des XIX. Jahrhunderts: I. Briefwechsel zwischen E. Erlenmeyer und A. M. Butlerow (von 1862 bis 1876)," *Physis* 8 (1966): 185–198, on 190–191.

44. On Beilstein's biography, see Michael D. Gordin, "Beilstein Unbound: The Pedagogical Unraveling of a Man and His *Handbuch*," in David Kaiser, ed., *Pedagogy and the Practice of Science: Historical and Contemporary Perspectives* (Cambridge, MA: MIT Press, 2005): 11–39.

45. Krätz, *Beilstein-Erlenmeyer*, 7.

46. Beilstein to Kekulé, 3 November 1865, quoted in Friedrich Richter, "K. F. Beilstein, sein Werk und seine Zeit: Zur Erinnerung an die 100. Wiederkehr seines Geburtstages," *Berichte der Deutschen Chemischen Gesellschaft* 71A (1938): 35–71, on 42. Emphasis in original.

47. Beilstein to Butlerov, 18/6 November 1866, reproduced in Bykow and Bekassowa, "Beiträge zur Geschichte der Chemie . . . II," 279.

48. Beilstein to Butlerov, 29/17 January 1865, reproduced in Bykow and Bekassowa, "Beiträge zur Geschichte der Chemie . . . II," 271. Emphasis in original.

49. Friedrich Beilstein, "O rabotakh chlenov Russkago fiziko-khimicheskago obshchestva po aromaticheskomu riadu," in *Russkoe khimicheskoe obshchestvo. XXV (1868–1893). Otdelenie khimii Russkago fiziko-khimicheskago obshchestva* (St. Petersburg: V. Demakov, 1894), 39–56, on 48; Beilstein, signed footnote in A. Engelhardt, "Ueber die Einwirkung der wasserfreien Schwefelsäure auf einige organische Verbindungen," *Zeitschrift für Chemie und Pharmacie* 7 (1864): 42–46 and 85–87, on 42n2.

50. Beilstein comment on D. Mendelejeff, "Ueber die Verbindung des Weingeistes mit Wasser," *Zeitschrift für Chemie*, N.S. 1 (1865): 257–264, on 264.

51. Beilstein to Erlenmeyer, 26/14 April 1871, reproduced in Krätz, *Beilstein-Erlenmeyer*, 16. See also Rudolph Fittig to Erlenmeyer, 2 January 1872, Tübingen, HBA.

52. Quoted in V. V. Kozlov and A. I. Lazarev, "Tri chetverti veka Russkogo Khimicheskogo Obshchestva (1869–1944)," in S. I. Vol'fkovich and V. S. Kiselev, eds., *75 let periodicheskogo zakona D. I. Mendeleeva i Russkogo Khimicheskogo Obshchestva* (Moscow: Izd. AN SSSR, 1947): 115–265, on 128.

53. N. A. Menshutkin and G. Shmidt, "Otchet o deiatel'nosti Russkago khimicheskago obshchestva v 1869 g.," *ZhRFKhO* 2 (1870): 3–6, on 5.

54. Minutes of Russian Chemical Society meeting of 4 March 1871 (O.S.), *ZhRFKhO* 3 (1871), 93.

55. V. V. Markovnikov to A. M. Butlerov, 9 October 1874 (O.S.), reproduced in Bykov, *Pis'ma russkikh khimikov k A. M. Butlerovu*, 272.

56. Minutes of Russian Chemical Society meeting of 3 April 1880 (O.S.), *ZhRFKhO* 12 (1880): 182–183.

57. Beilstein to Erlenmeyer, 23 September/5 October 1873, reproduced in Krätz, *Beilstein-Erlenmeyer*, 41.

58. A. N. Popov to A. M. Butlerov, 30 December 1871 (O.S.), reproduced in Bykov, *Pis'ma russkikh khimikov k A. M. Butlerovu*, 340.

59. V. V. Markovnikov to A. M. Butlerov, 13 January [1870] (O.S.), reproduced in Bykov, *Pis'ma russkikh khimikov k A. M. Butlerovu*, 259; M. D. L'vov to A. M. Butlerov, 22 July 1873 (O.S.), reproduced in ibid., 200.

60. V. V. Markovnikov to A. M. Butlerov, 17 January [1868] (O.S.), reproduced in Bykov, *Pis'ma russkikh khimikov k A. M. Butlerovu*, 252.

61. Beilstein to Erlenmeyer, 29 April/11 May 1872, reproduced in Krätz, *Beilstein-Erlenmeyer*, 26. Emphasis in original.

62. Draft of Erlenmeyer to Beilstein, 19 May 1872, reproduced in Krätz, *Beilstein-Erlenmeyer*, 33–34.

63. Johanna Meyer (née Volkmann) and her children, Tübingen, 12 April 1895, ADIM I-V-27-1-26.

64. Lothar Meyer, ed., *Die Anfänge des Systems der chemischen Elemente: Abhandlungen von J. W. Doebereiner 1829 und Max Pettenkofer 1850 nebst einer geschichtlichen Uebersicht der Weiterentwicklung der Lehre von den Triaden der Elemente* (Leipzig: W. Engelmann, 1895).

65. Lothar Meyer to Mendeleev, 16 August 1893, Tübingen, ADIM I-V-63-1-70.

66. Karl Seubert, ed., *Das natürliche System der chemischen Elemente: Abhandlungen von Lothar Meyer 1864–1869 und D. Mendelejeff 1869–1871* (Leipzig: W. Engelmann, 1895). The date span in the title represents a subtle priority claim in itself.

67. http://royalsociety.org/Content.aspx?id=3277. (Accessed 20 August 2012.)

68. Editorial comments in Seubert, ed., *Das natürliche System der chemischen Elemente*, 122–123. See also Nikolai A. Menshutkin, *Ocherk razvitiia khimicheskikh vozzrenii* (St. Petersburg: V. Demakov, 1888), 319.

69. Menshutkin in minutes of Russian Chemical Society meeting of 13 April 1895 (O.S.), *ZhRFKhO* 27 (1895): 197; and Butlerov, "Istoricheskii ocherk razvitiia khimii v poslednie 40 let," stenograph of lectures from 1879–1880, reproduced in Butlerov, *Sochineniia*, 3 v. (Moscow: AN SSSR, 1953–1958), III: 280.

70. P. Phillips Bedson, "Lothar Meyer Memorial Lecture," *Journal of the Chemical Society* 69 (1896): 1403–1439, on 1409.

71. See, for example, F. P. Venable, *The Development of the Periodic Law* (Easton, PA: Chemical Publishing Co., 1896), 95.

72. Minutes of Russian Chemical Society meeting of 3 April 1875 (O.S.), *ZhRFKhO* 7 (1875): 177.

73. In *Russkoe khimicheskoe obshchestvo*, 4.

74. In *Russkoe khimicheskoe obshchestvo*, 2.

75. D. Mendeléeff, "Comment j'ai trouvé le système périodique des éléments," *Revue générale de chimie pure et appliquée* 4 (1901): 533–546, on 546.

76. V. I. Modestov, *Russkaia nauka v posledniia dvadtsat' piat' let* (Odessa: Ekonomicheskaia tip., 1890), 9. Emphasis in original.

Chapter Three

1. F. M. Dostoevskii, *Polnoe sobranie sochinenii v tridtsati tomakh*, 30 v. (Leningrad: Nauka, 1972–1990), XXI: 121.

2. As of 1971, Russian speakers numbered just under half of all Slavic speakers. G. S. Vinokur, *The Russian Language: A Brief History*, tr. Mary A. Forsyth (Cambridge: Cambridge University Press, 1971), 1. That proportion is almost certainly higher today.

3. Vinokur, *The Russian Language*, 1.

4. W. K. Matthews, *The Structure and Development of Russian* (Cambridge: Cambridge University Press, 1953), 111.

5. Vinokur, *The Russian Language*, 22–23, quotation on 23. See also Lawrence L. Thomas, introduction to V. V. Vinogradov, *The History of the Russian Literary Language from the Seventeenth Century to the Nineteenth*, tr. and ed. Lawrence L. Thomas (Madison: University of Wisconsin Press, 1969), xii.

6. B. O. Unbegaun, "Colloquial and Literary Russian," *Oxford Slavonic Papers* 1 (1950): 26–36, on 26–27.

7. Vinokur, *The Russian Language*, 32.

8. Matthews, *The Structure and Development of Russian*, 14.

9. Vinokur, *The Russian Language*, 71; Matthews, *The Structure and Development of Russian*, 140.

10. W. K. Matthews, *Russian Historical Grammar* (London: Athlone Press, 1960), 63; Vinogradov, *The History of the Russian Literary Language*, 13–14.

11. Gerta Hüttl Worth, *Foreign Words in Russian: A Historical Sketch, 1550–1800* (Berkeley: University of California Press, 1963), 1–2.

12. Vinogradov, *The History of the Russian Literary Language*, 33. For a magisterial survey of transformations of Russian in this period, see V. M. Zhivov, *Iazyk i kul'tura v Rossii XVIII veka* (Moscow: Shkola "Iazyki russkoi kul'tury," 1996).

13. Unbegaun, "Colloquial and Literary Russian," 29; Vinogradov, *The History of the Russian Literary Language*, 31. On the context of Peter's translation movement, see Matthews, *The Structure and Development of Russian*, 156; and Dennis Ward, *The Russian Language Today: System and Anomaly* (London: Hutchinson University Library, 1965), 114.

14. Christopher D. Buck, "The Russian Language Question in the Imperial Academy of Sciences, 1724–1770," in Riccardo Picchio and Harvey Goldblatt, eds., *Aspects of the Slavic Language Question*, 2 vol. (New Haven: Yale Concilium on International and Area Studies, 1984), II: 187–233, on 188–189, 194.

15. Buck, "The Russian Language Question in the Imperial Academy of Sciences," 198–199.

16. Hans Rogger, *National Consciousness in Eighteenth-Century Russia* (Cambridge, MA: Harvard University Press, 1960), 109.

17. E. Lenz to council of St. Petersburg University, 10 October 1859 (O.S.), TsGIASPb, f. 14, op. 1, d. 6039, l. 1.

18. On Lomonosov's *Rossiiskaia grammatika* (1755) and *O pol'ze knig tserkovnykh v Rossiiskom iazyke* (1758) and their influence on Russian stylistics, see Vinokur, *The Russian Language*, 101; Unbegaun, "Colloquial and Literary Russian," 30; and Vinogradov, *The History of the Russian Literary Language*, 72–73. On French as a template for later Russian syntax, see ibid., 66.

19. Vinogradov, *The History of the Russian Literary Language*, 243.

20. On the history of the German language in Russia, see Alfons Höcherl, "Kulturelle und wissenschaftliche deutsche Einflüsse in Russland im historischen Überblick," in Ulrich Ammon and Dirk Kemper, eds., *Die deutsche Sprache in Russland: Geschichte, Gegenwart, Zukunftsperspektiven* (Munich: Iudicium, 2011), 23–40.

21. Edv. Hjelt, "Friedrich Konrad Beilstein," *Berichte der Deutschen Chemischen Gesellschaft* 40 (1907): 5041–5078, on 5069.

22. Petition to the Vice-President of the Academy of Sciences, 9 October 1854 (O.S.), PFARAN f. 5, op. 1(1854), d. 513, l. 2.

23. K. K. Klaus to A. M. Butlerov, 15 April 1853 (O.S.), reproduced in G. V. Bykov, ed., *Pis'ma russkikh khimikov k A. M. Butlerovu, Nauchnoe nasledstvo*, v. 4 (Moscow: Izd. AN SSSR, 1961), 161.

24. Klaus to Butlerov, 12 May 1857 (O.S.), reproduced in Bykov, *Pis'ma russkikh khimikov k A. M. Butlerovu*, 166.

25. Klaus to Butlerov, 11 August 1853 (O.S.), reproduced in Bykov, *Pis'ma russkikh khimikov k A. M. Butlerovu*, 164.

26. A. A. Inostrantsev, *Vospominaniia (Avtobiografiia)*, eds. V. A. Prozorovskii and I. L. Tikhonov (St. Petersburg: Peterburgskoe vostokovedenie, 1998), 95.

27. Borodin to M. A. Balakirev, [22–30 January 1867, O.S.], reproduced in A. P. Borodin, *Pis'ma: Polnoe sobranie, kriticheski sverennoe s podlinnymi tekstami*, 4 v., ed. S. A. Dianin (Moscow: Gos. muzykal'nyoe izd., 1927–1950), I: 94.

28. Minutes of the Russian Chemical Society meeting of 12 September 1902 (O.S.), *ZhRFKhO* 34 (1902): 637.

29. I. M. Sechenov, *Avtobiograficheskie zapiski* (Moscow: Izd. AN SSSR, 1945), 101n1.

30. Kablukov's autobiography, PFARAN, f. 474, op. 1, d. 201, quoted in Iu. I. Solov'ev, M. I. Kablukova, and E. V. Kolesnikov, *Ivan Alekseevich Kablukov* (Moscow: Izd. AN SSSR, 1957), 20–22.

31. V. V. Markovnikov to A. M. Butlerov, 22 July/3 August [1865], reproduced in Bykov, *Pis'ma russkikh khimikov k A. M. Butlerovu*, 215.

32. Markovnikov to Butlerov, 7 August [1865], reproduced in Bykov, *Pis'ma russkikh khimikov k A. M. Butlerovu*, 216.

33. A. Bulginskii to Emil Erlenmeyer, 29 October 1866, HBA.

34. Mendeleev's library is described in R. B. Dobrotin and N. G. Karpilo, *Biblioteka D. I. Mendeleeva* (Leningrad: Nauka, 1980).

35. See Mendeleev's diary entries of 1 and 6 January 1861, reproduced in D. I. Mendeleev, "Dnevnik 1861 g.," *Nauchnoe nasledstvo* 2 (1951): 111–212, on 112 and 114.

36. Letter of 25 November 1886 (O.S.), quoted in Iu. I. Solov'ev, *Istoriia khimii v Rossii: Nauchnye tsentry i osnovnye pravleniia issledovaniia* (Moscow: Nauka, 1985), 355.

37. Mendeleev to Kekulé, 28 June/10 July 1883, Boblovo, quoted in Richard Anschütz, *August Kekulé*, 2 v. (Berlin: Verlag Chemie, 1929), I: 692, 694.

38. Mendeleev to Erlenmeyer, 24 August/3 September [1870?], in Otto Krätz, "Zwei Briefe Dmitri Iwanowitsch Mendelejeffs an Emil Erlenmeyer," *Physis* 12 (1970): 347–352, on 350.

39. Mendeleev, *Dva londonskikh chteniia*, reproduced in Mendeleev, *Izbrannye sochineniia*, v. 2. (Leningrad: ONTI, 1934), 342.

40. Mendeleev to Menshutkin, 23 July 1889, Boblovo, reproduced in B. N. Menshutkin, *Zhizn' i deiatel'nost' Nikolaia Aleksandrovicha Menshutkina* (St. Petersburg: M. Frolova, 1908), 109.

41. Crum Brown to Mendeleev, 29 March 1884, reproduced in V. E. Tishchenko and M. N. Mladentsev, *Dmitrii Ivanovich Mendeleev, ego zhizn' i deiatel'nost': Universitetskii period, 1861–1890 gg., Nauchnoe nasledstvo*, v. 21 (Moscow: Nauka, 1993), 117. Presumably, this letter was originally written in German, but I was only able to find it in Russian translation.

42. William Ramsay to Mendeleev, 22 September 1889, ADIM Alb. 2/280.

43. William Ramsay to Mendeleev, undated, ADIM Alb. 2/201.

44. William Ramsay to Mendeleev, 6 January 1892, ADIM Alb. 3/500.

45. William Ramsay to Mendeleev, 20 January 1892, ADIM Alb. 3/501.

46. Minutes of the Russian Chemical Society meeting of 5 November 1870 (O.S.), *ZhRFKhO* 2 (1870), 290n1.

47. Otto Jespersen, "Nature and Art in Language," *American Speech* 5 (1929): 89–103, on 90–91. On similar layering in Russian chemical nomenclature, see David Kraus, "Sources of Scientific Russian," *Slavic and East European Journal* 5, no. 2 (Summer 1961): 123–131, on 128–129.

48. Viktor A. Kritsman, "Die Entstehung der russischen chemischen Nomenklatur im europäischen Kontext: Die Frühgeschichte," in Bernhard Fritscher and Gerhard Brey, eds., *Cosmographica et Geographica: Festschrift für Heribert M. Nobis zum 70. Geburtstag* (Munich: Institut für Geschichte der Naturwissenschaften, 1994): 199–218, on 18; Victor A. Kritsman and Briggite Hoppe, "The Study of Lavoisier's Works by Russian Scientists," *Revue d'histoire des sciences* 48 (1995): 133–142, esp. 135–136; Solov'ev, *Istoriia khimii v Rossii*, 58–60; Nikolai A. Menshutkin, *Ocherk razvitiia khimicheskikh vozzrenii* (St. Petersburg: V. Demakov, 1888), 31n1.

49. F. Savchenkov, "Istoricheskie materialy po russkoi khimicheskoi nomenklature," *ZhRFKhO* 2 (1870): 205–212, on 205.

50. Iakov D. Zakharov, "Razsuzhdenie o rossiiskom khimicheskom slovoznachenii," *Umozritel'nyia izsledovaniia Imperatorskoi Sanktpeterburgskoi Akademii nauk* 2 (1810): 332–354, on 332–333.

51. M. F. Solov'ev, S. Ia. Nechaev, P. G. Sobolevskii, and G. I. Gess, "Kratkii obzor khimicheskago imenosloviia," *Gornyi zhurnal* 2, no. 6 (1836): 457–463, on 457.

52. Frankland to Kolbe, 3 December 1871, reproduced in Rita Meyer, "Emil Erlenmeyer (1825–1909) als Chemietheoretiker und sein Beitrag zur Entwicklung der Strukturchemie" (Dissertation, Medical Faculty of Ludwig-Maximilians-Universität in Munich, 1984), on 344–345.

53. On the tortured history of the Geneva nomenclature, see the dissertation in progress by Evan Hepler-Smith at Princeton University, entitled "Nominally Rational: Systematic Nomenclature and the Structure of Organic Chemistry, 1889–1940."

54. See Michael D. Gordin, "Beilstein Unbound: The Pedagogical Unraveling of a Man and His *Handbuch*," in David Kaiser, ed., *Pedagogy and the Practice of Science: Historical and Contemporary Perspectives* (Cambridge, MA: MIT Press, 2005): 11–39.

55. Minutes of the Russian Chemical Society meeting of 8 October 1892 (O.S.), *ZhRFKhO* 24 (1892): 542–544. See also Beilstein's early comments on the Geneva nomenclature and Menshutkin's discussion of translations of the French rules into Russian: Minutes of the Russian Chemical Society meeting of 13 September 1890 (O.S.), *ZhRFKhO* 22 (1890): 480; and N. Menshutkin, "K voprosu o khimicheskoi nomenklature: Sostavlenie nazvanii organicheskikh kislot," *ZhRFKhO* 25 (1893): 10.

56. Solov'ev, *Istoriia khimii v Rossii*, 82, 86.

57. K. Ia. Parmenov, *Khimiia kak uchebnyi predmet v dorevoliutsionnoi i sovetskoi shkole* (Moscow: Akademiia pedagogicheskikh nauk RSFSR, 1963), 30.

58. C. G. Lehmann, *Handbuch der physiologischen Chemie* (Leipzig: W. Engelmann, 1854); Justus von Liebig, *Die organische Chemie in ihrer Anwendung auf Physiologie und Pathologie* (Braunschweig: F. Vieweg und Sohn, 1842).

59. On the issue of credit, see Alan J. Rocke, "Kekulé, Butlerov, and the Historiography of the Theory of Chemical Structure," *British Journal for the History of Science* 14 (1981): 27–57.

60. Much of the information in this section on the composition of the textbook is derived from G. V. Bykov, "Materialy k istorii trekh pervykh izdanii 'Vvedeniia k polnomu izucheniiu organicheskoi khimii' A. M. Butlerova," *Trudy Instituta istorii estestvoznaniia i tekhniki* 6 (1955): 243–291.

61. Karl Schmidt to A. M. Butlerov, 1/13 April 1865, reproduced in Bykov, *Pis'ma russkikh khimikov k A. M. Butlerovu*, 402. Emphasis in original.

62. Wurtz to Butlerov, 5 October 1864, reproduced in G. V. Bykov and J. Jacques, "Deux pionniers de la chimie moderne, Adolphe Wurtz et Alexandre M. Boutlerov, d'après une correspondance inédite," *Revue d'historie des sciences* 13 (1960): 115–134, on 126.

63. Markovnikov to Butlerov, 22 July/3 August [1865], reproduced in Bykov, *Pis'ma russkikh khimikov k A. M. Butlerovu*, 216.

64. P. P. Alekseev to Butlerov, 9 January 1867, reproduced in Bykov, *Pis'ma russkikh khimikov k A. M. Butlerovu*, 19.

65. Butlerov to Erlenmeyer, 23 July/4 August 1864, reproduced in G. W. Bykow and L. M. Bekassowa, "Beiträge zur Geschichte der Chemie der 60-er Jahre des XIX. Jahrhunderts: I. Briefwechsel zwischen E. Erlenmeyer und A. M. Butlerow (von 1862 bis 1876)," *Physis* 8 (1966): 185–198, on 193.

66. N. A. Golovkinskii to Butlerov, 11 March 1864, reproduced in Bykov, *Pis'ma russkikh khimikov k A. M. Butlerovu*, 97.

67. On Beilstein's assistance with many stages of the process, see Beilstein to Butlerov, 15/27 October 1867, reproduced in Elena Roussanova, *Friedrich Konrad Beilstein, Chemiker zweier Nationen: Sein Leben und Werk sowie einige Aspekte der deutsch-russischen Wissenschaftsbeziehungen in der zweiten Hälfte des 19. Jahrhunderts im Spiegel seines brieflichen Nachlasses*, vol. 2 (Hamburg: Norderstedt, 2007), 236.

68. Markovnikov to Butlerov, 15/27 January [1867], reproduced in Bykov, *Pis'ma russkikh khimikov k A. M. Butlerovu*, 240.

69. Beilstein to Erlenmeyer, 27 March 1861, reproduced in Roussanova, *Friedrich Konrad Beilstein*, 85. See also Beilstein to Erlenmeyer, 10 November 1861, reproduced in ibid., 78.

70. Friedrich Beilstein, review of D. Mendelejeff's *Organische Chemie*, *Zeitschrift für Chemie und Pharmacie* 5 (1862): 271–276, on 271.

71. Butlerov, *Vvedenie k polnomu izucheniiu organicheskoi khimii*, in Butlerov, *Sochineniia*, II: 12.

72. Paul Walden, "Ocherk istorii khimii v Rossii," in A. Ladenburg, *Lektsii po istorii razvitiia khimii do nashego vremeni*, tr. from 4th ed. by E. S. El'chaninov (Odessa: Mathesis, 1917): 361–654, on 421.

73. Markovnikov, "Sovremennaia khimiia i russkaia khimicheskaia promyshlennost'" (1879), in Markovnikov, *Izbrannye trudy*, ed. A. F. Plate and G. V. Bykov (Moscow: Izd. AN SSSR, 1955), 648. See also idem, "Moskovskaia rech' o Butlerove," ed. Iu. S. Musabekov, *Trudy Instituta istorii estestvoznaniia i tekhniki* 12 (1956): 135–181, on 161.

Chapter Four

1. From L. Zamenhof, ed., *Fundamenta Krestomatio de la lingvo Esperanto*, 18th ed. (Rotterdam: Universala Esperanto-Asocio, 1992 [1903]), 181.

2. Roland G. Kent, "The Scientist and an International Language," *Proceedings of the American Philosophical Society* 63 (1924): 162–170, on 163. Kent advocated the revival of Latin.

3. Luther H. Dyer, *The Problem of an International Auxiliary Language and Its Solution in Ido* (London: Putnam, 1923), 6–7.

4. Leopold Pfaundler, "The Need for a Common Scientific Language," in L. Couturat, O. Jespersen, R. Lorenz, W. Ostwald, and L. Pfaundler, *International Language and Science: Considerations on the Introduction of an International Language into Science*, tr. F. G. Donnan (London: Constable & Company, 1910): 1–10, on 2.

5. L. Couturat and L. Leau, *Histoire de la Langue Universelle* (Paris: Librairie Hachette, 1903), ix.

6. Louis Couturat, *A Plea for an International Language* (London: George J. Henderson, 1905), 8.

7. Otto Jespersen, *An International Language* (London: George Allen & Unwin, 1928), 14.

8. "Stated Meeting, January 6, 1888," *Proceedings of the American Philosophical Society* 25, no. 127 (1888): 1–18, on 4.

9. Jespersen, "Nature and Art in Language," *American Speech* 5 (1929): 89–103, on 89.

10. Detlev Blanke, "The Term 'Planned Language,'" in Humphrey Tonkin, ed., *Esperanto, Interlinguistics, and Planned Language* (Lanham, MD: University Press of America, 1997): 1–20; and Alicja Sakaguchi, "Towards a Clarification of the Function and Status of International Planned Languages," in Ulrich Ammon, ed., *Status and Function of Languages and Language Varieties* (Berlin: Walter de Gruyter, 1989): 399–440. This terminology originated in a scholarly monograph dedicated to standardizing technological nomenclature: Eugen Wüster, *Internationale Sprachnormung in der Technik, besonders in der Elektrotechnik (Die nationale Sprachnormung und ihre Verallgemeinerung)*, 2d ed. (Bonn: H. Bouvier u. Co. Verlag, 1966).

11. Louis Couturat, "Sur la langue internationale," *Revue des questions scientifiques* 52 (1902): 213–223.

12. Louis Couturat, "Autour d'une Langue internationale," *La Revue* 87 (1910): 381–385, on 382.

13. W. A. Oldfather, "Latin as an International Language," *Classical Journal* 16 (1921): 195–206.

14. Albert Léon Guérard, *A Short History of the International Language Movement* (London: T. Fisher Unwin, 1922), 169.

15. Giuseppe Peano, "De Latino sine flexione: Lingua auxiliare internationale," *Revista de mathematica* 8 (1903): 74–83, on 74. Peano even replicated, in his own language, a statement highly reminiscent of the Babel-rousers quoted at the beginning of this chapter: "Conoscentia de tres aut quatuor lingua principale suffice ut nos lege, in originale aut in versione omne libro jam celebre. Sed hodie Russo, Polacco, Rumeno, Japonico, . . . publica in suo lingua libro originale, et non solo libro scholastico" (p. 79). Ellipses in original.

16. M. Monnerot-Dumain, *Précis d'interlinguistique générale et spéciale* (Paris: Librairie Maloine, 1960), 512.

17. Couturat and Leau, *Histoire de la Langue Universelle*, 37. For more on Solresol, see Andrew Large, *The Artificial Language Movement* (Oxford: Basil Blackwell, 1985), 63.

18. Richard Lorenz, "The Relationship of the International Language to Science," in L. Couturat et al., *International Language and Science* (1910): 53–60, on 57.

19. Jespersen, *An International Language*, 27.

20. Karl Brugmann and August Leskien, *Zur Kritik der künstlichen Weltsprachen* (Straßburg: Karl J. Trübner, 1907), 19. For a response to this pamphlet from a linguist, see J. Baudouin de Courtenay, *Zur Kritik der künstlichen Weltsprachen* (Leipzig: Verlag von Veit, 1908).

21. Andrew Drummond, *A Hand-Book of Volapük: And an Elementary Manual of its Grammar and Vocabulary, Prepared from the Gathered Papers of Gemmell Hunter Ibidem Justice, together with an Account of Events Relating to the Annual General Meeting of 1891 of the Edinburgh Society for the Propagation of a Universal Language: Edited for the First Time by Dr. Charles Cordiner* (Edinburgh: Polygon, 2006). Today the term "volapuk" also refers to using Latin characters and Arabic numerals to render Cyrillic letters through their resemblances, so that ш would be written w, and б by the number 6. This was once quite common in text messaging, and is unrelated to the language discussed here.

22. Johann Martin Schleyer, *Volapük (Weltsprache): Grammatik der Universalsprache für alle gebildete Erdbewohner*, 4th. ed. (Überlingen am Bodensee: August Feyel, 1884), iii. Ellipses in original.

23. For example: Schleyer, *Grammar with Vocabularies of Volapük (The Language of the World) for all Speakers of the English Language*, 2d. rev. ed., tr. W. A. Seret (Glasgow: Thomas Murray & Son, 1887); and G. Krause, *The Volapük Commercial Correspondent* (London: Swan Sonnenschein & Co., 1889).

24. Frederick Bodmer, *The Loom of Language* (New York: W. W. Norton, 1944), 460; Couturat and Leau, *Histoire de la Langue Universelle*, 141. For more on Volapük's success in Paris, see Natasha Staller, "Babel: Hermetic Languages, Universal Languages, and Anti-Languages in Fin de Siècle Parisian Culture," *Art Bulletin* 76 (1994): 331–354.

25. Guérard, *A Short History of the International Language Movement*, 103n.

26. Constitution of the Volapükaklub Nolümelopik, [1890?], Volapük Exhibit, APSL.

27. Quoted in Guérard, *A Short History of the International Language Movement*, 98.

28. Alfred Kirchhoff, *Volapük, or Universal Language: A Short Grammatical Course*, 3d ed. (London: Swan Sonnenschein & Co., 1888), 25.

29. This point is emphasized in Guérard, *A Short History of the International Language Movement*, 105.

30. Couturat and Leau, *Histoire de la Langue Universelle*, 142; Guérard, *A Short History of the International Language Movement*, 97; Peter G. Forster, *The Esperanto Movement* (The Hague: Mouton, 1982), 46–47.

31. Large, *The Artificial Language Movement*, 69–70.

32. Guérard, *A Short History of the International Language Movement*, 103.

33. Quoted in Monnerot-Dumaine, *Précis d'interlinguistique générale et spéciale*, 88–90.

34. Ludwig Zamponi, *Zur Frage der Einführung einer internationalen Verkehrssprache* (Graz: Leykam, 1904); Arie De Jong, *Wörterbuch der Weltsprache: Vödabuk Volapüka pro Deutänapükans* (Leiden: E. J. Brill, 1931); and idem, *Gramat Volapüka: Dabükot Balid Pelautöl Nämätü e Zepü Kadäm Volapüka* (Leiden: E. J. Brill, 1931). For Wikipedia, see: http://vo.wikipedia.org/wiki/Cifapad. Ironically, most Volapükology and preservation of fragile and rare materials in the language takes place today through the medium of Esperanto. See Bernard Golden, "Conservation of the Heritage of Volapük," in Tonkin, ed., *Esperanto, Interlinguistics, and Planned Language* (1997): 183–189.

35. "Stated Meeting, January 6, 1888," 10, 12, 16. This report had a large impact on L. L. Zamenhof himself. See Edmond Privat, *Historio de la Lingvo Esperanto*, vol. 1: *Deveno kaj Komenco, 1887–1900* (Leipzig: Ferdinand Hirt & Sohn, 1923), 38.

36. Lawrence A. Sharpe, "Language Projects," *South Atlantic Bulletin* 27 (1961): 1–6, on 4.

37. For his own account, see L. L. Zamenhof, *The Birth of Esperanto: Extract of a Private Letter of Dr. L. L. Zamenhof to N. Borovko*, tr. Henry W. Hetzel (Fort Lee, NJ: Esperanto Association of North America, [1931]). For biographies, see René Centassi and Henri Masson, *L'homme qui a défié Babel: Ludwik Lejzer Zamenhof* (Paris: Éditions Ramsay, 1995); Marjorie Boulton, *Zamenhof: Creator of Esperanto* (London: Routledge and Kegan Paul, 1960); and Edmond Privat, *The Life of Zamenhof*, tr. Ralph Eliott (Oakville, ON: Esperanto Press, 1963 [1920]).

38. Privat, *Life of Zamenhof*, 48–49.

39. Privat, *Historio de la Lingvo Esperanto*, I: 43; Forster, *The Esperanto Movement*, 57.

40. Centassi and Masson, *L'homme qui a défié Babel*, 219; E. Drezen, *Historio de la Mondolinguo: Tra Jarcentoj da Serĉado*, 2d ed., tr. N. Hohlov and N. Nekrasov (Leipzig: Ekrelo, 1931), 181. Exact numbers are quite difficult to calculate due to the unsystematic quality of statistics collection by Esperanto organizations, as discussed in Forster, *The Esperanto Movement*, 18.

41. Boulton, *Zamenhof*, 62. The same kinds of criticisms of de Beaufront can be found in Centassi and Masson, *L'homme qui a défié Babel*, 211; and Privat, *Life of Zamenhof*, 85.

42. W. J. Clark, *International Language: Past, Present & Future* (London: J. M. Dent, 1907), 109. Even the great Esperantist Edmond Privat, who despised the man, gave de Beaufront credit for his propagation of Esperanto in France: Privat, *Historio de la Lingvo Esperanto*, I: 63.

43. Privat, *Historio de la Lingvo Esperanto*, I: 59n1.

44. Boulton, *Zamenhof*, 60; Privat, *Historio de la Lingvo Esperanto*, I: 59.

45. Forster, *The Esperanto Movement*, 75–76.

46. Zamenhof's views on these matters can be found in all the biographies and also Lazare Louis Zamenhof, *Le Hillélisme: Projet de solution de la question juive*, tr. Pierre Janton (Clermont-Ferrand: Association des publications de la Faculté des Lettres et Sciences Humaines, 1995). On neutrality, whereby "toutes les questions religieuses, politiques et sociales seraient rigoureusement exclues des séances publiques du Congrès," see L. Couturat and L. Leau, *Les Nouvelles Langues Internationales: Suite à L'histoire de la Langue Universelle* (Paris: M. L. Couturat, [1907]), 40.

47. Quoted in Forster, *The Esperanto Movement*, 90.

48. Dr. Esperanto [L. L. Zamenhof], *Mezhdunarodnyi iazyk: Predislovie i polnyi uchebnik* (Warsaw: Kh. Kel'ter, 1887), 28.

49. Zamenhof, *Mezhdunarodnyi iazyk*, 29.

50. Forster, *The Esperanto Movement*, 62.

51. Lorenz, "The Relationship of the International Language to Science," 53.

52. R. Mehmke, "Nüns Gletavik (Fovot 2[id])," *Nunel Valemik*, no. 2 (1889). Lorenz cited translations of Miess's *Craniology* and Winkler's *Petrification of Fishes*, but I have not been able to track these down. See Lorenz, "The Relationship of the International Language to Science," 54.

53. Zamenhof, *Mezhdunarodnyi iazyk*, 4–5.

54. Editors of *Internacia Scienca Revuo*, "Nia celo," *Internacia Scienca Revuo* 1, no. 1 (January 1904): 1.

55. Iv. Chetverikov to Mendeleev, 30 November 1904, ADIM II-V-24-Ch. The translation appeared as D. Mendelejev, "Provo de kemia kompreno de l'monda etero," tr. Iĉet-Verikov, *Internacia Scienca Revuo* 1, no. 6 (June 1904): 161–167; no. 7 (July 1904): 202–208; no. 8 (August 1904): 225–231. For more on Mendeleev's ether project, see Michael D. Gordin, *A Well-Ordered Thing: Dmitrii Mendeleev and the Shadow of the Periodic Table* (New York: Basic Books, 2004), chapter 8.

56. Paul Fruictier, "Unu jaro," *Internacia Scienca Revuo* 2, no. 13 (January 1905): 1–4, on 1.

57. Karl F. Kellerman, "The Advance of International Language," *Science* N.S. 30, no. 780 (10 December 1909): 843–844.

58. Clark, *International Language*, 111.

59. R. van Melckebeke and Th. Renard, "Projekto de kemia nomaro esperanta," *Internacia Scienca Revuo* 1, no. 1 (January 1904): 22–25, on 22.

60. Melckebeke and Renard, "Projekto de kemia nomaro esperanta," 23.

61. In "Korespondado," *Internacia Scienca Revuo* 1, no. 3 (March 1904): 92–95, on 92.

62. In "Korespondado," (March 1904), 93.

63. In "Korespondado," *Internacia Scienca Revuo* 1, no. 4 (April 1904): 123–128, on 123.

64. In "Korespondado," (April 1904), 125.

65. In "Korespondado," *Internacia Scienca Revuo* 1, no. 5 (May 1904): 156–158, on 156.

66. Amiko, "Terminaro de l'neorganika kemio sub vidpunkto de esperantisto," *Internacia Scienca Revuo* 1, no. 4 (April 1904): 120–122.

67. Quoted in Richard Lorenz, "The 'Délégation pour l'adoption d'une langue auxiliare internationale," in L. Couturat et al., *International Language and Science* (1910): 11–26, on 13.

68. See letter #1 reproduced in Karl Hansel and Fritz Wollenberg, eds., *Aus dem Briefwechsel Wilhelm Ostwalds zur Einführung einer Weltsprache, Mitteilungen der Wilhelm-Ostwald-Gesellschaft zu Großbothen e.V., Sonderheft* 6 (1999), 25.

69. Couturat and Leau, *Histoire de la Langue Universelle*, xii.

70. Clark, *International Language*, 30.

Chapter Five

1. Otto Jespersen, "Finala Diskurso," *Progreso* 7, no. 1 (73) (15 January 1914): 1–6, on 4. This was Jespersen's final lecture of a small series on the problem of a world lan-

guage at the University of Copenhagen in 1913. He delivered it entirely in Ido, without preparing the students beforehand.

2. Wilhelm Ostwald, *Lebenslinien: Eine Selbstbiographie*, 3 vol. (Berlin: Klasing & Co., 1927), III: 146–147. Chapter 5 concerns "Die Weltsprache."

3. Ostwald, *Lebenslinien*, III:141.

4. Albert Léon Guérard, *A Short History of the International Language Movement* (London: T. Fisher Unwin, 1922), 177. For detailed information on Ostwald's involvement with constructed languages, see Günter Anton, "Die Tätigkeit Professor Wilhelm Ostwalds für die internationale Sprache IDO," *Mitteilungen der Wilhelm-Ostwald-Gesellschaft zu Großbothen e.V.* 8, no. 4 (2003): 16–26.

5. Wilhelm Ostwald, "Pri la problemo de la helpa lingvo. II. La mondlingvo," tr. Dr. Helte, *Internacia Scienca Revuo* 1, no. 10 (October 1904): 289–295.

6. Ostwald, *Lebenslinien*, III:151.

7. Reproduced in Hans-Günther Körber, ed., *Aus dem wissenschaftlichen Briefwechsel Wilhelm Ostwalds, II. Teil: Briefwechsel mit Svante Arrhenius und Jacobus Hendricus Van't Hoff* (Berlin: Akademie Verlag, 1969), 196.

8. Letter of 24 March 1907, reproduced in Karl Hansel and Fritz Wollenberg, eds., *Aus dem Briefwechsel Wilhelm Ostwalds zur Einführung einer Weltsprache, Mitteilungen der Wilhelm-Ostwald-Gesellschaft zu Großbothen e.V., Sonderheft* 6 (1999), 42.

9. From the William James Papers, Houghton Library, Harvard University, as quoted in Niles R. Holt, "Wilhelm Ostwald's 'The Bridge,'" *British Journal for the History of Science* 10 (1977): 146–150, on 149n12.

10. Reproduced in Ostwald, *Lebenslinien*, III:143–144.

11. Wilhelm Ostwald, *Sprache und Verkehr* (Leipzig: Akademische Verlagsgesellschaft, 1911), 8.

12. Ostwald paraphrased in Edwin E. Slosson, *Major Prophets of To-Day* (Boston: Little, Brown, and Company, 1916), 222.

13. Wilhelm Ostwald, "Die Weltsprache," reproduced in Hansel and Wollenberg, *Aus dem Briefwechsel Wilhelm Ostwalds*: 4–14.

14. For surveys of the events chronicled in this chapter, see for example Guérard, *A Short History of the International Language Movement*, chapter 7; E. Drezen, *Historio de la Mondolinguo: Tra Jarcentoj da Serĉado*, 2d ed., tr. N. Hohlov and N. Nekrasov (Leipzig: Ekrelo, 1931), chapter 2 (from the Esperantist point of view); and Ward Nichols, "The Decision of the Delegation/La Decido Di La Delegitaro," *Internationalist* 2, no. 2 (6) (April-May 1910): 18–19 (from the Idist side). A comprehensive bibliography of Ido publications can be found in Tazio Carlevaro and Reinhard Hauptenthal, *Bibliografio di Ido* (Bellinzona: Hans Dubois, 1999).

15. Guérard, *A Short History of the International Language Movement*, 146–147.

16. Ostwald, *Lebenslinien*, III:165.

17. Peter G. Forster, *The Esperanto Movement* (The Hague: Mouton, 1982), 121–122.

18. Otto Jespersen, *A Linguist's Life*, eds. Arne Juul, Hans F. Nielsen, and Jørgen Erik Nielsen, tr. David Stoner (Odense: Odense University Press, 1995 [1938]), 149. This is the translation of his Danish autobiography: *En Sprogmands Levned* (Copenhagen: Nordisk Forlag, 1938).

19. Jespersen, *A Linguist's Life*, 150.

20. Guérard, *A Short History of the International Language Movement*, 136–140.

21. Otto Jespersen, "The Linguistic Principles Necessary for the Construction of

an International Auxiliary Language, with Appendix: Criticism of Esperanto," in L. Couturat, O. Jespersen, R. Lorenz, W. Ostwald, and L. Pfaundler, *International Language and Science: Considerations on the Introduction of an International Language into Science*, tr. F. G. Donnan (London: Constable & Company, 1910): 27–41, on 30.

22. W. J. Clark, *International Language: Past, Present & Future* (London: J. M. Dent, 1907), 99.

23. Reproduced in Hansel and Wollenberg, *Aus dem Briefwechsel Wilhelm Ostwalds*, 34–35.

24. Ostwald, *Lebenslinien*, III:155.

25. Ostwald to Carlo Bourlet, 16 June 1907, reproduced in Hansel and Wollenberg, *Aus dem Briefwechsel Wilhelm Ostwalds*, 47.

26. Ostwald, *Lebenslinien*, III:167; Jespersen, *A Linguist's Life*, 150.

27. Jespersen, *A Linguist's Life*, 150–151. On "middle course," see Otto Jespersen, *An International Language* (London: George Allen & Uwin, 1928), 42.

28. Quoted in Louis Couturat, "Le choix d'une langue internationale," *Revue du Mois* 7 (January-June 1909): 708–724, on 709.

29. Friedrich Schneeberger, "Pri la nomo di nia linguo," *Progresso* 2, no. 4 (16) (June 1909): 229.

30. Couturat, "Averto," *Progresso* 2, no. 4 (16) (June 1909): 2. Emphasis in original.

31. Max Talmey, "The Auxiliary Language Question," *Modern Language Journal* 23 (1938): 172–186, on 177; Jespersen, *An International Language*, 43.

32. Otto Jespersen, "International Language," *Science* N.S. 30, no. 776 (12 November 1909): 677.

33. That program also had plurals in *i*, invariable adjectives, and no accusative, but in addition eliminated the definite article and had verbal endings of *en*, *in*, and *on*. M. Monnerot-Dumain, *Précis d'interlinguistique générale et spéciale* (Paris: Librairie Maloine, 1960), 103.

34. Couturat, "Le choix d'une langue internationale," 720.

35. Louis Couturat, "Des rapports de la logique et de la linguistique dans le problème de la langue internationale," *Revue de Métaphysique et de Morale* 19 (1911): 509–516, on 512. Emphasis in original. Other Idists believed that credit had been "unjustly ascribed to Ostwald" (Max Talmey, "Word Derivation in a Logical Language," *Modern Language Journal* 24 [1940]: 617–628, on 620), and indeed one can find references to this notion in many nineteenth-century projects.

36. Wilhelm Ostwald, "Chemische Weltliteratur," *Zeitschrift für physikalische Chemie* 76 (January 1911): 1–20, on 5. Emphasis in original.

37. Otto Jespersen, "International Language," *Science* N.S. 31, no. 786 (21 January 1910): 109–112, on 109–110.

38. Esperanto was vulnerable to the same criticism. See Louvan E. Nolting, "The Deficiency of Esperanto as a World Language," *Federal Linguist* 5, no. 1–2 (1973): 18–22.

39. Jespersen, "The Linguistic Principles Necessary for the Construction," 28, 31. Emphasis in original.

40. Louis Couturat, "Pour la langue auxiliaire neutre," *Revue internationale de l'enseignement* 58 (1909): 255–259, on 256; Luther H. Dyer, *The Problem of an International Auxiliary Language and Its Solution in Ido* (London: Putnam, 1923), 115–116; Jespersen, "The Linguistic Principles Necessary for the Construction," 32.

41. Couturat, "D'une application de la logique au problème de la langue internationale," *Revue de Métaphysique et de Morale* 16 (1908): 761–769, on 768.

42. Couturat, "D'une application de la logique au problème de la langue internationale," 764. Emphasis in original.

43. Clark, *International Language*, 3n.

44. Reproduced in Hansel and Wollenberg, *Aus dem Briefwechsel Wilhelm Ostwalds*, 55.

45. Reproduced in Hansel and Wollenberg, *Aus dem Briefwechsel Wilhelm Ostwalds*, 57. Emphasis in original.

46. Guérard, *A Short History of the International Language Movement*, 149.

47. Louis Couturat and Léopold Leau, *Histoire de la Langue Universelle* (Paris: Librairie Hachette, 1903), 152.

48. Ostwald, *Sprache und Verkehr*, 23. See also Jespersen, *An International Language*, 36; Richard Lorenz, "The 'Delégation pour l'adoption d'une langue auxiliare internationale,'" in L. Couturat et al., *International Language and Science* (1910): 11–26, on 16; and P. Ahlberg, "A Few Statistics," *Internationalist* 2, no. 4 (8) (July 1910): 63–65, on 65.

49. Reproduced in Hansel and Wollenberg, *Aus dem Briefwechsel Wilhelm Ostwalds*, 55–56.

50. Reproduced in Hansel and Wollenberg, *Aus dem Briefwechsel Wilhelm Ostwalds*, 65. Emphasis in original.

51. Quoted in "Pri la alvoko 'Al la Delegitaro, al la Esperantistaro,'" *Internacia Scienca Revuo* 5, no. 49 (January 1908): 15–18, on 17. Emphasis in original. This letter was written in German and translated by the editors into Esperanto.

52. Ernest Naville et al., "Al la Delegitaro, al la Esperantistaro," *Internacia Scienca Revuo* 4, no. 48 (December 1907): 389–393, on 392.

53. Naville et al., "Al la Esperantistaro," *Internacia Scienca Revuo* 5, no. 49 (January 1908): 3–6, on 5.

54. See the very detailed letter by Baudoin de Courtenay, who had been a member of the Delegation committee, reproduced in Edmond Privat, *Historio de la Lingvo Esperanto*, vol. 2: *La Movado, 1900–1927* (Leipzig: Ferdinand Hirt & Sohn, 1927), 62–66.

55. Quoted in René Centassi and Henri Masson, *L'homme qui a défié Babel: Ludwik Lejzer Zamenhof* (Paris: Éditions Ramsay, 1995), 249.

56. Walter B. Sterrett, "To Esperantists and Idists; Importance of Mutual Good Will," *Internationalist* 1, no. 2 (July 1909): 5–6, on 5; Walther Borgius, *Warum ich Esperanto verließ: Eine Studie über die gegenwärtige Krisis und die Zukunft der Weltsprachen-Bewegung* (Berlin: Liebheit & Thiesen, 1908).

57. David K. Jordan, "Esperanto and Esperantism: Symbols and Motivations in a Movement for Linguistic Equality," in Humphrey Tonkin, ed., *Esperanto, Interlinguistics, and Planned Language* (Lanham, MD: University Press of America, 1997), 39–65, on 43; Forster, *The Esperanto Movement*, 135–136.

58. Reproduced in Hansel and Wollenberg, *Aus dem Briefwechsel Wilhelm Ostwalds*, 66.

59. Reproduced in Marjorie Boulton, *Zamenhof: Creator of Esperanto* (London: Routledge and Kegan Paul, 1960), 126.

60. Jespersen, *A Linguist's Life*, 153.

61. Louis de Beaufront, "Déclaration de Ido/Deklaro de Ido," *L'Esperantiste* 11 (May 1908): 97–100, on 97 and 99.

62. See, for example, Edmond Privat, *The Life of Zamenhof*, tr. Ralph Eliott (Oakville, ON: Esperanto Press, 1963 [1920]), 82–83.

63. Jespersen, *A Linguist's Life*, 149. The same phrasing is used in Jespersen, *An International Language*, 42.

64. Drezen, *Historio de la Mondolinguo*, 185; Privat, *Historio de la Lingvo Esperanto*, II: 58.

65. Boulton, *Zamenhof*, 131.

66. Forster, *The Esperanto Movement*, 130 (quotation); Monnerot-Dumaine, *Précis d'interlinguistique générale et spéciale*, 41.

67. Ostwald, *Sprache und Verkehr*, 29. On the symbolism provided by de Beaufront's authorship, see Privat, *Historio de la Lingvo Esperanto*, II: 57.

68. G. Aymonier and L. Couturat, "Ido et Esperanto," *Revue du Mois* 9 (January-June 1910): 219–229; Louis Couturat, "Entre l'Ido et l'Esperanto," *Revue mondiale* (April 1912): 381–392; idem, "Ido ou français," *La grande revue* (25 February 1910): 791–793; idem, "Entre Idistes et Espérantistes," *La Revue* 78 (1909): 110–113.

69. Louis Couturat, "'Wait until Zamenhof is Dead'/'Expektez la Morto di Zamenhof!,'" *Internationalist* 2, no. 5–6 (August-September 1910): 77–81, on 80. Emphasis in original. See also idem, "Makiavelatra Taktiko," *Progreso* 2, no. 8 (20) (October 1909): 449–452, on 450.

70. Couturat, "L'Ido devant la science: Lettre ouverte à M. A. Cotton, Professeur à la Sorbonne," *La Langue Auxiliaire* 3 (1910): 21–27, on 22.

71. Couturat, "Ido contre Esperanto," *La coopération des idées* (1912): 444–449, on 446.

72. Letter of 4 November 1907, reproduced in Hansel and Wollenberg, *Aus dem Briefwechsel Wilhelm Ostwalds*, 57.

73. Jordan, "Esperanto and Esperantism," 43.

74. Drezen, *Historio de la Mondolinguo*, 186.

75. Couturat, "Pri Nia Metodo," *Progreso* 2, no. 10 (22) (December 1909): 579–582, on 580–581.

76. Couturat, "Le choix d'une langue internationale," 722.

77. Lorenz, "The 'Délégation pour l'adoption d'une langue auxiliare internationale,'" 20.

78. Jespersen, "The Linguistic Principles Necessary for the Construction," 37. In 1924, in a general linguistic text, he would repeatedly invoke Ido as an example of the rational development of language: Otto Jespersen, *The Philosophy of Grammar* (Chicago: University of Chicago Press, 1992 [1924]), 41, 60, 136, 208n1, 232, 321n1. For Couturat, see for example his "On the Application of Logic to the Problem of an International Language," in L. Couturat et al., *International Language and Science* (1910): 41–52.

79. Couturat, "Le choix d'une langue internationale," 723.

80. Leopold Pfaundler, "The Need for a Common Scientific Language," in L. Couturat et al., *International Language and Science* (1910): 1–10, on 6.

81. Couturat, "Entre Idistes et Espérantistes," 112.

82. Jespersen, "International Language [1910]," 112.

83. Nichols, "Hear the Other Side," *Internationalist*, no. 3–4 (February 1910): 8–9,

on 9. For Couturat's interest in Lamarck, see his "La Stabileso di la Vivo," *Progreso* 5, no 3 (51) (May 1912): 140–141. This issue of *Progreso* contains several articles on Gregor Mendel, on Lamarck, and on heredity in general—all translated into Ido by Couturat.

84. Couturat, "Autour d'une Langue internationale," *La Revue* 87 (1910): 381–385, on 383–384. Emphasis in original.

85. See E. H. MacPike, "La Praktikal Utileso di Ido/The Practical Usefulness of Ido," *Internationalist*, no. 3–4 (February 1910): 11–12; C. S. Pearson, "Kiropraktiko," *Internationalist* 2, no. 2 (6) (April-May 1910): 28; and Louis Couturat, *Internaciona matematikal lexiko en Ido, Germana, Angla, Franca, e Italiana* (Jena: Gustav Fischer, 1910).

86. Reproduced in Körber, ed., *Aus dem wissenschaftlichen Briefwechsel Wilhelm Ostwalds*, 320.

87. Reproduced in Hansel and Wollenberg, *Aus dem Briefwechsel Wilhelm Ostwalds*, 113.

88. Quoted in "Prof. William Ostwald and International Language," *Internationalist*, no. 3–4 (February 1910): 12–13, on 13. On the donation, see also Grete Ostwald, *Wilhelm Ostwald: Mein Vater* (Stuttgart: Berliner Union, 1953), 110; Slosson, *Major Prophets of To-Day*, 222n1; and Eugen Wüster, *Internationale Sprachnormung in der Technik, besonders in der Elektrotechnik (Die nationale Sprachnormung und ihre Verallgemeinerung)*, 2d ed. (Bonn: H. Bouvier u. Co. Verlag, 1966), 335.

89. Slosson, *Major Prophets of To-Day*, 223.

90. Wilhelm Ostwald, "The Question of Nomenclature," in L. Couturat et al., *International Language and Science* (1910): 61–68, on 61, 67.

91. Letter of 4 January 1911, reproduced in Körber, ed., *Aus dem wissenschaftlichen Briefwechsel Wilhelm Ostwalds*, 322.

92. Reproduced in Körber, ed., *Aus dem wissenschaftlichen Briefwechsel Wilhelm Ostwalds*, 323.

93. See the letters to Ostwald of 24 and 31 December 1910, reproduced in Hansel and Wollenberg, *Aus dem Briefwechsel Wilhelm Ostwalds*, 124.

94. Reproduced in Körber, ed., *Aus dem wissenschaftlichen Briefwechsel Wilhelm Ostwalds*, 323–325, quotation on 325.

95. See the letter of 25 January 1901, reproduced in Hansel and Wollenberg, *Aus dem Briefwechsel Wilhelm Ostwalds*, 129.

96. Ostwald, *Lebenslinien*, III:176.

97. Ostwald, "Chemische Weltliteratur," 1n1.

98. Ostwald, "Chemische Weltliteratur," 2–3.

99. Ostwald, "Chemische Weltliteratur," 7.

100. Ostwald, "Memorial on the Foundation of an International Chemical Institute," *Science* N.S. 40, no. 1022 (31 July 1914): 147–158, on 155.

101. Alexander Batek, "Pri la ĥemia nomigado," *Internacia Scienca Revuo* 6, nos. 68–69 (August-September 1909): 264–266, on 265. For Couturat's accusations of plagiarism, see *Progreso* 5, no. 3 (51) (May 1912): 162.

102. Maurice Rollet de l'Isle, "Konsilaro por la farado de la sciencaj kaj teknikaj vortoj," *Internacia Sciencia Revuo* 7, no. 83 (November 1910): 279–295; no. 84 (December 1910): 311–334; 8, no. 85 (January 1911): 1–21; no. 86 (February 1911): 33–49.

103. Ostwald, "Memorial on the Foundation of an International Chemical Institute," 147, 155. On the Bridge, see Holt, "Wilhelm Ostwald's 'The Bridge,'" although at points this piece is unreliable. For example, on p. 146 Holt claims that the universal language promoted by Ostwald was Esperanto.

104. Otto Jespersen, "Grava Propozo," *Internationalist* 2, no. 3 (7) (June 1910): 45–46, on 45 (Jespersen) and 46 (Couturat). Couturat had earlier been a fierce defender of eternal experimentation: Couturat, "Entre l'Ido et l'Esperanto," 390–391.

105. F. Schneeberger and L. Couturat, "Pri la periodo di stabileso," *Progreso* 5, no. 3 (51) (May 1912): 191. On Solothurn, see *Progreso* 7, no. 4 (76) (April 1914): 197. In his history of constructed languages, Esperantist Ernest Drezen was convinced that the period of stability had damaged Ido by restricting innovation and dampening advocates' enthusiasm: Drezen, *Historio de la Mondolinguo*, 187.

106. Guérard, *A Short History of the International Language Movement*, 122–123.

107. Wüster, *Internationale Sprachnormung in der Technik*, 335; Drezen, *Historio de la Mondolinguo*, 198–199.

Chapter Six

1. Franz Thierfelder, *Die deutsche Sprache im Ausland*, 2 v. (Hamburg: R. v. Decker, 1956–1957), I:18.

2. Quoted in Günter Anton, "Die Tätigkeit Professor Wilhelm Ostwalds für die internationale Sprache IDO," *Mitteilungen der Wilhelm-Ostwald-Gesellschaft zu Großbothen e.V.* 8, no. 4 (2003): 16–26, on 22. Ellipses in original.

3. Wilhelm Ostwald, "Weltdeutsch," *Monistische Sonntagspredigten*, no. 36 (31 October 1915): 545–559, on 553.

4. Ostwald, "Weltdeutsch," 555–556.

5. Ostwald, "Weltdeutsch," 557. J. A. Large attributes the program of Weltdeutsch to a Professor Baumann in Munich, also in 1915. Large, *The Foreign-Language Barrier: Problems in Scientific Communication* (London: André Deutsch, 1983), 148. I have not been able to demonstrate a connection between Baumann and Ostwald.

6. Leopold Pfaundler to Wilhelm Ostwald, 30 December 1915, reproduced in Karl Hansel and Fritz Wollenberg, eds., *Aus dem Briefwechsel Wilhelm Ostwalds zur Einführung einer Weltsprache, Mitteilungen der Wilhelm-Ostwald-Gesellschaft zu Großbothen e.V., Sonderheft* 6 (1999), 147.

7. Ostwald to Pfaundler, 12 January 1916, reproduced in Hansel and Wollenberg, *Aus dem Briefwechsel Wilhelm Ostwalds*, 148.

8. Otto Jespersen to Franz Boas, 4 December 1914, Franz Boas Papers, Mss.B.B61, APSL, Folder: "Jespersen, Otto."

9. A. Meillet, *Les langues dans l'Europe nouvelle* (Paris: Payot & Cie, 1918), 292.

10. Félix Henneguy, "Du rôle de l'Allemagne dans l'évolution des sciences biologiques," *Revue scientifique* 53 (27 February-6 March 1915): 70–74, on 71.

11. W. F. Twaddell, "Standard German," *Anthropological Linguistics* 1, no. 3 (1959): 1–7, on 1; Ulrich Ammon, *Die internationale Stellung der deutschen Sprache* (Berlin: Walter de Gruyter, 1991), 27.

12. W. B. Lockwood, *An Informal History of the German Language* (Cambridge: W. Heffer and Sons, 1965), 116.

13. Ruth H. Sanders, *German: Biography of a Language* (New York: Oxford University Press, 2010), 98; R. E. Keller, *The German Language* ([Atlantic Highlands], NJ: Humanities Press, 1978), 237.

14. Twaddell, "Standard German," 1, 3; Lockwood, *An Informal History of the German Language*, 109; Keller, *The German Language*, 338–339.

15. Jeanne Pfeiffer, "La création d'une langue mathématique allemande par Albrecht Dürer. Les raisons de sa non réception," in Roger Chartier and Pietro Corsi, eds., *Sciences et langues en Europe* (Paris: European Communities, 2000 [1994]): 77–90.

16. Quoted in Adolf Bach, *Geschichte der deutschen Sprache*, 8th ed. (Heidelberg: Quelle & Meyer, 1965 [1938]), 331.

17. Ulrich Ricken, "Zum Thema Christian Wolff und die Wissenschaftssprache der deutschen Aufklärung," in Heinz L. Kretzenbacher and Harald Weinrich, eds., *Linguistik der Wissenschaftssprache* (Berlin: Walter de Gruyter, 1995): 41–90; Eric A. Blackall, *The Emergence of German as a Literary Language, 1700–1775*, 2d ed. (Ithaca: Cornell University Press, 1978 [1959]).

18. Lockwood, *An Informal History of the German Language*, 129–130.

19. Such figures are standard in any history of German. See, for example: Bach, *Geschichte der deutschen Sprache*, 309; W. Walker Chambers and John R. Wilkie, *A Short History of the German Language* (London: Methuen & Co, 1970), 46; Keller, *The German Language*, 360, 485.

20. Denise Phillips, *Acolytes of Nature: Defining Natural Science in Germany, 1770–1850* (Chicago: University of Chicago Press, 2012), 75, 109–111.

21. Quoted in Keller, *The German Language*, 487.

22. Keller, *The German Language*, 487; see also Richard Games Brunt, *The Influence of the French Language on the German Vocabulary (1649–1735)* (Berlin: Walter de Gruyter, 1983).

23. There are many excellent studies that track these developments. For the social transformation, see Peter Borscheid, *Naturwissenschaft, Staat und Industrie in Baden (1848–1914)* (Stuttgart: Ernst Klett, 1976); for the intellectual developments, see Alan J. Rocke, *The Quiet Revolution: Hermann Kolbe and the Science of Organic Chemistry* (Berkeley: University of California Press, 1993).

24. Owen Hannaway, "The German Model of Chemical Education in America: Ira Remsen at Johns Hopkins (1876–1913)," *Ambix* 23 (1976): 145–164. While Remsen was studying chemistry at Tübingen, he ran courses for English-speaking students to get them up to snuff on the language so they could sustain their oral examinations.

25. Jeffrey Allan Johnson, *The Kaiser's Chemists: Science and Modernization in Imperial Germany* (Chapel Hill: University of North Carolina Press, 1990), 29–30.

26. William Coleman, ed., *French Views of German Science* (New York: Arno Press, 1981); Robert Fox, "The View over the Rhine: Perceptions of German Science and Technology in France, 1860–1914," in Yves Cohen and Klaus Manfrass, eds., *Frankreich und Deutschland: Forschung, Technologie und industrielle Entwicklung im 19. und 20. Jahrhundert* (Munich: C. H. Beck, 1990): 14–24; Maurice Crosland, *Science under Control: The French Academy of Sciences, 1795–1914* (Cambridge: Cambridge University Press, 1992); and Harry W. Paul, *The Sorcerer's Apprentice: The French Scientist's Image of German Science, 1840–1919* (Gainesville: University of Florida Press, 1972).

27. Henneguy, "Du rôle de l'Allemagne dans l'évolution des sciences biologiques," 71.

28. Adolphe Würtz, *Dictionnaire de chimie pure et appliquée* (Paris: Hachette, 1869), I: i.

29. Jakob Volhard, "Die Begrundung der Chemie durch Lavoisier," *Journal für praktische Chemie* 110, N.F. 2 (1870): 1–47. On the tradition of anti-Lavoisier polemics, see Hans-Georg Schneider, "The 'Fatherland of Chemistry': Early Nationalistic Currents in Late Eighteenth Century German Chemistry," *Ambix* 36 (1989): 14–21.

30. Hermann Kolbe, "Ueber den Zustand der Chemie in Frankreich," *Journal für praktische Chemie* 110, N.F. 2 (1870): 173–183, on 177.

31. Rudolf Fittig to Emil Erlenmeyer, 5 March 1872, HBA.

32. N. Zinin, A. Butlerow, D. Mendelejew, and A. Engelhardt, "[Letter to the Editor]," *St.-Petersburger Zeitung* 9, no. 271 (21 October 1870): 4. The Russian Chemical Society seconded these sentiments: Minutes of meeting of 8 October 1870, *ZhRF-KhO* 2 (1870): 253.

33. Wurtz to Butlerov, 1 January 1874, reproduced in G. V. Bykov and J. Jacques, "Deux pionniers de la chimie moderne, Adolphe Wurtz et Alexandre M. Boutlerov, d'après une correspondance inédite," *Revue d'Historie des Sciences* 13 (1960): 115–134, on 127.

34. Jakob Volhard, "Berichtigung," *Journal für praktische Chemie* 110, N.F. 2 (1870): 381–384, on 384.

35. Liebig to Kolbe, 19 November 1870, reproduced in Alan J. Rocke and Emil Heuser, eds., *Justus von Liebig und Hermann Kolbe in ihren Briefen, 1846–1873* (Mannheim: Bionomica, 1994), 119.

36. Rocke, *The Quiet Revolution*, 346; idem, *Nationalizing Science: Adolphe Wurtz and the Battle for French Chemistry* (Cambridge, MA: MIT Press, 2001), 333–334.

37. Helmholtz to his wife, 14 September 1853, in Richard L. Kremer, ed., *Letters of Hermann von Helmholtz to His Wife, 1847–1859* (Stuttgart: Franz Steiner Verlag, 1990), 134. See also 105 and 120. I would like to thank Rich Kremer for bringing these letters to my attention.

38. Helmholtz to his wife, 16 August 1851, in Kremer, *Letters of Hermann von Helmholtz to His Wife*, 64.

39. Rocke, *The Quiet Revolution*, 69 and 341.

40. Bach, *Geschichte der deutschen Sprache*, 409.

41. Rainald Von Gizycki, "Centre and Periphery in the International Scientific Community: Germany, France and Great Britain in the 19th Century," *Minerva* 11 (1973): 474–494, on 477.

42. Meillet, *Les langues dans l'Europe nouvelle*, 295.

43. Brigitte Schröder-Gudehus, "Deutsche Wissenschaft und internationale Zusammenarbeit, 1914–1928: Ein Beitrag zum Studium kultureller Beziehungen in politischen Krisenzeiten" (PhD dissertation, University of Geneva, 1966), 42–43. See also idem, ed., "Les Congrès Scientifiques Internationaux," *Relations internationales* 62 (1990): 111–211.

44. The manifesto is reproduced in Jürgen von Ungern-Sternberg and Wolfgang von Ungern-Sternberg, *Der Aufruf "An die Kulturwelt!": Das Manifest der 93 und die Anfänge der Kriegspropaganda im Ersten Weltkrieg* (Stuttgart: Franz Steiner Verlag, 1996), 144–157, quotation on 144.

45. For Max Planck's later defense (in a letter to Dutch physicist Hendrik A. Lorentz), see Armin Hermann, ed., *Max Planck in Selbstzeugnissen und Bilddokumenten* (Hamburg: Rohwolt, 1973), 54.

46. For a survey of German chemical warfare in World War I, and especially Haber's role, see L. F. Haber, *The Poisonous Cloud: Chemical Warfare in the First World War* (New York: Oxford University Press, 1986).

47. Roy MacLeod, "Der wissenschaftliche Internationalismus in der Krise: Die Akademien der Alliierten und ihre Reaktion auf den Ersten Weltkrieg," tr. Peter Jaschner, in *Die Preussische Akademie der Wissenschaften zu Berlin, 1914–1945* (Berlin: Wolfram Fischer, 2000): 317–349.

48. Henneguy, "Du rôle de l'Allemagne dans l'évolution des sciences biologiques," 73.

49. Carolyn N. Biltoft, "Speaking the Peace: Language, World Politics and the League of Nations, 1918–1935" (PhD dissertation, Princeton University, 2010), 28–29; Ammon, *Die internationale Stellung der deutschen Sprache*, 289.

50. Brigitte Schroeder-Gudehus, *Les scientifiques et la paix: La communauté scientifique internationale au cours des années 20* (Montreal: Presses de l'Université de Montréal, 1978), 131.

51. Schröder-Gudehus, "Deutsche Wissenschaft und internationale Zusammenarbeit"; idem, "Challenge to Transnational Loyalties: International Scientific Organizations after the First World War," *Science Studies* 3, no. 2 (April 1973): 93–118; Daniel J. Kevles, "'Into Hostile Political Camps': The Reorganization of International Science in World War I," *Isis* 62, no. 1 (Spring 1971): 47–60; Lawrence Badash, "British and American Views of the German Menace in World War I," *Notes and Records of the Royal Society of London* 34 (1979): 91–121. For an interesting Marxist account from East Germany, see Siegfried Grundmann, "Zum Boykott der deutschen Wissenschaft nach dem ersten Weltkrieg," *Wissenschaftliche Zeitschrift der Technischen Universität Dresden* 14, no. 3 (1965): 799–805.

52. A. G. Cock, "Chauvinism and Internationalism in Science: The International Research Council, 1919–1926," *Notes and Records of the Royal Society of London* 37, no. 2 (March 1983): 249–288, on 249.

53. Roswitha Reinbothe, *Deutsch als internationale Wissenschaftssprache und der Boykott nach dem Ersten Weltkrieg* (Frankfurt am Main: Peter Lang, 2006), 150–151.

54. Reinbothe, *Deutsch als internationale Wissenschaftssprache und der Boykott*, 204–205; Ulrich Ammon, "German as an International Language of the Sciences—Recent Past and Present," in Andreas Gardt and Bernd Hüppauf, eds., *Globalization and the Future of German* (Berlin: Mouton de Gruyter, 2004): 157–172, on 163.

55. See Arnold Sommerfeld to Albert Einstein, 4 July 1921, reproduced in Armin Hermann, *Albert Einstein/Arnold Sommerfeld Briefwechsel: Sechzig Briefe aus dem goldenen Zeitalter der modernen Physik* (Basel/Stuttgart: Schwabe & Co., 1968), 81.

56. Reinbothe, *Deutsch als internationale Wissenschaftssprache und der Boykott*, 288, 342; Schroeder-Gudehus, *Les scientifiques et la paix*, 244–245.

57. Ludwig Aschoff, "Der Geist von Locarno und die Wissenschaft," *Frankfurter Zeitung* (1 May 1926).

58. Schröder-Gudehus, "Deutsche Wissenschaft und internationale Zusammenarbeit," 191.

59. Schroeder-Gudehus, *Les scientifiques et la paix*, 123.

60. Robert Marc Friedman, *The Politics of Excellence: Behind the Nobel Prize in Science* (New York: W. H. Freeman, 2001), 77.

61. Friedman, *The Politics of Excellence*, 187.

62. Susan Gross Solomon, "Introduction: Germany, Russia, and Medical Cooperation between the Wars," in Solomon, ed., *Doing Medicine Together: Germany and Russia between the Wars* (Toronto: University of Toronto Press, 2006): 3–31, on 3–6; Elizabeth Hachten, "How to Win Friends and Influence People: Heinz Zeiss, Boundary Objects, and the Pursuit of Cross-National Scientific Collaboration in Microbiology," in ibid.: 159–198, on 166; Ljudmila Nosdrina, "Deutsch als internationale Wissenschaftssprache in Russland," in Ulrich Ammon and Dirk Kemper, eds., *Die deutsche Sprache in Russland: Geschichte, Gegenwart, Zukunftsperspektiven* (Munich: Iudicium, 2011): 106–111, on 108. Of course, Soviet scientists, especially mathematicians, also published in French. See Jean-Michel Kantor, "Mathematics East and West, Theory and Practice: The Example of Distributions," *Mathematical Intelligencer* 26, no. 1 (2004): 39–50, on 41.

63. Austin M. Patterson, "International Chemistry," *Science* N.S. 69, no. 1795 (24 May 1929): 531–536, on 532; Roger Fennell, *History of IUPAC, 1919–1987* (Oxford: Blackwell Science, 1994), 3–4. In 1912, Ido was put forward to be the language for this chemical association. The proposal was rejected (ibid., 8).

64. Edwin Bidwell Wilson, "Insidious Scientific Control," *Science* 48, no. 1246 (15 November 1918): 491–493, on 491.

65. Wilson, "Insidious Scientific Control," 492; "The Co-ordination of Scientific Publication," *Nature* 101, no. 2533 (16 May 1918): 213.

66. Paul Weindling, "The League of Nations and International Medical Communication in Europe between the First and Second World Wars," in Chartier and Corsi, *Sciences et langues en Europe* (2000): 201–211. On language politics at the League of Nations, see Biltoft, "Speaking the Peace."

67. Patterson, "International Chemistry," 535.

68. Reinbothe, *Deutsch als internationale Wissenschaftssprache und der Boykott*, 409.

69. Schroeder-Gudehus, "Challenge to Transnational Loyalties," 98.

70. Reinbothe, *Deutsch als internationale Wissenschaftssprache und der Boykott*, 415.

71. Heinz Kloss, *The American Bilingual Tradition* (Rowley, MA: Newbury House, 1977), 26–27.

72. Edwin H. Zeydel, "The Teaching of German in the United States from Colonial Times to the Present," *German Quarterly* 37 (September 1964): 315–392, on 333.

73. Zeydel, "The Teaching of German in the United States from Colonial Times to the Present," 355–356. On French education, often associated with the Northeast and with girls, as opposed to the masculine, German Midwest, see George B. Watts, "The Teaching of French in the United States: A History," *French Review* 37, no. 1, pt. 2 (October 1963): 11–165.

74. Kloss, *The American Bilingual Tradition*, 52, 60–61; William R. Parker, *The National Interest and Foreign Languages* (Washington, DC: Government Printing Office, 1954), 77; and Paul Finkelman, "The War on German Language and Culture, 1917–1925," in Hans-Jürgen Schröder, ed., *Confrontation and Cooperation: Germany and the United States in the Era of World War I, 1900–1924*, vol. 2 (Providence,

RI: Berg, 1993): 177–205, on 183, 188. Some of this sentiment was a holdover from late nineteenth-century resentment of parochial schools, especially Catholic and Lutheran ones. See ibid., 185–186.

75. United States Supreme Court, *Meyer v. State of Nebraska*, 262 U.S. 390 (1923), available at http://laws.findlaw.com/us/262/390.html. See also Finkelman, "The War on German Language and Culture," 191–193.

76. Holmes's dissent in United States Supreme Court, *Bartels v. State of Iowa*, 262 U.S. 404, 43 S. Ct. 628 (1923), *412.

77. These figures are drawn from Zeydel, "The Teaching of German in the United States from Colonial Times to the Present," 362, 368; Franz Thierfelder, *Deutsch als Weltsprache. 1. Band: Die Grundlagen der deutschen Sprachgeltung in Europa* (Berlin: Verlag für Volkstum, Wehr und Wirtschaft Hans Kurzeja, 1938), 39; and Parker, *The National Interest and Foreign Languages*, 56, 76–77. On the collapse of French, see Watts, "The Teaching of French in the United States," 42.

78. G. A. Miller, "Scientific Activity and the War," *Science* 48, no. 1231 (2 August 1918): 117–118, on 117.

79. Fritz Stern, *Five Germanys I Have Known* (New York: Farrar, Straus & Giroux, 2006), 62.

80. Georg Karo, *Der geistige Krieg gegen Deutschland*, 2d expanded ed. (Halle: Wilhelm Knapp, 1926), 10; Schroeder-Gudehus, *Les scientifiques et la paix*, 136–137.

81. Reinbothe, *Deutsch als internationale Wissenschaftssprache und der Boykott*, 391–395; Cock, "Chauvinism and Internationalism in Science," 267–268.

Chapter Seven

1. Dolf Sternberger, Gerhard Storz, and W. E. Süskind. *Aus dem Wörterbuch des Unmenschen* (Hamburg: Claassen, 1957), 9.

2. E. J. Crane, "Growth of Chemical Literature: Contributions of Certain Nations and the Effects of War," *Chemical & Engineering News* 22, no. 17 (10 September 1944): 1478–1481, 1496, on 1481; Ruth H. Sanders, *German: Biography of a Language* (New York: Oxford University Press, 2010), 180; Michael Clyne, *The German Language in a Changing Europe* (Cambridge: Cambridge University Press, 1995), 6; Paul Weindling, "The League of Nations and International Medical Communication in Europe between the First and Second World Wars," in Roger Chartier and Pietro Corsi, eds., *Sciences et langues en Europe* (Paris: European Communities, 2000 [1994]): 201–211, on 211; and Hans Joachim Meyer, "Global English—a New Lingua Franca or a New Imperial Culture?," in Andreas Gardt and Bernd Hüppauf, eds., *Globalization and the Future of German* (Berlin: Mouton de Gruyter, 2004): 65–84, on 68.

3. Alan D. Beyerchen, *Scientists under Hitler: Politics and the Physics Community in the Third Reich* (New Haven: Yale University Press, 1977), 14, 40. On later estimates, which place the numbers possibly as "low" as 15.5%, see Mitchell G. Ash, "Scientific Changes in Germany 1933, 1945, 1990: Towards a Comparison," *Minerva* 37 (1999): 329–354, on 332; and Ute Deichmann, *Biologists under Hitler*, tr. Thomas Dunlap (Cambridge, MA: Harvard University Press, 1996 [1992]), 26.

4. Ute Deichmann and Benno Müller-Hill, "Biological Research at Universities and Kaiser Wilhelm Institutes in Nazi Germany," in Monika Renneberg and Mark

Walker, eds., *Science, Technology and National Socialism* (Cambridge: Cambridge University Press, 1994): 160–183, on 161, 165; Deichmann, *Biologists under Hitler*, 5.

5. Florian Schmaltz, "Chemical Weapons Research in National Socialism: The Collaboration of the Kaiser Wilhelm Institutes with the Military and Industry," in Susanne Heim, Carola Sachse, and Mark Walker, eds., *The Kaiser Wilhelm Society under National Socialism* (Cambridge: Cambridge University Press, 2009): 312–338, on 321; Ute Deichmann, "'To the Duce, the Tenno and Our Führer: A Threefold Sieg Heil': The German Chemical Society and the Association of German Chemists during the Nazi Era," in Dieter Hoffmann and Mark Walker, eds., *The German Physical Society in the Third Reich: Physicists between Autonomy and Accommodation*, tr. Ann M. Hentschel (Cambridge: Cambridge University Press, 2012 [2007]), 280–316, on 280. On the Haber Institute, see Jeremiah James, Thomas Steinhauser, Dieter Hoffmann, and Bretislav Friedrich, *One Hundred Years at the Intersection of Chemistry and Physics: The Fritz Haber Institute of the Max Planck Society, 1911–2011* (Berlin: De Gruyter, 2011).

6. Reinhard Siegmund-Schultze, *Mathematicians Fleeing Nazi Germany: Individual Fates and Global Impact* (Princeton: Princeton University Press, 2009), 102–103; Charles Weiner, "A New Site for the Seminar: The Refugees and American Physics in the Thirties," in Donald Fleming and Bernard Bailyn, eds., *The Intellectual Migration: Europe and America, 1930–1960* (Cambridge, MA: Belknap, 1969), 190–234, on 214, 217.

7. See Carola Tischler, "Crossing Over: The Emigration of German-Jewish Physicians to the Soviet Union after 1933," in Susan Gross Solomon, ed., *Doing Medicine Together: Germany and Russia between the Wars* (Toronto: University of Toronto Press, 2006), 462–500, on 476.

8. Yoshiyuki Kikuchi, "World War I, International Participation and Reorganisation of the Japanese Chemical Community," *Ambix* 58, no. 2 (July 2011): 136–149; and Scott L. Montgomery, *Science in Translation: Movements of Knowledge through Cultures and Time* (Chicago: University of Chicago Press, 2000), 221.

9. The Depression also decreased the number of foreign members of German scientific societies. Beyerchen, *Scientists under Hitler*, 73.

10. Mark Walker, *Nazi Science: Myth, Truth, and the German Atomic Bomb* (Cambridge, MA: Perseus, 1995), 125–126, 140–141; Volker R. Remmert, "The German Mathematical Association during the Third Reich: Professional Policy within the Web of National Socialist Ideology," in Hoffmann and Walker, eds., *The German Physical Society in the Third Reich* (2012): 246–279, on 270.

11. Beyerchen, *Scientists under Hitler*, 72.

12. E. J. Gumbel, "Arische Naturwissenschaft?" in Gumbel, ed., *Freie Wissenschaft: Ein Sammelbuch aus der deutschen Emigration* (Strasbourg: Sebastian Brant, 1938): 246–262, on 252. On the "normality" of German science publications in this period, Gerhard Simonsohn, "The German Physical Society and Research," in Hoffmann and Walker, eds., *The German Physical Society in the Third Reich* (2012): 187–245, on 206.

13. Beyerchen, *Scientists under Hitler*, 76.

14. Hildegard Brücher and Clemens Münster, "Deutsche Forschung in Gefahr?," *Frankfurter Hefte* 4 (1949): 333–344, on 335.

15. Christopher M. Hutton, *Linguistics and the Third Reich: Mother-Tongue Fascism, Race and the Science of Language* (London: Routledge, 1999), 4; Claus Ahlz-

weig, "Die deutsche Nation und ihre Muttersprache," in Konrad Ehlich, ed., *Sprache im Faschismus* (Frankfurt am Main: Suhrkamp, 1989): 35–57, on 36–37.

16. Hutton, *Linguistics and the Third Reich*, 5. On Esperanto and Yiddish, see ibid., 202.

17. Leo Weisgerber, "Die deutsche Sprache im Aufbau des deutschen Volkslebens," *Von deutscher Art in Sprache und Dichtung*, v. 1 (1941): 3–41, on 3.

18. Weisgerber, "Die deutsche Sprache im Aufbau des deutschen Volkslebens," 3.

19. Weisgerber, "Die deutsche Sprache im Aufbau des deutschen Volkslebens," 12.

20. Hermann M. Flasdieck, "England und die Sprachwissenschaft: Englische Sprachforschung als Ausdruck völkischen Denkstils," *Germanisch-Romanische Monatsschrift* 31 (1943): 169–184, on 183.

21. Konrad Ehlich, "Über den Faschismus sprechen—Analyse und Diskurs," in Ehlich, ed., *Sprache im Faschismus* (1989): 7–34.

22. Clyne, *The German Language in a Changing Europe*, 175.

23. R. E. Keller, *The German Language* ([Atlantic Highlands], NJ: Humanities Press, 1978), 604, 607.

24. Eugen Seidel and Ingeborg Seidel-Slotty, *Sprachwandel im Dritten Reich: Eine kritische Untersuchung faschistischer Einflüsse* (Halle: VEB Verlag Sprache und Literatur, 1961), 18. On militarism, see p. vii, on nominalization, see p. 31.

25. Seidel and Seidel-Slotty, *Sprachwandel im Dritten Reich*, 136.

26. Victor Klemperer, *LTI: Notizbuch eines Philologen* (Leipzig: Verlag Philipp Reclam jun., [1966]), 20.

27. Klemperer, *LTI*, 24. See also Sternberger et al., *Aus dem Wörterbuch des Unmenschen*.

28. Lothar G. Tirala, "Nordische Rasse und Naturwissenschaft," in August Becker, ed., *Naturforschung im Aufbruch: Reden und Vorträge zur Einweihungsfeier des Philipp Lenard-Instituts der Universität Heidelberg am 13. und 14. Dezember 1935* (Munich: J. F. Lehmanns Verlag, 1936): 27–38, on 28.

29. Edmund Andrews, *A History of Scientific English: The Story of Its Evolution Based on a Study of Biomedical Terminology* (New York: Richard R. Smith, 1947), 298.

30. Johannes Stark, *Nationalsozialismus und Wissenschaft* (Munich: Zentralverlag der NSDAP, 1934), 10.

31. Deichmann, *Biologists under Hitler*, 89.

32. Mark Walker, "The German Physical Society under National Socialism in Context," in Hoffmann and Walker, eds., *The German Physical Society in the Third Reich* (2012): 1–21, on 8.

33. Philipp Lenard, *Deutsche Physik*, 4 v., 3d ed. (Munich: J. S. Lehmanns Verlag, 1942 [1936]), ix.

34. Philipp Lenard, *England und Deutschland zur Zeit des großen Krieges* (Heidelberg: Carl Winters Universitätsbuchhandlung, 1914), 5.

35. Jan Wirrer, "Dialekt und Standardsprache im Nationalsozialismus—am Beispiel des Niederdeutschen," in Ehlich, ed., *Sprache im Faschismus* (1989): 87–103; Franz Thierfelder, *Die deutsche Sprache im Ausland*, 2 v. (Hamburg: R. v. Decker, 1956–1957), II:58; Dirk Scholten, *Sprachverbreitung des nationalsozialistischen Deutschlands* (Frankfurt a/M: Lang, 2000).

36. Theodore Huebener, "What Our Enemies Think of the Value of Foreign Languages in the 'Air Age,'" *Hispania* 26 (1943): 193–194, on 194.

37. See document #91 in Klaus Hentschel and Ann M. Hentschel, eds., *Physics and National Socialism: An Anthology of Primary Sources*, tr. Ann M. Hentschel (Basel: Birkhäuser, 1996), 281–283.

38. Hannah Arendt and Günter Gaus, "Was bleibt? Es bleibt die Muttersprache," in Günter Gaus, *Zur Person* (Munich: Deutscher Taschenbuch Verlag, 1965): 11–30, on 21–22.

39. Arendt in Arendt and Gaus, "Was bleibt?," 21.

40. Theodor W. Adorno, "Auf die Frage: Was ist deutsch," in Adorno, *Stichworte: Kritische Modelle 2*. Frankfurt am Main: Suhrkamp, 1969): 102–112, on 110.

41. Adorno, "Auf die Frage: was ist deutsch," 111. On his English competence, see Adorno, "Scientific Experiences of a European Scholar in America," tr. Donald Fleming, in Fleming and Bailyn, eds., *The Intellectual Migration* (1969), 338–370, on 340.

42. Julius Schaxel, "Faschistische Verfälschung der Biologie," in Gumbel, *Freie Wissenschaft*: 229–245, on 234. Emphasis in original.

43. Siegmund-Schultze, *Mathematicians Fleeing Nazi Germany*, 153. See also ibid., 38, 142; Weiner, "A New Site for the Seminar," 220–221; and Beyerchen, *Scientists under Hitler*, 36, on the linguistic and general psychological difficulties of emigration.

44. Jean Matter Mandler and George Mandler, "The Diaspora of Experimental Psychology: The Gestaltists and Others," in Fleming and Bailyn, eds., *The Intellectual Migration* (1969): 371–419, on 417.

45. Constance Reid, *Courant in Göttingen and New York: The Story of an Improbable Mathematician* (New York: Springer-Verlag, 1976), 157.

46. Einstein to Lanczos, 11 September 1935, AEDA, Box 18, Folder: "C. Lanczos, 1919–1940," 15–246. An account of this correspondence can be found in Stefan L. Wolff, "Marginalization and Expulsion of Physicists under National Socialism: What Was the German Physical Society's Role?," in Hoffmann and Walker, eds., *The German Physical Society in the Third Reich* (2012): 50–95, on 77.

47. C. Lanczos to Einstein, 14 September 1935, AEDA, Box 18, Folder: "C. Lanczos, 1919–1940," 15-248-1.

48. On French, see Fritz Stern, *Einstein's German World* (Princeton: Princeton University Press, 1999), 93. On Italian, see Einstein to Besso, 20 October [1921], in Albert Einstein and Michele Besso, *Correspondance, 1903–1955*, tr. and ed. Pierre Speziali (Paris: Hermann, 1972), 170.

49. Einstein to Besso, [late 1913], in Einstein and Besso, *Correspondance*, 50. There is only one sentence in English in their entire correspondence, written by Besso: "I have not yet found a practicable way to confront the results of the theory with experimental evidence." Besso to Einstein, 12 July 1954, in ibid., 523.

50. Einstein to Max Born, 7 September 1944, in Albert Einstein, Hedwig Born, und Max Born, *Briefwechsel, 1916–1955*, ed. Max Born (Munich: Nymphenburger Verlagshandlung, 1969), 202.

51. Einstein to Hahn, 28 January 1948, quoted in Klaus Hentschel, *Die Mentalität deutscher Physiker in der frühen Nachkriegszeit (1945–1949)* (Heidelberg: Synchron, 2005), 159. See also Einstein to Sommerfeld, 14 December 1946, in Armin Hermann, ed., *Albert Einstein/Arnold Sommerfeld Briefwechsel: Sechzig Briefe aus dem goldenen Zeitalter der modernen Physik* (Basel/Stuttgart: Schwabe & Co., 1968), 121.

52. Max Born, *My Life & My Views* (New York: Charles Scribner's Sons, 1968), 16. This memoir was originally composed in English.

53. Max Born, *Problems of Atomic Dynamics: Two Series of Lectures* (Cambridge, MA: MIT, 1926); and Born, *Probleme der Atomdynamik* (Berlin: Julius Springer, 1926).

54. Max Born to Einstein, 2 June 1933, in Einstein, Born, and Born, *Briefwechsel*, 163–164.

55. Born, *My Life & My Views*, 38.

56. Max Born commentary on letter to Einstein, 10 April 1940, in Einstein, Born, and Born, *Briefwechsel*, 192–193.

57. Born to Einstein, 26 September 1953, in Einstein, Born, and Born, *Briefwechsel*, 265.

58. Einstein to Born, 12 October 1953, in Einstein, Born, and Born, *Briefwechsel*, 266.

59. Robert Marc Friedman, *The Politics of Excellence: Behind the Nobel Prize in Science* (New York: W. H. Freeman, 2001), 232–250.

60. On their facility with languages, see Jost Lemmerich, ed., *Lise Meitner zum 125. Geburtstag: Ausstellungskatalog* (Berlin: ERS-Verlag, [2003]), 12, 17.

61. Ruth Lewin Sime, *Lise Meitner: A Life in Physics* (Berkeley: University of California Press, 1996), 214, 258.

62. See, for example, Meitner to von Laue, 19 November 1940, in Jost Lemmerich, ed., *Lise Meitner—Max von Laue: Briefwechsel 1938–1948* (Berlin: ERS Verlag, 1998), 104.

63. Meitner to von Laue, 23 December 1940, in Lemmerich, *Lise Meitner—Max von Laue*, 108.

64. Max von Laue, "Mein physikalischer Werdegang," from Hans Hartmann, ed., *Schöpfer des neuen Weltbildes* (Hamburg: Deutsche Hausbücherei, 1952), 178–207, reproduced in von Laue, *Gesammelte Schriften und Vorträge*, 3 v. (Braunschweig: Freidrich Vieweg & Sohn, 1961), III: xi.

65. Von Laue, "Mein physikalischer Werdegang," xi–xii.

66. Meitner to Eva von Bahr-Bergius, 21 June 1944, reproduced in Lemmerich, *Lise Meitner zum 125. Geburtstag*, 112.

67. Sime, *Lise Meitner*, 358.

68. Meitner to von Laue, 28 June 1946, in Lemmerich, *Lise Meitner—Max von Laue*, 452–453.

69. See Meitner to Hahn, 27 June 1945, reproduced in Lemmerich, *Lise Meitner zum 125. Geburtstag*, 116–117.

70. Meitner to von Laue, 12 November 1946, in Lemmerich, *Lise Meitner—Max von Laue*, 470–471.

71. P. W. Bridgman, "'Manifesto' by a Physicist," *Science* 89, no. 2304 (24 February 1939): 179.

72. Warren Weaver, diary entry of 25 May 1947, Rockefeller Archive Center, RG 12.1, p. 45, reproduced in Klaus Hentschel, *The Mental Aftermath: The Mentality of German Physicists, 1945–1949*, tr. Ann M. Hentschel (New York: Oxford University Press, 2007), 24.

73. "The Fate of German Science: Impressions of a BIOS Officer," *Discovery* (August 1947): 239–243.

74. Gerhard Rammer, "'Cleanliness among Our Circle of Colleagues': The German Physical Society's Policy toward Its Past," in Hoffmann and Walker, eds., *The German Physical Society in the Third Reich* (2012): 367–421, on 377.

75. Niko Tinbergen to Margaret Nice, 23 June 1945, reproduced in Deichmann, *Biologists under Hitler*, 203.

76. Samuel A. Goudsmit, "Our Task in Germany," *Bulletin of the Atomic Scientists* 4, no. 4 (1948): 106.

77. For Meitner, see Meitner to von Laue, 12 November 1946, in Lemmerich, *Lise Meitner—Max von Laue*, 470; and Meitner to Hahn, 27 June 1945, reproduced in Lemmerich, *Lise Meitner zum 125. Geburtstag*, 117. On other émigrés, see Wolff, "Marginalization and Expulsion of Physicists under National Socialism," 61.

78. Henry J. Kellermann, *Cultural Relations as an Instrument of U.S. Foreign Policy: The Educational Exchange Program between the United States and Germany, 1945–1954* (Washington, DC: Department of State, 1978), 3.

79. Arendt in Arendt and Gaus, "Was bleibt?," 23.

Chapter Eight

1. D. Zaslavskii, "Velikii iazyk nashei epokhi," *Literaturnaia gazeta* (1 January 1949).

2. The full machine consisted of the 701 Analytic Control Unit, the 706 Electrostatic Storage Unit, the 711 Punched Card Reader, the 716 Alphabetical Printer, the 721 Punched Card Recorder, the 726 Magnetic Tape Readers and Recorders, the 731 Magnetic Drum Reader and Recorder, and a Power Supply and Distribution Box. On computer hardware in this period, see Mina Rees, "Computers: 1954," *Scientific Monthly* 79, no. 2 (August 1954): 118–124; and Cuthbert C. Hurd, "Computer Development at IBM," in N. Metropolis, J. Howlett, and Gian-Carlo Rota, eds., *A History of Computing in the Twentieth Century: A Collection of Essays* (New York: Academic Press, 1980): 389–418. For all Hurd's pride in the experiment, he got the name of his Georgetown collaborator wrong, naming him "Professor Dorot" (on 406).

3. J. B. Donnelly, IBM Press Release, 8 January 1954, GUA-SLL 1:1–6/1954.

4. Dostert quoted in W. Schweisheimer, "Language Translation by Electronic Computer," *Mechanical World* (December 1955): 534–535, on 534.

5. Quoted in Dorothy M. Bishop, "Breaking the Language Barrier," *Phi Delta Kappan* 35, no. 8 (May 1954): 315–317, 320, on 317.

6. Dostert quoted in Donnelly Press Release, 8 January 1954, GUA-SLL 1:1–6/1954.

7. See, for example: Jacob Ornstein, "Mechanical Translation: New Challenge to Communication," *Science* 122, no. 3173 (21 October 1955): 745–748; "Language Translation by the Electronic 'Brain,'" *Science News-Letter* 65, no. 4 (23 January 1954): 59; and William N. Locke, "Speech Typewriters and Translating Machines," *PMLA* 70, no. 2 (April 1955): 23–32, on 30. As W. John Hutchins has noted in his excellent history of the experiment, science was deemphasized in most mainstream media accounts as "the newspaper reporters tended to choose only non-chemistry examples, since these gave impressions of the quality of the translations which could be more readily appreciated by readers than the chemistry ones." Hutchins, "The First Public Demonstration of Machine Translation: The Georgetown-IBM System, 7th January 1954," (March 2006), available at http://hutchinsweb.me.uk/GU-IBM-2005.pdf, accessed 16 September 2011, on p. 12. Hutchins includes a sizeable bibliography of the press accounts.

8. Leon Dostert, "Report on Academic Developments, The Institute of Languages and Linguistics, 1952–53, Projected Activities, 1953–54," 7 October 1953, GUA-SLL 1:1953.

9. Leon E. Dostert to Edward B. Bunn, S.J., 24 November 1953, GUA-SLL 1:1953.

10. Memorandum of phone call, Robert Avery to Leon Dostert, 11 December 1953, 2 pm, GUA-SLL 1:1953.

11. N. W. Baklanoff, "Scientific Russian," *Modern Language Journal* 32 (1948): 190–194, on 191.

12. J. G. Tolpin, "Teaching of Scientific Russian," *American Slavic and East European Review* 4 (August 1945): 158–164, on 158; idem, "The Place of Russian Scientific Literature in Bibliographical Work," *Journal of Chemical Education* 21 (July 1944): 336–342, on 336; Kurt Gingold, "Translation Pools—Ideal and Reality," *Journal of Chemical Documentation* 1, no. 2 (1961): 14–19, on 14; and Joseph J. Gwirtsman, "Coverage of Russian Chemical Literature in Chemical Abstracts," *Journal of Chemical Documentation* 1, no. 2 (1961): 38–44, on 38.

13. Boris I. Gorokhoff, *Providing U.S. Scientists with Soviet Scientific Information*, rev. ed. (Washington, DC: Publications Office of the National Science Foundation, 1962), i.

14. E. J. Crane, "Growth of Chemical Literature: Contributions of Certain Nations and the Effects of War," *Chemical & Engineering News* 22, no. 17 (10 September 1944): 1478–1481, 1496, on 1478, 1481 (quotation). The number of Russian chemists grew by twenty-five times between 1875 and 1940, and they were increasingly productive, producing eighty times the quantity of publications over this period. I. I. Zaslavskii, "Rol' russkikh uchenykh v sozdanii mirovoi khimii," *Uspekhi khimii* 13, no. 4 (1944): 328–335, on 331.

15. Gingold, "Translation Pools," 14.

16. Advisory Panel on Scientific Information, "Minutes of the First Meeting," 12 October 1953, NSF Records, RG 307, Box 18, Folder: "Scientific Information Office: Advisory Panel on Scientific Information," p. 4.

17. Saul Herner, "American Use of Soviet Medical Research," *Science*, N.S. 128, no. 3314 (4 July 1958): 9–15, on 14.

18. J. G. Tolpin, "Surveying Russian Technical Publications: A Brief Course," *Science* 146, no. 3648 (27 November 1964): 1143–1144, on 1143.

19. Office of Scientific Information, "International Exchange of Scientific Information," 2 November 1955, NSF Records, RG 307, Box 18, Folder: "Scientific Information," page 19. The Russian language barrier had been cited in Vannevar Bush's original position paper calling for the creation of this organization: Bush, *Science: The Endless Frontier* (Washington, DC: Government Printing Office, 1945), 114.

20. George Alan Connor, Doris Tappan Connor, William Solzbacher, and J. B. Se-Tsien Kao, *Esperanto: The World Interlanguage*, 2d rev. ed. (South Brunswick: Thomas Yoseloff, 1966 [1948]), 4, 32.

21. For a complete list of members of the IALA, see "Practical World Language," *Science News-Letter* 66, no. 3 (17 July 1954): 34; and Watson Davis, "Practical World Language," *Science News-Letter* 62 (5 July 1952): 10–11.

22. Gode in IALA, *Interlingua-English: A Dictionary of the International Language* (New York: Storm, 1951), xxi.

23. Alexander Gode and Hugh E. Blair, *Interlingua: A Grammar of the International Language* (New York: Storm Publishers, 1951), v.

24. Alexander Gode, *Interlingua: A Prime Vista* (New York: Storm Publishers, 1954).

25. Andrew Large, *The Artificial Language Movement* (Oxford: Basil Blackwell, 1985), 151.

26. Arika Okrent, *In the Land of Invented Languages: Esperanto Rock Stars, Klingon Poets, Loglan Lovers, and the Mad Dreamers Who Tried to Build a Perfect Language* (New York: Spiegel & Grau, 2009), 210.

27. E. Glyn Lewis, *Multilingualism in the Soviet Union: Aspects of Language Policy and Its Implementation* (The Hague: Mouton, 1972), 50. On propaganda, see Selig S. Harrison, *The Most Dangerous Decades: An Introduction to the Comparative Study of Language Policy in Multi-Lingual States* (New York: Language and Communication Research Center, Columbia University, 1957), 24–27.

28. The exceptions were Georgian, Armenian, and Abkhaz in the Caucasus, which used their traditional scripts; Estonian, Lithuanian, Latvian, and Karelian in the Baltic region, which kept Latin scripts; and Yiddish, which retained Hebrew letters. (Yiddish was strongly repressed in the postwar period.) For the information on Soviet language policy in this section, see Barbara A. Anderson and Brian D. Silver, "Equality, Efficiency, and Politics in Soviet Bilingual Education Policy, 1934–1980," *American Political Science Review* 78, no. 4 (December 1984): 1019–1039; Uriel Weinreich, "The Russification of Soviet Minority Languages," *Problems of Communism* 2, no. 6 (1953): 46–57; and Michael Kirkwood, ed., *Language Planning in the Soviet Union* (Basingstoke: Macmillan, 1989).

29. Nicholas DeWitt, *Soviet Professional Manpower: Its Education, Training, and Supply* (Washington, DC: NSF, 1955), 55. The situation became even more sharply Russophone, especially in technical publication, by the 1970s. Iu. D. Desheriev, *Razvitie obshchestvennykh funktsii literaturnykh iazykov* (Moscow: Nauka, 1976), 11–12, 17.

30. Tatjana Kryuchkova, "English as a Language of Science in Russia," in Ulrich Ammon, ed., *The Dominance of English as a Language of Science: Effects on Other Languages and Language Communities* (Berlin: Mouton de Gruyter, 2001): 405–423, on 407.

31. T. P. Lomtev, "I. V. Stalin o razvitii natsional'nykh iazykov v epokhu sotsializma," *Voprosy filosofii* 1(6) (1949): 131–141, on 136.

32. J. G. Garrard, "The Teaching of Foreign Languages in the Soviet Union," *Modern Language Journal* 46 (1962): 71–74; Jacob Ornstein, "Foreign Language Training in the Soviet Union—A Qualitative View," *Modern Language Journal* 42 (1958): 382–392, on 388.

33. Alexander G. Korol, *Soviet Education for Science and Technology* (New York: Technology Press and John Wiley & Sons, 1957), 230. Emphasis in original. See also pages 121, 185–186, and 228.

34. For the contrasting positions, see P. V. Terent'ev, "Obiazatel'noe uslovie uspeshnoi podgotovki kadrov," *Vestnik vysshei shkoly*, no. 12 (December 1955): 50–51; and A. F. Plate, "Kakoi inostrannyi iazyk v pervuiu ochered' neobkhodim sovetskomu khimiku," *Uspekhi khimii* 9, no. 5 (1940): 589–595, on 593.

35. Dale Lockard Barker, "Characteristics of the Scientific Literature Cited by

Chemists of the Soviet Union" (PhD dissertation, University of Illinois-Urbana, 1966). For the 80% figure, see Department of Scientific and Industrial Research, *Scientific and Technical Information in the Soviet Union: Report of the D.S.I.R.–Aslib Delegation to Moscow and Leningrad, 7th–24th June, 1963* (London: Department of Scientific and Industrial Research, 1964), 17.

36. J. G. Tolpin, J. Danaczko Jr., R. A. Liewald, et al., "The Scientific Literature Cited by Russian Organic Chemists," *Journal of Chemical Education* 28 (May 1951): 254–258, on 254.

37. Ornstein, "Foreign Language Training in the Soviet Union," 392.

38. See Nils Roll-Hansen, *The Lysenko Effect: The Politics of Science* (Amherst, NY: Humanity Books, 2005) and David Joravsky, *The Lysenko Affair* (Cambridge, MA: Harvard University Press, 1970).

39. The theory is complicated and not entirely internally consistent. For English-language accounts, see Ethan Pollock, *Stalin and the Soviet Science Wars* (Princeton: Princeton University Press, 2006), chapter 5; Jeffrey Ellis and Robert W. Davies, "The Crisis in Soviet Linguistics," *Soviet Studies* 2, no. 3 (January 1951): 209–264; W. K. Matthews, "The Japhetic Theory," *Slavonic and East European Review* 27, no. 68 (December 1948): 172–192; M. Miller, "Marr, Stalin, and the Theory of Language," *Soviet Studies* 2, no. 4 (April 1951): 364–371.

40. Jindrich Kucera, "Soviet Nationality Policy: The Linguistic Controversy," *Problems of Communism* 3, no. 2 (March-April 1954): 24–29, on 26.

41. Quoted in E. Drezen, *Osnovy iazykoznaniia: Teorii i istorii mezhdunarodnogo iazyka* (Moscow: Izd. TsK SESR, 1932), 26.

42. J. G. Tolpin, "Searching the Russian Technical Literature," *Journal of Chemical Education* 23 (October 1946): 485–489, on 488.

43. Quoted in Nikolai Krementsov, *The Cure: A Story of Cancer and Politics from the Annals of the Cold War* (Chicago: University of Chicago Press, 2002), 131.

44. John Turkevich, "Science," in Harold H. Fisher, ed., *American Research on Russia* (Bloomington: Indiana University Press, 1959): 103–112, on 105–106.

45. Theodosius Dobzhansky, "Russian Genetics," in Ruth C. Christman, ed., *Soviet Science* (Washington, DC: AAAS, 1952): 1–7, on 6; J. R. Kline, "Soviet Mathematics," in ibid.: 80–84, on 82–83.

46. Albert Parry, *America Learns Russian: A History of the Teaching of the Russian Language in the United States* (Syracuse: Syracuse University Press, 1967), 50–58; Rudolf Sturm, "The Changing Aspects of Teaching Russian," in Joseph S. Roucek, ed., *The Study of Foreign Languages* (New York: Philosophical Library, 1968): 170–183, on 175.

47. Sturm, "The Changing Aspects of Teaching Russian," 173.

48. Jacob Ornstein, "Structurally Oriented Texts and Teaching Methods since World War II: A Survey and Appraisal," *Modern Language Journal* 40, no. 5 (May 1956): 213–222; idem, *Slavic and East European Studies: Their Development and Status in the Western Hemisphere* (Washington, DC: Department of State, Office of Intelligence Research, 1957), 9; Arthur Prudden Coleman, "The Teaching of Russian in the United States," *Russian Review* 4, no. 1 (Autumn 1944): 83–88; Parry, *America Learns Russian*, 107.

49. Parry, *America Learns Russian*, 112, 130; Jacob Ornstein, "A Decade of Russian Teaching: Notes on Methodology and Textbooks," *Modern Language Journal* 35 (1951): 263–279, on 263; Sturm, "The Changing Aspects of Teaching Russian," 177.

50. Parry, *America Learns Russian*, 152; Fan Parker, "Report of the National Information Center on the Status of Russian in Secondary Schools," *Slavic and East European Journal* 3, no. 1 (Spring 1959): 55–61.

51. L. E. Dostert, Frederick D. Eddy, W. P. Lehmann, and Albert H. Markwardt, "Tradition and Innovation in Language Teaching," *Modern Language Journal* 44, no. 5 (May 1960): 220–226, on 220.

52. William B. Edgerton, "A Modest Proposal: The Teaching of Russian in America," *Slavic and East European Journal* 6, no. 4 (Winter 1962): 354–372, on 357. On rising enrollments, see Sturm, "The Changing Aspects of Teaching Russian," 179.

53. Alan T. Waterman, "Research and the Scholar," remarks at the dedication of the Price Gilbert Library at the Georgia Institute of Technology, 20 November 1953, Waterman Papers, LOC, Box 57, Folder: "1953 Apr.8-Dec.29," p. 14.

54. Jacob Chaitkin, "The Challenge of Scientific Russian," *Scientific Monthly* 60, no. 4 (April 1945): 301–306, on 301.

55. George A. Znamensky, *Elementary Scientific Russian Reader* (New York: Pitman, 1944), v.

56. V. A. Pertzoff, *Translation of Scientific Russian* (New York: Exposition-University Books, [1964]), 41.

57. Tolpin, "Teaching of Scientific Russian," 160. There were, of course, others who claimed that the presence of false cognates and convoluted sentence structure made translating scientific Russian in many ways *harder* than "standard" Russian. See, for example, Vijay Pandit, "'Misleading' Words in Scientific Translation from Russian into English," *Babel* 25 (1979): 148–151; and John Turkevich, "Scientific Russian," *American Scientist* 34, no. 3 (July 1946): 464–466, 470, on 464.

58. Lorraine T. Kapitanoff, "The Teaching of Technical Russian," *Slavic and East European Journal* 7, no. 1 (Spring 1963): 51–56, on 52.

59. David Kaiser, "The Physics of Spin: Sputnik Politics and American Physicists in the 1950s," *Social Research* 73, no. 4 (Winter 2006): 1225–1252, on 1239–1240; Parry, *America Learns Russian*, 146; Arthur Prudden Coleman, *A Report on the Status of* Russian and Other Slavic and East European Languages *in the Educational Institutions of the United States, Its Territories, Possessions and Mandates, with Additional Data on Similar Studies in Canada and Latin America* (New York: American Association of Teachers of Slavic and East European Languages, 1948).

60. J. G. Tolpin, "The Present Status of Teaching Russian for Scientists," *Modern Language Journal* 33, no. 1 (January 1949): 27–30, on 27. For self-training, see James W. Perry, "Chemical Russian, Self-Taught," *Journal of Chemical Education* (August 1944): 393–398.

61. Kapitanoff, "The Teaching of Technical Russian," 51.

62. E. F. Langley, "New Course in Russian," 8 September 1942, MIT Archives, AC359, Box 1, Folder 1.

63. Baklanoff, "Scientific Russian"; R. D. Burke, "Some Unique Problems in the Development of Qualified Translators of Scientific Russian," RAND Report P-1698 (12 May 1959); Tolpin, "Surveying Russian Technical Publications," 1143; and George E. Condoyannis, "Russian for Scientists," *Modern Language Journal* 32, no. 5 (May 1948): 392–393.

64. Ornstein, *Slavic and East European Studies*, 21–22.

65. Burke, "Some Unique Problems in the Development of Qualified Translators,"

7. See also C. R. Buxton and H. Sheldon Jackson, *Russian for Scientists: A Grammar and Reader* (New York: Interscience, 1960), 4, 73.

66. James W. Perry, "Translation of Russian Technical Literature by Machine: Notes on Preliminary Experiments," *Mechanical Translation* 2, no. 1 (July 1955): 15–26, on 15.

67. For example, James W. Perry, *Scientific Russian: A Textbook for Classes and Self-Study* (New York: Interscience Publishers, 1950); idem, *Chemical Russian, Self-Taught* (Easton, PA: Journal of Chemical Education, 1948); Noah D. Gershevsky, ed., *Scientific Russian Reader: Selected Modern Readings in Chemistry and Physics* (New York: Pitman Publishing, 1948); George E. Condoyannis, *Scientific Russian: A Concise Description of the Structural Elements of Scientific and Technical Russian* (New York: John Wiley & Sons, 1959); John Turkevich and Ludmilla B. Turkevich, *Russian for the Scientist* (Princeton: D. Van Nostrand, 1959); Buxton and Jackson, *Russian for Scientists*; and Pertzoff, *Translation of Scientific Russian.*

68. Parry, *America Learns Russian*, 149. On teaching by radio in the United States and Great Britain, respectively, see Coleman, "The Teaching of Russian in the United States," 87–88; and Ariadne Nicolaeff, "Problems of Teaching Russian by Radio," in C. V. James, ed., *On Teaching Russian* (Oxford: Pergamon, 1963), 64–75.

69. Biographical information is drawn from: R. Ross Macdonald, "Léon Dostert," in William M. Austin, ed., *Papers in Linguistics in Honor of Léon Dostert* (The Hague: Mouton, 1967): 9–14; "New Institute of Linguistics," *Georgetown University Alumni Bulletin* (Fall 1949): 5, 16; *The Lado Years, 1960–1973* (Washington, DC: Georgetown University Press, 1973), 1–2; and Muriel Vasconcellos, "The Georgetown Project and Léon Dostert: Recollections of a Young Assistant," in W. John Hutchins, ed., *Early Years in Machine Translation: Memoirs and Biographies of Pioneers* (Amsterdam: John Benjamins, 2000): 87–96.

70. This does not mean that everyone liked him. Michael Zarechnak, one of his employees at Georgetown on developing MT, once wrote tactfully that "Dostert was a person toward whom an observer could not display a neutral attitude." Michael Zarechnak, "The History of Machine Translation," in Bożena Henisz-Dostert, R. Ross Macdonald, and Michael Zarechnak, *Machine Translation* (The Hague: Mouton, 1979): 1–87, on 21. Some of the translators in Nuremberg dubbed him "le Petit Napoléon." Francesca Gaiba, *The Origins of Simultaneous Interpretation: The Nuremberg Trial* (Ottowa: Ottowa University Press, 1998), 134.

71. Léon Dostert, *France and the War*, No. 24 *America in a World at War* (New York: Oxford University Press, 1942).

72. See Gaiba, *The Origins of Simultaneous Interpretation.* On the relationship with Watson, see Vasconcellos, "The Georgetown Project and Léon Dostert," 87n.

73. L. E. Dostert, "The Georgetown Institute Language Program," *PMLA* 68, no. 2 (April 1953): 3–12. See also William M. Austin's dedicatory foreword in *Papers in Linguistics in Honor of Léon Dostert*, 5.

74. L. E. Dostert, "Languages in Preparedness: Link or Obstacle?," *Armor* (May-June 1951): 12–14, on 12–13.

75. The memorandum is reprinted as Warren Weaver, "Translation," in William N. Locke and A. Donald Booth, eds., *Machine Translation of Languages: Fourteen Essays* (Cambridge, MA: MIT Press, 1955), 15–23, quotation on 18. On Booth's early impact and claims to priority, see Andrew D. Booth, "Mechanical Translation," *Computers*

and Automation 2, no. 4 (May 1953): 6–8; and idem, ed., *Machine Translation* (Amsterdam: North-Holland, 1967), vi. Priority is tricky here, however, since there were two active proposals in the 1930s for machine translation: one in France by George Artsrouni, and a more elaborate one in the Soviet Union by Petr Smirnov-Troianskii. Both were largely ignored until the self-historicization of the field in the late 1950s. On Artsrouni, see W. J. Hutchins, *Machine Translation: Past, Present, Future* (Chichester: Ellis Horwood, 1986), 22–23. On Troianskii, John Hutchins and Evgenii Lovtskii, "Petr Petrovich Troyanskii (1894–1950): A Forgotten Pioneer of Mechanical Translation," *Machine Translation* 15 (2000): 187–221.

76. Quoted in Weaver, "Translation," 18. Despite the take-off of MT, consensus has sided with Wiener, since cryptography, being a map within the same language, is a poor analog to translation. See, for example, W. John Hutchins, "From First Conception to First Demonstration: The Nascent Years of Machine Translation, 1947–1954. A Chronology," *Machine Translation* 12, no. 3 (1997): 195–252, on 208; and Émile Delavenay, *La machine à traduire* (Paris: Presses universitaires de France, 1963), 16–17.

77. Quoted in A. Donald Booth and William N. Locke, "Historical Introduction," in Locke and Booth, eds., *Machine Translation of Languages* (1955), 1–14, on 4.

78. For the program, see "M.I.T. Conference on Mechanical Translation, June 17—June 20, 1952 Program," MIT Archives, AC359, Box 2, Folder: "Machine Translation Conf.-1952." This conference immediately followed one on electronic speech analysis, the other technological-linguistic hybrid that occupied MIT's marginalized Department of Modern Languages during the 1950s.

79. Bar-Hillel, "Some Linguistic Problems Connected with Machine Translation," *Philosophy of Science* 20, no. 3 (July 1953): 217–225; idem, "Machine Translation," *Computers and Automation* 2, no. 5 (July 1953): 1–6; idem, "Can Translation Be Mechanized?," *American Scientist* 42, no. 2 (April 1954): 248–260; and idem, "The Present State of Research on Mechanical Translation," *American Documentation* 2, no. 4 (1951): 229–237.

80. Léon Dostert, "Development Plan for the Institute of Languages and Linguistics, 1953–1958," 31 December 1952, GUA-SLL 1:1952.

81. Rev. Cyprian Towney to Dostert, 4 January 1955, GUA-SLL 1:1955; Robert Emmett Curran, *A History of Georgetown University*, 3 v. (Washington, DC: Georgetown University Press, 2010), II: 349–350.

82. A. C. Reynolds, Jr., "The Conference on Mechanical Translation Held at M.I.T., June 17–20, 1952," *Mechanical Translation* 1, no. 3 (December 1954): 47–55, on 48.

83. Dostert, "The Georgetown-I.B.M. Experiment," in Locke and Booth, eds., *Machine Translation of Languages* (1955): 124–135, on 125.

84. Erwin Reifler, "The First Conference on Mechanical Translation," *Mechanical Translation* 1, no. 2 (August 1954): 23–32, on 27, 31 (quotation).

85. Erwin Reifler, "The Mechanical Determination of Meaning," in Locke and Booth, eds., *Machine Translation of Languages* (1955): 136–164, on 136.

86. Victor A. Oswald, Jr., and Stuart L. Fletcher, Jr., "Proposals for the Mechanical Resolution of German Syntax Patterns," *Modern Languages Forum* 36, no. 3–4 (1951): 81–104.

87. Kenneth Harper, "The Mechanical Translation of Russian: A Preliminary Study," *Modern Language Forum* 38, no. 3–4 (1953): 12–29, on 12. For the Soviet view

of the same issue, see I. K. Belskaja, "Machine Translation of Languages," *Research* 10 (1957): 383–389, on 383.

88. Kenneth E. Harper, "A Preliminary Study of Russian," in Locke and Booth, eds., *Machine Translation of Languages* (1955): 66–85, on 67 and 69.

89. Regrettably, we have no detailed description of how the experiment actually worked, much less a copy of the computer code. The closest we come is Paul Garvin's April 1953 articulation of his plan of attack: "Statement of Opinion Concerning Machine Translation," 14 April 1953, GUA-MTP. The lack of public specificity was a source of frustration to contemporaries who wished to replicate the experiment or at least compare the results with other ventures, as described in L. Brandwood, "Previous Experiments in Mechanical Translation," *Babel* 2, no. 3 (October 1956): 125–127. Historical accounts must rely on the rather vague presentation in Dostert, "The Georgetown-I.B.M. Experiment," and the more technical but retrospective 1967 presentation by Garvin, "The Georgetown-IBM Experiment of 1954: An Evaluation in Retrospect," in Austin, ed., *Papers in Linguistics in Honor of Léon Dostert* (1967), 46–56. The best available reconstruction (although still speculative) is Hutchins, "The First Public Demonstration of Machine Translation."

90. Dostert, "An Experiment in Mechanical Translation: Aspects of the General Problem," August 1954, GUA-SLL 1:7–12/1954, p. 7. A slightly amended version of this description can be found in "The Georgetown-I.B.M. Experiment," 127.

91. Peter Sheridan, "Research in Language Translation on the IBM Type 701," *IBM Technical Newsletter* 9 (1955): 5–24, on 5. See also Garvin, "The Georgetown-IBM Experiment of 1954," 50.

92. For an explanation and survey of the various approaches in this period through the 1970s, see Jonathan Slocum, "A Survey of Machine Translation: Its History, Current Status, and Future Prospects," *Computational Linguistics* 11, no. 1 (January-March 1985): 1–17.

93. Sheridan, "Research in Language Translation on the IBM Type 701," 17.

94. Michael Zarechnak, "The Early Days of GAT-SLC," in Hutchins, ed., *Early Years in Machine Translation* (2000): 111–128, on 112; and Zarechnak, "The History of Machine Translation," 24.

95. Undated 2-page typescript, entitled "A Sample of Russian Sentences translated by the IBM Type-701 Data Processing Machines, together with the English translations," GUA-SLL 1:1–6/1954.

96. Léon E. Dostert to Director of the Mathematical Sciences Division at the office of the Chief of Naval Research, 25 May 1954, GUA-SLL 1:1–6/1954; Rear Admiral and Navy Chief of Staff L. H. Frost to Edward B. Bunn SJ, 21 July 1954, GUA-SLL 1:7–12/1954.

97. L. E. Dostert, "Outline for Extension of Research on Mechanical Translation," 6 July 1954, GUA-SLL 1:7–12/1954.

98. Paul W. Howerton to Dostert, 26 January 1954, GUA-MTP.

99. W. John Hutchins, "The Evolution of Machine Translation Systems," in Veronica Lawson, ed., *Practical Experience of Machine Translation: Proceedings of a Conference, London, 5–6 November 1981* (Amsterdam: North-Holland, 1982), 21–37, on 22 (quotation); and Hutchins, "The First Public Demonstration of Machine Translation," 26.

100. Anthony G. Oettinger, "A Study for the Design of an Automatic Dictionary" (PhD dissertation, Harvard University, 1954).

Chapter Nine

1. Evgenii Zamiatin, "O literature, revoliutsii, entropii i o prochem [1923]," *Izbrannye proizvedeniia* (Moscow: Sovetskaia Rossiia, 1990), 434.

2. V. P. Berkov and B. A. Ershov, "O popytkakh mashinnogo perevoda," *Voprosy iazykoznaniia*, no. 6 (November-December 1955): 145–148. On the visit to New York, see Cuthbert C. Hurd, "Computer Development at IBM," in N. Metropolis, J. Howlett, and Gian-Carlo Rota, eds., *A History of Computing in the Twentieth Century: A Collection of Essays* (New York: Academic Press, 1980): 389–418, on 406. On the work of Liapunov's group, see Olga S. Kulagina, "Pioneering MT in the Soviet Union," in W. John Hutchins, ed., *Early Years in Machine Translation: Memoirs and Biographies of Pioneers* (Amsterdam: John Benjamins, 2000): 197–204, on 197. On Soviet cybernetics and Liapunov, see Slava Gerovitch, *From Newspeak to Cyberspeak: A History of Soviet Cybernetics* (Cambridge, MA: MIT Press, 2002).

3. D. Panov, I. Mukhin, and I. Bel'skaia, "Mashina perevodit s odnogo iazyka na drugoi," *Pravda*, no. 22 (13685) (22 January 1956): 4.

4. Yehoshua Bar-Hillel, *Language and Information: Selected Essays on Their Theory and Application* (Reading, MA: Addison-Wesley, 1964), 181. For surveys of Soviet research in this period, see I. K. Belskaja, "Machine Translation of Languages," *Research* 10 (1957): 383–389; Anthony G. Oettinger, "A Survey of Soviet Work on Automatic Translation," *Mechanical Translation* 5, no. 3 (December 1958): 101–110; V. Yu. Rozentsveig, "The Work on Machine Translation in the Soviet Union: Fourth International Congress of Slavicists Reports, Sept. 1958," tr. Lew R. Micklesen, *Machine Translation* 5, no. 1 (July 1958): 95–100; and W. J. Hutchins, *Machine Translation: Past, Present, Future* (Chichester: Ellis Horwood, 1986), chapter 6. The Soviets preferred calling the field "automatic translation," as discussed in I. A. Mel'chuk and R. D. Ravich, *Avtomaticheskii perevod, 1949–1963: Kritiko-bibliograficheskii spravochnik* (Moscow: VINITI, 1967), 5.

5. G. P. Zelenkevich, L. N. Korolev, and S. P. Razumovskii, "Opyty avtomaticheskogo perevoda na elektronnoi vychislitel'noi mashine BESM," *Priroda*, no. 8 (1956): 81–85, on 82. Three sentences are given as examples of the input and output of English-to-Russian translation using this program. They are suspiciously good for such complicated input sentences.

6. Sylvie Archaimbault and Jacqueline Léon, "La langue intermédiaire dans la traduction automatique en URSS (1954–1960)," *Histoire Épistémologie Langage* 19, no. 2 (1997): 105–132.

7. L. E. Dostert, "Certain Aspects and Objectives of Research in Machine Translation," undated [1957], GUA-MTP, pp. 2–3. See also Leon Dostert, "Brief History of Machine Translation Research," in *Eighth Annual Round Table Meeting on Linguistics and Language Studies* (Washington, DC: Georgetown University, 1957): 3–10, on 4.

8. Oettinger, "Machine Translation at Harvard," in Hutchins, ed., *Early Years in Machine Translation* (2000): 73–86, on 80.

9. For example, Erwin Reifler, "Machine Translation," *Bulletin of the Medical*

Library Association 50, no. 3 (1962): 473–480, on 475; and William N. Locke, "Machine Translation," in Eleanor D. Dym, *Subject and Information Analysis* (New York: Marcel Dekker, 1985): 124–153, on 134.

10. Michael Zarechnak, "The Early Days of GAT-SLC," in Hutchins, ed., *Early Years in Machine Translation* (2000): 111–128, on 113.

11. L. E. Dostert to Alberto Thompson, 17 April 1956, GUA-SLL 1:1–4/1956. Emphasis in original.

12. James Clayton, "Device to Translate Foreign Science Data," *Washington Post and Times Herald* (30 July 1956): 19.

13. Muriel Vasconcellos, "The Georgetown Project and Léon Dostert: Recollections of a Young Assistant," in Hutchins, ed., *Early Years in Machine Translation* (2000): 87–96, on 88. See the year-by-year breakdown by funding agency and scholarly institution in Automatic Language Processing Advisory Committee (hereafter ALPAC), *Language and Machines: Computers in Translation and Linguistics* (Washington, DC: National Academy of Sciences, National Research Council, 1966), 107–110; and R. Ross Macdonald, *General Report, 1952–1963* (Washington, DC: Georgetown University Machine Translation Research Project, 1963), v. On publicity, see "Machine Translation Tests Are Sponsored by CIA," *Georgetown Record* 11, no. 4 (January 1962): 1.

14. US House of Representatives, Committee on Science and Astronautics, *Research on Mechanical Translation*, 86th Congress, second session, Serial d, House Report No. 2021 (Washington, DC: Government Printing Office, 1960), 11–12.

15. For informative descriptions of Georgetown's characteristic approach to MT research under Dostert, see Anthony F. R. Brown, "Machine Translation: Just a Question of Finding the Right Programming Language?," in Hutchins, ed., *Early Years in Machine Translation* (2000): 129–134; and R. Ross Macdonald, "The Problem of Machine Translation," in Bożena Henisz-Dostert, R. Ross Macdonald, and Michael Zarechnak, *Machine Translation* (The Hague: Mouton, 1979): 89–145, on 133–135.

16. Philip H. Smith, Jr., "Machine Analysis of Russian Lexical Terms in Organic Chemistry," Georgetown University Occasional Papers on Machine Translation No. 24, August 1959, GUA-MTP, p. 1. Lawrence Summers, a graduate student in chemistry, was hired for two years as a consultant to the project.

17. Macdonald, *General Report*, 9; Group III, "Machine Translation Glossary: Complete List of Forms Occurring in the Corpus of the Journal of General Chemistry Vol. XXII, 1952, Moscow," MT Seminar Work Paper, Series B No 2, 1958, GUA-MTP.

18. Dr. S. Glazer, "Article Requirements of Plural Nouns in Russian Chemistry Texts," Seminar Work Paper MT-42, 1957, GUA-MTP.

19. Dostert, "Georgetown University Machine Translation Project: Proposal for Extension and Expansion of Research, 1961–1962," 1 December 1959, GUA-SLL 1:1959; Macdonald, *General Report*, 212; Brown, "Machine Translation," 131–132.

20. Martin Kay, "Automatic Translation of Natural Languages," *Daedalus* 102, no. 3 (Summer 1973): 217–230, on 218–219.

21. Léon Dostert, "The Georgetown-I.B.M. Experiment," in William N. Locke and A. Donald Booth, eds., *Machine Translation of Languages: Fourteen Essays* (Cambridge, MA: MIT Press, 1955): 124–135, on 124n.

22. Oettinger, "Machine Translation at Harvard," 79.

23. Andrew D. Booth and Kathleen H. V. Booth, "The Beginnings of MT," in Hutchins, ed., *Early Years in Machine Translation* (2000): 253–261, on 255.

24. Peter Toma, "From SERNA to SYSTRAN," in Hutchins, ed., *Early Years in Machine Translation* (2000): 135–145, on 138. For Lehmann, see the quotation in Vasconcellos, "The Georgetown Project and Léon Dostert," 91.

25. Section 901 of Public Law 85–864, quoted in US House of Representatives, *Research on Mechanical Translation*, 3.

26. Victor H. Yngve, "Early Research at M.I.T.: In Search of Adequate Theory," in Hutchins, ed., *Early Years in Machine Translation* (2000): 39–72, on 50–51.

27. Dostert's preface to Macdonald, *General Report*, xiii; and Andrew D. Booth, "The Nature of a Translating Machine," *Engineering* 182 (7 September 1956): 302–304, on 302.

28. I. I. Revzin and V. Iu. Rozentsveig, *Osnovy obshchego i mashinnogo perevoda* (Moscow: Vysshaia shkola, 1964), 46–51; Bar-Hillel, *Language and Information*, 186.

29. Wallace R. Brode's testimony to US Senate, Committee on Government Operations, *Science and Technology Act of 1958, on S. 3126*, 85th Congress, 2nd Session (Washington: U.S. Government Printing Office, 1958), 30. On the scale of *Chemical Abstracts*, along with time-series data, see Edward P. Donnell, "Growth and Change of the World's Chemical Literature as Reflected in *Chemical Abstracts*," *Publishing Research Quarterly* (Winter 1994/1995): 38–46.

30. Boris I. Gorokhoff, *Publishing in the U.S.S.R.* (Washington, DC: Council on Library Resources, 1959), 135–136.

31. Department of Scientific and Industrial Research, *Scientific and Technical Information in the Soviet Union: Report of the D.S.I.R.–Aslib Delegation to Moscow and Leningrad, 7th–24th June, 1963* (London: DSIR, 1964), 4. On VINITI's history and activities, see A. I. Mikhailov, A. I. Chernyi, and R. S. Giliarevskii, *Osnovy nauchnoi informatsii* (Moscow: Nauka, 1965), 7; A. I. Mikhailov, ed., *Izdaniia VINITI za 15 let, 1952–1966: Bibliograficheskii ukazatel'* (Moscow: VINITI, 1969); and A. I. Chernyi, *Vserossiiskii institut nauchnoi i tekhnicheskoi informatsii: 50 let sluzheniia nauke* (Moscow: VINITI, 2005). On the offset prints, see Boris I. Gorokhoff, *Providing U.S. Scientists with Soviet Scientific Information*, rev. ed. (Washington, DC: Publications Office of the NSF, 1962), 11; and Emilio Segrè, "What's Science Like in Russia?," *Chemical & Engineering News* (20 August 1956): 4058–4066, on 4060.

32. Gorokhoff, *Publishing in the U.S.S.R.*, 140; Mordecai Hoseh, "Scientific and Technical Literature of the USSR," *Advances in Chemistry* 30 (1961): 144–171, on 151–152.

33. Gordon E. Randall, "Research in Libraries and Information Centers," in John P. Binnington, ed., *Mutual Exchange in the Scientific Library and Technical Information Center Fields: A Report from the Special Libraries Association Delegation to the Soviet Union, 1966* (New York: Special Libraries Association, 1966): 43–50, on 48.

34. Ruggero Giliarevskii, "Soviet Scientific and Technical Information System: Its Principles, Development, Accomplishments, and Defects," in Mary Ellen Bowden, Trudi Bellardo Hahn, and Robert V. Williams, eds., *Proceedings of the 1998 Conference on the History and Heritage of Science Information Systems* (Medford, NJ: Information Today, Inc., 1999): 195–205, on 201.

35. John Turkevich, "Science," in Harold H. Fisher, ed., *American Research on Russia* (Bloomington: Indiana University Press, 1959): 103–112, on 110.

36. Gorokhoff, *Providing U.S. Scientists with Soviet Scientific Information*, 24.

37. H. H. Goldsmith to Henry A. Barton, 16 June 1948, and Henry A. Barton to J. R. Klein [*sic*: Kline], 2 July 1948, both in AIP, Barton Records, 78:3; J. R. Kline, "Soviet Mathematics," in Ruth C. Christman, ed., *Soviet Science* (Washington, DC: AAAS, 1952), 80–84, on 83.

38. NSF, *The Third Annual Report: Year Ending June 30, 1953* (Washington, DC: US Government Printing Office, 1953), 56–57; UNESCO, *Scientific and Technical Translating: And Other Aspects of the Language Problem* (Geneva: UNESCO, 1957), 129; Kurt Gingold, "Translation Pools—Ideal and Reality," *Journal of Chemical Documentation* 1, no. 2 (1961): 14–19, on 15.

39. Elmer Hutchisson, "Preliminary Statistics Relating to American Institute of Physics Translation Survey, Sponsored by the National Science Foundation," 30 June 1954 (revised 7 August 1954), AIP, Barton Records 78:4.

40. Earl Maxwell Coleman, "The Mass Production of Translation—For a Limited Market," *Publishing Research Quarterly* 10, no. 4 (Winter 1994–1995): 22–29. Emphasis in original.

41. Coleman, "The Mass Production of Translation," 25. Emphasis in original.

42. "Hurdling the Language Barrier," *Chemical & Engineering News* 32, no. 52 (27 December 1954): 5158–5159.

43. A. Tybulewicz, "Cover-to-Cover Translations of Soviet Scientific Journals," *Aslib Proceedings* 22, no. 2 (1970): 55–62, on 58–59.

44. "Hurdling the Language Barrier," 5158.

45. Paul A. Kiefer, Jr., "On Translating Chemical Russian," *Journal of Chemical Documentation* 10, no. 2 (1970): 119–124.

46. G. I. Braz, "The Reaction of Ethylene Sulfide with Amines," *Journal of General Chemistry of the USSR* 21, no. 4 (April 1951): 757–762, on 757.

47. G. I. Braz, "Vzaimodeistvie etilensul'fida s aminami," *Zhurnal obshchei khimii* 21, no. 4 (1951): 688–693, on 688.

48. Braz, "The Reaction of Ethylene Sulfide with Amines," 758.

49. Braz, "Vzaimodeistvie etilensul'fida s aminami," 689.

50. Karl F. Heumann and Peter M. Bernays, "Fifty Foreign Languages at *Chemical Abstracts*," *Journal of Chemical Education* 36, no. 10 (October 1959): 478–482, on 481.

51. David Kaiser, "The Physics of Spin: Sputnik Politics and American Physicists in the 1950s," *Social Research* 73, no. 4 (Winter 2006): 1225–1252; V. Ambegaokar, "The Landau School and the American Institute of Physics Translation Program," *Physics—Uspekhi* 51, no. 12 (2008): 1287–1290.

52. Dwight E. Gray, "American Institute of Physics Translation Survey: Final Report on Part II," 23 February 1955, AIP, Barton Records, 78:4, p. 1.

53. Elmer Hutchisson, "Preliminary Statistics Relating to American Institute of Physics Translation Survey."

54. Elmer Hutchisson, "Final Report, American Institute of Physics Translation Survey Sponsored by the National Science Foundation," 6 October 1954, AIP, Hutchisson Records 13:35.

55. Ralph E. O'Dette, "Russian Translation," *Science* 125 (29 March 1957): 579–585, on 582. For the translator's lot, see the account by Freeman Dyson, quoted in Ambegaokar, "The Landau School and the American Institute of Physics," 1289.

56. Dwight Gray, notes on a conversation with E. J. Crane, 9 November 1954, AIP, Barton Records 78:4.

57. O'Dette, "Russian Translation," 582.

58. NSF, *Sixth Annual Report for the Fiscal Year Ended June 30, 1956* (Washington, DC: U.S. Government Printing Office, 1956), 81.

59. For lists, see Gorokhoff, *Providing U.S. Scientists with Soviet Scientific Information*, 16–19; O'Dette, "Russian Translation," 580.

60. C. E. Sunderlin to Verner W. Clapp, 22 September 1955, AIP, Barton Records, Box 59, Folder: "JETP Dr. R. T. Beyer." On foreign copyright in the Soviet Union, see Gorokhoff, *Publishing in the U.S.S.R.*, 68.

61. The archives are littered with discussions of this question. See, for example, Barton to Beyer, 16 December 1955; Barton to Beyer, D. Gray, M. Hamermesh, E. Hutchisson, V. Rojansky, W. Waterfall, and V. Weisskopf, 11 October 1955; V. Rojansky to Beyer, 12 October 1955; Morton Hamermesh to Beyer, 14 October 1955; Victor Weisskopf to Beyer, 13 October 1955; Beyer to Henry A. Barton, 7 October 1955, all in AIP, Barton Records, Box 59, Folder: "JETP Dr. R. T. Beyer"; Henry A. Barton, memorandum on "Soviet Physics—JETP," 23 September 1955, AIP, Hutchisson Records 13:34; and Ralph A. Sawyer to Alan T. Waterman, 11 October 1960, AIP, Hutchisson Records 13:38.

62. "Greetings to Comrade J. V. Stalin from the Academy of Sciences of the USSR," *Journal of General Chemistry of the USSR* 20, no. 1 (January 1950): 1–2. The original is "Privetstvie tovarishchu I. V. Stalinu ot Akademii Nauk SSSR," *Zhurnal obshchei khimii* 20, no. 1 (1950): 3–4.

63. Hutchisson to Beyer, 2 September 1955, AIP, Hutchisson Records 13:34. For heated discussions in the periodical literature, see O'Dette, "Russian Translation," 584; Gregory Razran, "Transliteration of Russian," *Science* 129, no. 3356 (24 April 1959): 1111–1113; and Razran, Eric P. Hamp, A. C. Fabergé, Miriam B. London, Ivan D. London, and David T. Ray, "Russian-English Transliteration," *Science* 130, no. 3374 (28 August 1959): 482–488.

64. Beyer, "Report of Soviet Physics JETP for 1956," 25 January 1957, AIP, Barton Records, Box 59, Folder: "JETP Dr. R. T. Beyer."

65. E. J. Crane to Henry Barton, 5 June 1948, AIP, Barton Records 78:3; Alberto Thompson to Alan Waterman, 1 March 1956, NSF Records, RG 307, Box 18, Folder: "Scientific Information."

66. Moira Phillips, "The Translation Problem in Science," *Revue de la documentation* 28, no. 2 (1961): 52–55, on 54; and D. N. Wood, "Chemical Literature and the Foreign-Language Problem," *Chemistry in Britain* 2 (1966): 346–350, on 348.

67. Robert T. Beyer, "Hurdling the Language Barrier," *Physics Today* (January 1965): 46–52, on 48.

68. V. K. Rangra, "A Study of Cover to Cover English Translations of Russian Scientific and Technical Journals," *Annals of Library Science and Documentation* 15 (1968): 7–23, on 9; Melville J. Ruggles, "Translations of Soviet Publications," *College & Research Libraries* 20 (September 1959): 347–352, on 348.

69. Eugene Garfield, "'Cover-to-Cover' Translations of Soviet Journals—A Wrong 'Solution' of the Wrong Problem," in Garfield, *Essays of an Information Scientist*, 3 v. (Philadelphia: ISI Press, 1977–1980), I:334–335, on 334.

70. Phillips, "The Translation Problem in Science," 54.

71. Dostert, "Brief History of Machine Translation Research," 7.

72. Bar-Hillel, "The Present Status of Automatic Translation of Languages," *Advances in Computing* 1 (1960): 91–163, quotations on 109, 94, 135.

73. Bar-Hillel, "Machine Translation," *Computers and Automation* 2, no. 5 (July 1953): 1–6, on 1; idem, "The Present State of Research on Mechanical Translation," *American Documentation* 2, no. 4 (1951): 229–237, on 230.

74. Jonathan Slocum, "A Survey of Machine Translation: Its History, Current Status, and Future Prospects," *Computational Linguistics* 11, no. 1 (January-March 1985): 1–17, on 4.

75. W. J. Hutchins, "Machine Translation and Machine-Aided Translation," *Journal of Documentation* 34, no. 2 (June 1978): 119–159, on 126.

76. The evaluator's verdict: "In brief, the translation can be understood by a chemist thoroughly familiar with the subject under discussion. A competent chemist can, from a study of the chemical formulas, reconstruct many of the chemical terms which appear to be a cross between a translation and a transliteration." Quoted in US House of Representatives, *Research on Mechanical Translation*, 48.

77. Vasconcellos, "The Georgetown Project and Léon Dostert," 94–95.

78. Ross Macdonald assumed control of the Georgetown program, and it was reintegrated with the Institute of Languages and Literatures. "MT Research at Georgetown," *Finite String* 2, no. 6 (June 1965): 4.

79. Léon Dostert, "Machine Translation and Automatic Data Processing," in Paul W. Howerton and David C. Weeks, eds., *Vistas in Information Handling, Volume I: The Augmentation of Man's Intellect by Machine* (Washington, DC: Spartan Books, 1963): 92–110, on 94. Emphasis in original.

80. ALPAC, *Language and Machines*, quotations on iii, 16, and 23–24.

81. Kay, "Automatic Translation of Natural Languages"; Slocum, "A Survey of Machine Translation," 1. See also Michael Zarechnak, "The History of Machine Translation," in Henisz-Dostert, Macdonald, and Zarechnak, *Machine Translation* (1979): 1–87, on 57, 86.

82. Locke, "Machine Translation," 137.

83. Mel'čuk, "Machine Translation and Formal Linguistics in the USSR," in Hutchins, ed., *Early Years in Machine Translation* (2000): 204–226, on 221. On the negative international ramifications of ALPAC, which seemed to exclude only France, see Hutchins, *Machine Translation*, 167.

84. Elmer Hutchisson to Members of the Advisory Committee on Russian Translation, 27 December 1955, AIP, Hutchisson Records 13:34.

85. Wallace Waterfall to Robert Beyer, 9 August 1956, AIP, Hutchisson Records 13:30.

86. Coleman, "The Mass Production of Translation," 26–27.

87. Beyer to Waterfall, 15 November 1965, AIP, Koch Records 51:32.

88. Beyer to Waterfall, 16 February 1966, AIP, Koch Records 51:32; Rangra, "A Study of Cover to Cover English Translations of Russian Scientific and Technical Journals," 9.

89. Tybulewicz, "Cover-to-Cover Translations of Soviet Scientific Journals," 57.

90. Rangra, "A Study of Cover to Cover English Translations of Russian Scientific and Technical Journals," 8; Alice Frank, "Translations of Russian Scientific and Tech-

nical Literature in Western Countries," *Revue de la documentation* 28, no. 2 (1961): 47–51, on 48.

91. NSF, *Tenth Annual Report for the Fiscal Year Ended June 30, 1960* (Washington, DC: US Government Printing Office, 1960), 120–121.

92. Quoted in "Hurdling the Language Barrier," 5158.

93. Beyer, "Hurdling the Language Barrier," 48, 50.

94. Vannevar Bush, *Science: The Endless Frontier* (Washington, DC: Government Printing Office, 1945), 114.

Chapter Ten

1. Hermann Hesse, *Das Glasperlenspiel* (Frankfurt: Suhrkamp Verlag, 1963 [1943]), 324.

2. Franz Thierfelder, *Die deutsche Sprache im Ausland*, 2 v. (Hamburg: R. v. Decker, 1956–1957), I:114.

3. Michael Balfour, "Four-Power Control in Germany 1945–1946," in Balfour and John Mair, *Four-Power Control in Germany and Austria, 1945–1946* (London: Oxford University Press, 1956), 46.

4. Norman M. Naimark, *The Russians in Germany: A History of the Soviet Zone of Occupation, 1945–1949* (Cambridge, MA: Belknap, 1995), 20–21.

5. Balfour, "Four-Power Control in Germany 1945–1946," 74–75.

6. Balfour, "Four-Power Control in Germany 1945–1946," 109–110.

7. Naimark, *The Russians in Germany*, 39–41.

8. Max von Laue, "A Report on the State of Physics in Germany," *American Journal of Physics* 17 (1949): 137–141.

9. R. C. Evans, "Naturforschung in Deutschland," *Physikalische Blätter* 3, no. 1 (1947): 12–14, on 12.

10. Klaus Hentschel, *The Mental Aftermath: The Mentality of German Physicists, 1945–1949*, tr. Ann M. Hentschel (New York: Oxford University Press, 2007).

11. J. G. Tolpin, J. Danaczko Jr., R. A. Liewald, et al., "The Scientific Literature Cited by Russian Organic Chemists," *Journal of Chemical Education* 28 (May 1951): 254–258, on 256.

12. Thierfelder, *Die deutsche Sprache im Ausland*, I:7, II:37. Bulgaria was still using German as a leading language of science well into the Cold War. See ibid.: II:22.

13. Ulrich Ammon, "German as an International Language," *International Journal of the Sociology of Language* 83 (1990): 135–170, on 144.

14. James F. Tent, *Mission on the Rhine: Reeducation and Denazification in American-Occupied Germany* (Chicago: University of Chicago Press, 1982), 40.

15. John Krige, *American Hegemony and the Postwar Reconstruction of Science in Europe* (Cambridge, MA: MIT Press, 2006), passim, but especially 35, 65, and 109–112.

16. Tent, *Mission on the Rhine*, 98; Mitchell G. Ash, "Scientific Changes in Germany 1933, 1945, 1990: Towards a Comparison," *Minerva* 37 (1999): 329–354, on 333–334.

17. Naimark, *The Russians in Germany*, 441–443; John Connelly, *Captive University: The Sovietization of East German, Czech, and Polish Higher Education, 1945–1956* (Chapel Hill: University of North Carolina Press, 2000), 203.

18. Falk Pingel, "Attempts at University Reform in the British Zone," tr. David Phillips and Alison Goodall, in David Phillips, ed., *German Universities after the Sur-*

render: British Occupation Policy and the Control of Higher Education (Oxford: University of Oxford Department of Educational Studies, 1983), 20–27, on 22.

19. Burghard Weiss, "The 'Minerva' Project: The Accelerator Laboratory at the Kaiser Wilhelm Institute/Max Planck Institute of Chemistry: Continuity in Fundamental Research," in Monika Renneberg and Mark Walker, eds., *Science, Technology and National Socialism* (Cambridge: Cambridge University Press, 1994): 271–290.

20. David Phillips, "The Re-opening of Universities in the British Zone: The Problem of Nationalism and Student Admissions," in Phillips, ed., *German Universities after the Surrender* (1983): 4–19, on 4.

21. Tent, *Mission on the Rhine*, 238, 247–249, 299.

22. Minutes of the 97th Faculty Meeting, 21 November 1959, FUA, MatNatFak 2.6, Fak.-Protokolle (1959), 395.

23. "Promotionsordnung der Mathematisch-Naturwissenschaftlichen Fakultät der Freien Universität Berlin," 18 January 1953, FUA, MatNatFak 2.6, Fak.-Protokolle (1953), paragraph 12.

24. Edward Hartshorne to his wife, 28 September 1945, reproduced in James F. Tent, ed., *Academic Proconsul: Harvard Sociologist Edward Y. Hartshorne and the Reopening of German Universities 1945–1946* (Trier: Wissenschaftlicher Verlag Trier, 1998), 128.

25. Ernst Brücke, "Ost und West," *Physikalische Blätter*, no. 5 (1948): 224.

26. Mark Walker, "The German Physical Society under National Socialism in Context," in Dieter Hoffmann and Mark Walker, eds., *The German Physical Society in the Third Reich: Physicists between Autonomy and Accommodation*, tr. Ann M. Hentschel (Cambridge: Cambridge University Press, 2012 [2007]): 1–21, on 19.

27. Ute Deichmann, *Biologists under Hitler*, tr. Thomas Dunlap (Cambridge, MA: Harvard University Press, 1996 [1992]), 292.

28. Hentschel, *The Mental Aftermath*, 31. The Americans and British also heavily subsidized the purchase of German translation rights for English-language material, paid for using occupation currencies, which provided a substantial boost for West German publishing. See Balfour, "Four-Power Control in Germany 1945–1946," 224.

29. Gerhard Rammer, "'Cleanliness among Our Circle of Colleagues': The German Physical Society's Policy toward Its Past," in Hoffmann and Walker, eds., *The German Physical Society in the Third Reich* (2012): 367–421, on 383–384.

30. On the Soviets' conflicted relationship toward East German science, see Dolores L. Augustine, *Red Prometheus: Engineering and Dictatorship in East Germany, 1945–1990* (Cambridge, MA: MIT Press, 2007), 144.

31. Connelly, *Captive University*, 40.

32. Raymond G. Stokes, "Chemistry and the Chemical Industry under Socialism," in Kristie Macrakis and Dieter Hoffmann, eds., *Science under Socialism: East Germany in Comparative Perspective* (Cambridge, MA: Harvard University Press, 1999): 199–211.

33. On the upheavals in Soviet education, see Michael David-Fox, *Revolution of the Mind: Higher Learning among the Bolsheviks, 1918–1929* (Ithaca: Cornell University Press, 1997).

34. Connelly, *Captive University*, 62.

35. Naimark, *The Russians in Germany*, 234; Peter Nötzoldt, "From German

Academy of Sciences to Socialist Research Academy," tr. Thomas Dunlap, in Macrakis and Hoffmann, eds., *Science under Socialism* (1999): 140–157, figures on 140.

36. Eckart Förtsch, "Science, Higher Education, and Technology Policy," tr. Deborah Lucas Schneider, in Macrakis and Hoffmann, eds., *Science under Socialism* (1999): 25–43, on 32.

37. Connelly, *Captive University*, 42, 51.

38. Augustine, *Red Prometheus*, 174.

39. Kristie Macrakis, "Espionage and Technology Transfer in the Quest for Scientific-Technical Prowess," in Macrakis and Hoffmann, eds., *Science under Socialism* (1999): 82–121, on 83–84.

40. Ferenc Fodor and Sandrine Peluau, "Language Geostrategy in Eastern and Central Europe: Assessment and Perspectives," in Jacques Maurais and Michael A. Morris, eds., *Languages in a Globalising World* (Cambridge: Cambridge University Press, 2003): 85–98, on 86–87; Péter Medgyes and Mónika László, "The Foreign Language Competence of Hungarian Scholars: Ten Years Later," in Ulrich Ammon, ed., *The Dominance of English as a Language of Science: Effects on Other Languages and Language Communities* (Berlin: Mouton de Gruyter, 2001): 261–286, on 263. On Romania, see David C. Gordon, *The French Language and National Identity (1930–1975)* (The Hague: Mouton, 1978), 66.

41. Theme Plan for "Deutsch-Russisches Wörterbuch," to run from 1 September 1961 to 31 December 1970, BBAW, Zentralinstitut für Sprachwissenschaft 10, VA 4760/3, Appendix 3.

42. Akademie-Verlag, publication plan for 1954, BBAW, Bestand Klassen 59; Alice Frank, "Translations of Russian Scientific and Technical Literature in Western Countries," *Revue de la documentation* 28, no. 2 (1961): 47–51, on 49. On SVAG's activities, see Naimark, *The Russians in Germany*, 447.

43. "Bericht über die Entwicklung der Deutschen Akademie der Wissenschaften zu Berlin in der Periode zwischen dem V. und dem VI. Parteitag der Sozialistischen Einheitspartie Deutschlands," 16 December 1962, BBAW, Bestand Akademieleitung 594, p. 13.

44. I. Schilling to Rector of Humboldt University, 5 June 1954, HUA, Rektorat 373, ll.170–190, on 170.

45. Horst Dieter Schlosser, "Die Sprachentwicklung in der DDR im Vergleich zur Bundesrepublik Deutschland," in Manfred Hättich and Paul Dietmar Pfitzner, eds., *Nationalsprachen und die Europäische Gemeinschaft: Probleme am Beispiel der deutschen, französischen und englischen Sprache* (Munich: Olzog, 1989): 36–52.

46. Staatssekretar Prof. Dr. Harig, "Anweisung Nr. 91 des Staatssekretariats für Hochschulwesen über Ziele, Aufgaben und Organisation des Sprachunterrichts an den Universitäten und Hochschulen," 8 February 1957, HUA, Rektorat 373, ll.36–37ob., on 36.

47. For example, M. Lehnert to Rector W. Neye, 10 March 1956, HUA, Rektorat 373, ll.44–45.

48. Dr. Schilling, "Zur Vorlage an der Senat der Humboldt-Universität," May 1960, HUA, Rektorat 373, l.1.

49. Prof. Dr. Werner Hartke to Minister of Volksbildung and Staatssekretär für das Hochschulwesen, 16 December 1956, HUA, Rektorat 373, ll.101–108, on 101. Em-

phasis in the original. Mathematics and Chemistry were particularly emphatic about the importance of Russian for keeping abreast of the scholarly literature (ibid., 103).

50. Prof. Dr. Werner Hartke to Minister of Volksbildung and Staatssekretär für das Hochschulwesen, 16 December 1956, HUA, Rektorat 373, ll.101–108, on 101.

51. Prof. Dr. Werner Hartke to Minister of Volksbildung and Staatssekretär für das Hochschulwesen, 16 December 1956, HUA, Rektorat 373, ll.101–108, on 102. That said, a special "Russian for Physicians" course began to be offered at Humboldt after 1959. See I. Schilling to the Prorektorat für Forschungsangelegenheiten, 11 March 1959, HUA, Philosophische Fakultät (nach 1945) 0, l.2.

52. For earlier contexts and comparisons with more recent times, see Ann M. Blair, *Too Much to Know: Managing Scholarly Information before the Modern Age* (New Haven: Yale University Press, 2010). On the development of these "secondary" publications in chemistry, see Helen Schofield, "The Evolution of the Secondary Literature in Chemistry," in Mary Ellen Bowden, Trudi Bellardo Hahn, and Robert V. Williams, eds., *Proceedings of the 1998 Conference on the History and Heritage of Science Information Systems* (Medford, NJ: Information Today, 1999): 94–106.

53. Richard Willstätter, "Zur Hundertjahrfeier des Chemischen Zentralblattes," *Zeitschrift für angewandte Chemie* 42, no. 45 (9 November 1929): 1049–1052, on 1052. Throughout, I have drawn details of the general history of the *Zentralblatt* from Christian Weiske, "Das Chemische Zentralblatt—ein Nachruf," *Chemische Berichte* 106, no. 4 (1973): i–xvi; Maximilian Pflücke, "Das Chemische Zentralblatt 125 Jahre alt," *Angewandte Chemie* 66, no. 17/18 (1954): 537–541; and Dunja Langanke, "Das Chemische Zentralblatt im Wandel der Zeiten—der Weg von der gedruckten zur elektronischen Ausgabe," *Information: Wissenschaft & Praxis* 60, no. 3 (2009): 143–150.

54. Adolf Windaus to Herr Stille, 24 October 1946, BBAW, Akademieleitung—Gesellschaftwiss. Einricht. 101.

55. Fritz Pangritz et al. to Erich Thilo, 23 May 1947, BBAW, Bestand Adlershof C 494.

56. Georg Kurt Schauer to Akademie-Verlag, 1 October 1947, BBAW, Akademieleitung—Gesellschaftwiss. Einricht. 101.

57. Pangritz et al. to Thilo, 23 May 1947, BBAW, Bestand Adlershof C 494.

58. Kaeser to the Deutsche Verwaltung für Volksbildung in the Soviet Occupation Zone, Abteilung Verlagswesen, 9 November 1948, BBAW, Akademieleitung—Gesellschaftwiss. Einricht. 101.

59. Pflücke, "Das Chemische Zentralblatt 125 Jahre alt," 540; Langanke, "Das Chemische Zentralblatt im Wandel der Zeiten."

60. "Die Deutsche Akademie der Wissenschaften zu Berlin," May 1966, BBAW, Bestand Akademieleitung 594, p. 16.

61. Langanke, "Das Chemische Zentralblatt im Wandel der Zeiten," 146.

62. Weiske, "Das Chemische Zentralblatt."

63. Langanke, "Das Chemische Zentralblatt im Wandel der Zeiten," 147.

64. On English within West Germany, see Michael Clyne, *The German Language in a Changing Europe* (Cambridge: Cambridge University Press, 1995), 201; and Theodore Huebener, "The Teaching of Foreign Languages in the Schools of West Germany," *Modern Language Journal* 46 (1962): 69–70. On state promotion of the German language, see Ulrich Ammon, "The Federal Republic of Germany's Policy of Spreading German," *International Journal of the Sociology of Language* 95 (1992): 33–50.

65. For example, in 1959 West Germany established a Russian Scientific and Technical Literature Section at the Foreign Technical Information Centre of the Technical University in Hamburg. Frank, "Translations of Russian Scientific and Technical Literature in Western Countries," 49. See also Günther Reichardt, ed., *Sowjetische Literatur zur Naturwissenschaft und Technik: Bibliographischer Wegweiser*, 2d ed. (Wiesbaden: Franz Steiner Verlag, 1959 [1957]); and John Turkevich, "Science," in Harold H. Fisher, ed., *American Research on Russia* (Bloomington: Indiana University Press, 1959): 103–112, on 111. On the negative impact of Anglophone cover-to-cover journals on German scientists' competence in the Russian language, see Dr. Edmund Marsch (MPG) to Ambassador Dr. Wilhelm Haas (President of the Deutsche Gesellschaft für Osteuropakunde), 26 February 1965, AMPG, II. Abt., Rep. 1A, Nr. IM2/ - Wissenschaftl. Zusammenarbeit mit dem Auslande, UdSSR v. 1, p. 1.

66. Hentschel, *The Mental Aftermath*, 88.

67. "Kurzprotokoll über die Besprechung am 19. Juli 1967 im Bundesministerium für wissenschaftliche Forschung," 19 July 1967, AMPG, II. Abt., Rep. 1A, Nr. IM5/1 Abwanderung deutscher Wissenschaftl.

68. Marsch to President of the MPG, 24 June 1968, AMPG, II. Abt., Rep. 1A, Nr. IM5/1 Abwanderung deutscher Wissenschaftl.

69. Arnold Ebel to the Pressestelle of the MPG, 22 September 1969, AMPG, II. Abt., Rep. 1A, Nr. IM5/1 Abwanderung deutscher Wissenschaftl.; Roeske to Friedrich Helm, 25 November 1966, AMPG, II. Abt., Rep. 1A, Nr. IM5/1 Abwanderung deutscher Wissenschaftl.

70. For more on Beilstein's life and career in St. Petersburg, see Michael D. Gordin, "Beilstein Unbound: The Pedagogical Unraveling of a Man and His *Handbuch*," in David Kaiser, ed., *Pedagogy and the Practice of Science: Historical and Contemporary Perspectives* (Cambridge, MA: MIT Press, 2005): 11–39. For a compact general chronology of the development of the *Handbuch*, see *Einhundert Jahre Beilsteins Handbuch der Organischen Chemie* (Würzburg: H. Stürtz, 1981), 9–23.

71. Friedrich Beilstein, *Handbuch der Organischen Chemie*, 2 v. (Hamburg and Leipzig: Leopold Voss, 1883). On the expansion of pages and compounds, see Friedrich Richter, "Beilsteins Handbuch—75 Jahre organisch-chemischer Dokumentation," *Angewandte Chemie* 70 (1958): 279–284, on 280.

72. Beilstein to Heinrich Göppert, 6/18 September 1881, reproduced in Elena Roussanova, *Friedrich Konrad Beilstein, Chemiker zweier Nationen: Sein Leben und Werk sowie einige Aspekte der deutsch-russischen Wissenschaftsbeziehungen in der zweiten Hälfte des 19. Jahrhunderts im Spiegel seines brieflichen Nachlasses*, vol. 2 (Hamburg: Norderstedt, 2007), 299.

73. "Rundschreiben" from the protocol of the meeting of the German Chemical Society, 4 February 1896, in *Berichte der Deutschen Chemischen Gesellschaft* 29 (1896): 321–324.

74. Volhard to Erlenmeyer, 27 March 1896, reproduced in Otto Krätz, ed., *Beilstein-Erlenmeyer: Briefe zur Geschichte der chemischen Dokumentation und des chemischen Zeitschriftenwesens* (Munich: Werner Fritsch, 1972), 84.

75. Ernest Hamlin Huntress, "1938: The One Hundredth Anniversary of the Birth of Friedrich Konrad Beilstein (1838–1906)," *Journal of Chemical Education* 15 (1938): 303–309, on 309. The same point was reiterated as late as 1990: Gayle S. Baker and David C. Baker, "Experiences of Two Academic Users: Evaluation of Applications

in an Academic Environment," in Stephen R. Heller, ed., *The Beilstein Online Database: Implementation, Content, and Retrieval* (Washington, DC: American Chemical Society, 1990), 130–142, on 131.

76. Marion E. Sparks, "Chemical Literature and Its Use," *Science* 47, no. 1216 (19 April 1918): 377–381, on 377.

77. John Theodore Fotos and R. Norris Shreve, eds., *Advanced Readings in Chemical and Technical German from Practical Reference Books* (New York: John Wiley & Sons, 1940), ii.

78. Paul Jacobson, "Beilsteins Handbuch der Organischen Chemie, ein Spiegel ihrer Entwicklung," *Die Naturwissenschaften* 7 (1919): 222–225, on 225.

79. Ute Deichmann, "'To the Duce, the Tenno and Our Führer: A Threefold Sieg Heil': The German Chemical Society and the Association of German Chemists during the Nazi Era," in Hoffmann and Walker, eds., *The German Physical Society in the Third Reich* (2012): 280–316, on 291–293; Roger Adams, "Beilstein Marks 75th Anniversary," *Chemical and Engineering News* (1956): 6310, reprinted in Friedrich Richter, ed., *75 Jahre Beilsteins Handbuch der Organischen Chemie: Aufsätze und Reden* (Berlin: Springer Verlag, 1957): 14–16. For the number of collaborators, see Friedrich Richter, "How Beilstein Is Made," tr. Ralph E. Oesper, *Journal of Chemical Education* 15 (1938): 310–316, on 316.

80. F. Richter, "Jahresbericht des Vorstands des Beilstein-Instituts für das Geschäftsjahr 1951," [17 June 1952], AMPG, II. Abt., Rep. 1A, Nr. IL1/i-Beilstein, v. 1, pp. 1–3. See also Richter, "Beilsteins Handbuch—75 Jahre organisch-chemischer Dokumentation," 281.

81. Roger Fennell, *History of IUPAC, 1919–1987* (Oxford: Blackwell Science, 1994), 86.

82. Friedrich Richter, "75 Jahre Beilsteins Handbuch der Organischen Chemie: Ein Jubiläum der Wissenschaft," *Frankfurter Allgemeine Zeitung*, 13 December 1956, reprinted in Richter, ed., *75 Jahre Beilsteins Handbuch der Organischen Chemie* (1957): 5–10, on 10. On the financial state of *Beilstein* in the 1950s, see Protokoll über die Stiftungsratssitzung des Beilstein-Instituts, 25 March 1958, AMPG, II. Abt., Rep. 1A, Nr. IL1/-Beilstein, v. 2, p. 4.

83. Otto Hahn to Hinsch, 14 October 1952, AMPG, II. Abt., Rep. 1A, Nr. IL1/-Beilstein, v. 1, p. 1.

84. In German, see: B. Prager, D. Stern, and K. Ilberg, *System der organischen Verbindungen: Ein Leitfaden für die Benutzung von Beilsteins Handbuch der organischen Chemie* (Berlin: Julius Springer, 1929); and Friedrich Richter, ed., *Kurze Anleitung zur Orientierung in Beilsteins Handbuch der Organischen Chemie* (Berlin: Julius Springer, 1936). In English, see Oskar Weissbach, *The Beilstein Guide: A Manual for the Use of Beilsteins Handbuch der Organischen Chemie* (Berlin: Springer-Verlag, 1976); *How to Use Beilstein: Beilstein Handbook of Organic Chemistry* (Frankfurt/Main: Beilstein Institute, [1979]); *Stereochemical Conventions in the Beilstein Handbook of Organic Chemistry* (Berlin: Springer-Verlag, [198-]); Olaf Runquist, *A Programmed Guide to Beilstein's Handbuch* (Minneapolis: Burgess Publishing, [1966]). For a bilingual guide, see Friedo Giese, *Beilstein's Index: Trivial Names in Systematic Nomenclature of Organic Chemistry* (Berlin: Springer-Verlag, 1986).

85. Reiner Luckenbach, "Der Beilstein," *CHEMTECH* (October 1979): 612–621, on 613.

86. Ernest Hamlin Huntress, *A Brief Introduction to the Use of Beilstein's Handbuch der Organischen Chemie* (New York: John Wiley and Sons, 1930), iv.

87. Otto Hahn, minutes of meeting on Beilstein held in Frankfurt US headquarters, 1 February 1950, AMPG, II. Abt., Rep. 1A, Nr. IL1/-Beilstein, v. 1, p. 9. Emphasis in original.

88. R. Fraser, minutes of meeting on Beilstein held in Frankfurt US headquarters, 1 February 1950, AMPG, II. Abt., Rep. 1A, Nr. IL1/-Beilstein, v. 1, p. 10.

89. Protokol der Stiftungsrats-Sitzung des Beilstein-Instituts, 3 December 1954, AMPG, II. Abt., Rep. 1A, Nr. IL1/-Beilstein, v. 2, p. 6.

90. Dermot A. O'Sullivan, "Germany's Beilstein Will Change to English," *Chemical and Engineering News* 59 (18 May 1981): 21–22.

91. Reiner Luckenbach, "The Beilstein Handbook of Organic Chemistry: The First Hundred Years," *Journal of Chemical Information and Computer Sciences* 21 (1981): 82–83; Luckenbach and Josef Sunkel, "Das wissenschaftliche Handbuch: 100 Jahre Beilstein," *Die Naturwissenschaften* 68, no. 2 (February 1981): 53–55.

92. Ken Rouse and Roger Beckman, "Beilstein's CrossFire: A Milestone in Chemical Information and Interlibrary Cooperation Academia," in Stephen R. Heller, ed., *The Beilstein System: Strategies for Effective Searching* (Washington, DC: American Chemical Society, 1998): 133–148, on 134–135; Ieva O. Hartwell and Katharine A. Haglund, "An Overview of DIALOG," in Heller, ed., *The Beilstein Online Database* (1990): 42–63.

93. Stephen R. Heller, "The Beilstein System: An Introduction," in Heller, ed., *The Beilstein System* (1998): 1–12, on 1–2.

94. Heller, "The Beilstein System," 5.

95. See, for example, Clyne, *The German Language in a Changing Europe*, 73; Manfred W. Hellmann, "Zwei Gesellschaften—Zwei Sprachkulturen?: Acht Thesen zur öffentlichen Sprache in der Bundesrepublik Deutschland und in der Deutschen Demokratischen Republik," *Forum für interdisziplinäre Forschung* 2 (1989): 27–38; Gotthard Lerchner, "Zur Spezifik der Gebrauchsweise der deutschen Sprache in der DDR und ihrer gesellschaftlichen Determination," *Deutsch als Fremdsprache* 11 (1974): 259–265; Herbert Christ, "Die kulturelle und politische Funktion des Deutschen als 'Nationalsprache,'" in Hättich and Pfitzner, eds., *Nationalsprachen und die Europäische Gemeinschaft* (1989): 25–35; Schlosser, "Die Sprachentwicklung in der DDR im Vergleich zur Bundesrepublik Deutschland."

96. Clyne, *The German Language in a Changing Europe*, 6.

97. Dankwart Guratzsch, "Deutsch: Die dritte Weltsprache," *Lebende Sprachen* 22 (1977): 149–150.

98. Ammon, "German as an International Language," 152.

Chapter Eleven

1. James Joyce, *A Portrait of the Artist as a Young Man* (New York: Viking, 1964 [1916]), 189.

2. James S. Miller, "Flora, Now in English," *New York Times* (22 January 2012). On the Spanish delegation, see Geoffrey C. Bowker, "The Game of the Name: Nomenclatural Instability in the History of Botanical Informatics," in Mary Ellen Bowden, Trudi Bellardo Hahn, and Robert V. Williams, eds., *Proceedings of the 1998 Confer-*

ence on the History and Heritage of Science Information Systems (Medford, NJ: Information Today, 1999): 74–83, on 79. On Botanical Latin and its persistence, see William T. Stearn, *Botanical Latin: History, Grammar, Syntax, Terminology and Vocabulary* (London: Nelson, 1966); and Françoise Waquet, *Latin, or the Empire of a Sign: From the Sixteenth to the Twentieth Centuries*, tr. John Howe (London: Verso, 2001 [1998]), 93–94.

3. A. Tybulewicz, "Languages Used in Physics Papers," *Physics Bulletin* 20, no. 1 (1969): 19–20, on 20; Ulrich Ammon, "Deutsch als Publikationssprache der Wissenschaft: Zum Umfang seiner Verwendung im Vergleich mit anderen Sprachen," *Germanistische Mitteilungen* 28 (1988): 75–86, on 78; Uwe Pörksen, "Anglisierung—der dritte große Entlehnungsvorgang in der deutschen Sprachgeschichte: Zur Einführung," in Pörksen, ed., *Die Wissenschaft spricht Englisch? Versuch einer Standortbestimmung* (Göttingen: Wallstein, 2005): 9–16, on 10. For specific data on chemistry, see also Fletcher S. Boig and Paul W. Howerton, "History and Development of Chemical Periodicals in the Field of Organic Chemistry: 1877–1949," *Science*, N.S. 115, no. 2976 (11 January 1952): 25–31; and idem, "History and Development of Chemical Periodicals in the Field of Analytical Chemistry: 1877–1950," *Science*, N.S. 115, no. 2995 (23 May 1952): 555–560. There is also a danger in focusing too heavily on numbers. For certain subfields, such as tropical medicine, French is sometimes the more appropriate language for publication, since one's intended audience is largely Francophone. See Geoffrey Nunberg, "Les langues des sciences dans le discours électronique," in Roger Chartier and Pietro Corsi, eds., *Sciences et langues en Europe* (Paris: European Communities, 2000 [1994]): 239–247, on 240.

4. The crucial scholarly reference on this question remains Ulrich Ammon, ed., *The Dominance of English as a Language of Science: Effects on Other Languages and Language Communities* (Berlin: Mouton de Gruyter, 2001). On the globalization of English in general, see, for example: David Crystal, *English as a Global Language*, 2d ed. (Cambridge: Cambridge University Press, 2003 [1997]); Robert McCrum, *Globish: How the English Language Became the World's Language* (New York: W. W. Norton, 2010); Leslie Dunton-Downer, *The English Is Coming! How One Language is Sweeping the World* (New York: Touchstone, 2010); and Henry Hitchings, *The Secret Life of Words: How English Became English* (New York: Farrar, Straus & Giroux, 2008).

5. David Graddol, *The Future of English?* (London: British Council, 1997), 56. See also Richard W. Bailey and Jay L. Robinson, eds., *Varieties of Present-Day English* (New York: Macmillan, 1973); and Richard W. Bailey and Manfred Görlach, eds., *English as a World Language* (Ann Arbor: University of Michigan Press, 1982).

6. David Crystal, "The Past, Present, and Future of World English," in Andreas Gardt and Bernd Hüppauf, eds., *Globalization and the Future of German* (Berlin: Mouton de Gruyter, 2004): 27–45, on 39.

7. Interested readers will gain a good overview from David Crystal, *The Stories of English* (Woodstock, NY: Overlook Press, 2004).

8. Robert B. Kaplan, "English in the Language Policy of the Pacific Rim," *World Englishes* 6 (1987): 137–148, on 138.

9. Quoted in Otto Jespersen, *Growth and Structure of the English Language*, 4th ed. (New York: D. Appleton and Company, 1923), 250.

10. Jacob Ornstein, "English the Global Way," *Modern Language Journal* 46 (1962): 9–13, on 12.

11. Jespersen, *Growth and Structure of the English Language*, 251. On English as a

low priority for continental Europeans throughout the seventeenth century, see Peter Burke, *Languages and Communities in Early Modern Europe* (Cambridge: Cambridge University Press, 2004), 115.

12. Albert Léon Guérard, *A Short History of the International Language Movement* (London: T. Fisher Unwin, 1922), 42; Michael West, "English as a World Language," *American Speech* 9, no. 3 (October 1934): 163–174; Herbert Newhard Shenton, *Cosmopolitan Conversation: The Language Problems of International Conferences* (New York: Columbia University Press, 1933), 463.

13. Alexander Melville Bell, *World-English: The Universal Language* (New York: N. D. C. Hodges, 1886), 7.

14. C. K. Ogden and I. A. Richards, *The Meaning of Meaning: A Study of the Influence of Language upon Thought and of the Science of Symbolism* (San Diego: Harcourt Brace Jovanovich, 1989 [1923]).

15. I. A. Richards, *Basic English and Its Uses* (New York: W. W. Norton, [1943]), 26.

16. There were, of course, competitors with the same idea, who differed largely in the number and choice of words. H. E. Palmer, advisor to the Japanese Department of Education, developed a series of tools for teaching English at IRET (Institute for Research in English Teaching) in Tokyo, including the 1937 *Thousand-Word English* with A. S. Hornby. Elaine Swenson, head of the Language Research Institute at New York University, proffered "Swenson English," based on 900 words. On such contemporary efforts, see Janet Rankin Aiken, "English as the International Language," *American Speech* 9 (1934): 98–110, on 107–108.

17. C. K. Ogden, *Debabelization: With a Survey of Contemporary Opinion on the Problem of a Universal Language* (London: Kegan Paul, Trench, Trubner, 1931), 10.

18. I. A. Richards, "Basic English and Its Applications," *Journal of the Royal Society of Arts*, no. 4515 (2 June 1939): 733–747, reproduced in C. K. Ogden, *From Russell to Russo: Reviews and Commentaries*, ed. W. Terrence Gordon (London: Routledge/Thoemmes, 1994), 15.

19. Richards, *Basic English and Its Uses*, 24; C. K. Ogden, "A New Solution of the Universal Language Problem," *Psyche* 10 (October 1929), reproduced in Ogden, *From Significs to Orthology*, ed. W. Terrence Gordon (London: Routledge/Thoemmes Press, 1994), 106–107. See also Ogden, extracts from *Basic for Science* (London: Kegan Paul, Trench, Trubner, 1942), reproduced in Ogden, *From Bentham to Basic English*, ed. W. Terrence Gordon (London: Routledge/Thoemmes Press, 1994), 321–349.

20. Aiken, "English as the International Language," 108.

21. Richards, *Basic English and Its Uses*, 11.

22. On the War Cabinet, see Carolyn N. Biltoft, "Speaking the Peace: Language, World Politics and the League of Nations, 1918–1935" (PhD dissertation, Princeton University, 2010), on 120. On Basic in China, see Rodney Koeneke, *Empire of the Mind: I. A. Richards and Basic English in China, 1929–1979* (Stanford: Stanford University Press, 2004).

23. Lancelot T. Hogben, *Essential World English: Being a Preliminary Mnemotechnic Programme for Proficiency in English Self-Expression for International Use, Based on Semantic Principles* (New York: W. W. Norton, 1963), 9.

24. I. A. Richards, "Basic English and Its Applications," *Journal of the Royal Society of Arts*, no. 4515 (2 June 1939): 733–747, reproduced in Ogden, *From Russell to Russo*, 25.

25. See, for example, J. A. Large, *The Foreign-Language Barrier: Problems in Sci-*

entific Communication (London: André Deutsch, 1983), 147; Franz Thierfelder, *Die deutsche Sprache im Ausland*, 2 v. (Hamburg: R. v. Decker, 1956–1957), I:37; Aiken, "English as the International Language," 102; and UNESCO, *Scientific and Technical Translating: And Other Aspects of the Language Problem* (Geneva: UNESCO, 1957), 188.

26. Edmund Andrews, *A History of Scientific English: The Story of Its Evolution Based on a Study of Biomedical Terminology* (New York: Richard R. Smith, 1947), 281.

27. J. A. Lauwerys preface to Henry Jacob, *On the Choice of a Common Language* (London: Sir Isaac Pitman & Sons, 1946), vii. For the statistics, see "English Is Most Popular Scientific Language," *Science News-Letter* 56, no. 11 (10 September 1949): 166.

28. "Rapport de l'Académie des Sciences sur la langue française et le rayonnement de la science française," *Comptes Rendus* 295 (8 November 1982): 131–146, on 133.

29. Franck Ramus, "A quoi servent les revues scientifiques francophones?," available at http://www.lscp.net/persons/ramus/fr/revues.html, accessed 18 July 2012.

30. Crystal, *The Stories of English*, 454.

31. Hitchings, *The Secret Life of Words*, 182.

32. Robert B. Kaplan, "The Hegemony of English in Science and Technology," *Journal of Multilingual and Multicultural Development* 14 (1993): 151–172, on 157.

33. Robert Schoenfeld, *The Chemist's English*, 3d ed. (Weinheim: Wiley-VCH, 2001 [1985]), 143.

34. Ulrich Ammon, *Die internationale Stellung der deutschen Sprache* (Berlin: Walter de Gruyter, 1991), 265.

35. Rolf Tatje, "Fachsprachliche Kommunikation: Zum Status des Deutschen, Englischen und Französischen als Wissenschafts- und Publikationssprachen in der Mineralogie," in Theo Bungarten, ed., *Beiträge zur Fachsprachenforschung: Sprache in Wissenschaft und Technik, Wirtschaft und Rechtswesen* (Tostedt: Attikon Verlag, 1992): 73–90, on 76; Wolfgang Wickler, "Englisch als deutsche Wissenschaftssprache," in Hartwig Kalverkämper and Harald Weinrich, eds., *Deutsch als Wissenschaftssprache: 25. Konstanzer Literaturgespräch des Buchhandels, 1985* (Tübingen: Gunter Narr Verlag, 1986): 26–31, on 26; Robert A. Day, *Scientific English: A Guide for Scientists and Other Professionals* (Phoenix, AZ: Oryx Press, 1992), 9; W. Wayt Gibbs, "Lost Science in the Third World," *Scientific American* (August 1995): 92–99; Ulrich Ammon, "The International Standing of the German Language," in Jacques Maurais and Michael A. Morris, eds., *Languages in a Globalising World* (Cambridge: Cambridge University Press, 2003): 231–249, on 244; Thomas Karger, "Englisch als Wissenschaftssprache im Spiegel der Publikationsgeschichte," in Kalverkämper and Weinrich, eds., *Deutsch als Wissenschaftssprache* (1986): 48–50, on 49; Ralph A. Lewin and David K. Jordan, "The Predominance of English and the Potential Use of Esperanto for Abstracts of Scientific Articles," in M. Kageyama, K. Nakamura, T. Oshima, and T. Uchida, eds., *Science and Scientists: Essays by Biochemists, Biologists, and Chemists* (Tokyo: Japan Scientific Societies Press, 1981): 435–441, on 436; Ulrich Ammon and Grant McConnell, *English as an Academic Language in Europe: A Survey of Its Use in Teaching* (Frankfurt am Main: Peter Lang, 2002), 19; and Scott L. Montgomery, *Science in Translation: Movements of Knowledge through Cultures and Time* (Chicago: University of Chicago Press, 2000), 224.

36. All quoted in Werner Traxel, "Internationalität oder Provinzialismus? Über die Bedeutung der deutschen Sprache für deutschsprachige Psychologen," *Psychologische Beiträge* 17 (1975): 584–594, on 585–586. Ellipses in original.

37. Day, *Scientific English*. See also E. Sopher, "An Introductory Approach to the Teaching of Scientific English to Foreign Students," *English Language Teaching* 28 (July 1974): 353–359; Peter Strevens, "Technical, Technological, and Scientific English (TTSE)," *English Language Teaching* 27 (1973): 223–234.

38. Eija Ventola and Anna Mauranen, "Non-Native Writing and Native Revising of Scientific Articles," in Eija Ventola, ed., *Functional and Systemic Linguistics: Approaches and Uses* (Berlin: Mouton de Gruyter, 1991): 457–492. On the expense of translation, see Bonnie Lee La Madeleine, "Lost in Translation," *Nature* 445 (25 January 2007): 454–455, on 455.

39. Scott L. Montgomery, *Does Science Need a Global Language? English and the Future of Research* (Chicago: University of Chicago Press, 2013), 99.

40. Alan G. Gross, Joseph E. Harmon, and Michael Reidy, *Communicating Science: The Scientific Article from the 17th Century to the Present* (New York: Oxford University Press, 2002), 230.

41. Dietrich Voslamber, "Wissenschaftssprache am Scheideweg—Die Sprachenproblematik aus der Sicht eines Physikers," in Hermann Zabel, ed., *Deutsch als Wissenschaftssprache: Thesen und Kommentare zum Problemkreis "Denglisch"* (Paderborn: IFB Verlag, 2005): 87–95, on 91.

42. Joan K. Swinburne, "The Use of English as the International Language of Science: A Study of the Publications and Views of a Group of French Scientists," *Incorporated Linguist* 22 (1983): 129–132, on 131. On English at international conferences, see Sabine Skudlik, *Sprachen in den Wissenschaften: Deutsch und Englisch in der internationalen Kommunikation* (Tübingen: Gunter Narr, 1990), 113; and Tatje, "Fachsprachliche Kommunikation," 77.

43. La Madeleine, "Lost in Translation," 454; Florian Coulmas, "Les facteurs économiques et les langues de la communication scientifique: Le japonais et l'allemand," in Conseil de la langue française, Gouvernement du Québec, *Le français et les langues scientifiques de demain: Actes du colloque tenu à l'Université du Québec à Montréal du 19 au 21 mars 1996* (Quebec, Canada: Gouvernment du Québec, 1996): 69–79, on 75; Montgomery, *Science in Translation*, 201; Christian Galinski, "Information Technology and Documentation in Science and Technology in Japan," *Journal of Information Science* 5 (1982): 63–77; and C. F. Foo-Kune, "Japanese Scientific and Technical Periodicals: An Analysis of Their European Language Content," *Journal of Documentation* 26 (1970): 111–119.

44. Florian Vollmers, "Zu viel Englisch ist auch nicht gut," *Frankfurter Allgemeine Zeitung* (24 January 2012), accessed online at www.faz.net/-gyq-6wy7d.

45. Vollmers, "Zu viel Englisch ist auch nicht gut"; Ulrich Ammon, "Die Politik der deutschsprachigen Länder zur Förderung der deutschen Sprache in Russland," in Ammon and Dirk Kemper, eds., *Die deutsche Sprache in Russland: Geschichte, Gegenwart, Zukunftsperspektiven* (Munich: Iudicium, 2011): 327–343, on 337. On textbooks, see Wickler, "Englisch als deutsche Wissenschaftssprache," 28.

46. Wickler, "Englisch als deutsche Wissenschaftssprache," 27. On the general state of English in European academia, see Ammon and McConnell, *English as an Academic Language in Europe*.

47. Ralph Mocikat, Wolfgang Haße, and Hermann H. Dieter, "Sieben Thesen zur deutschen Sprache in der Wissenschaft," in Zabel, ed., *Deutsch als Wissenschaftssprache* (2005): 12–17, on 15.

48. Molina's biography, at: http://www.nobelprize.org/nobel_prizes/chemistry/laureates/1995/molina-bio.html.

49. Skou's biography, at: http://www.nobelprize.org/nobel_prizes/chemistry/laureates/1997/skou-bio.html. The other two are Yves Chauvin, born in 1930 in France and laureate in 2005, and Gerhard Ertl, born in 1936 in Bad Cannstatt, Germany, and laureate in 2007.

50. Tanaka's biography, at: http://www.nobelprize.org/nobel_prizes/chemistry/laureates/2002/tanaka-bio.html.

51. Negishi's biography, at: http://www.nobelprize.org/nobel_prizes/chemistry/laureates/2010/negishi-bio.html.

52. Zewail's biography, at: http://www.nobelprize.org/nobel_prizes/chemistry/laureates/1999/zewail-bio.html.

53. Skudlik, *Sprachen in den Wissenschaften*, 178–181. For excellent scholarly analyses of the Nobel population, see Robert Marc Friedman, *The Politics of Excellence: Behind the Nobel Prize in Science* (New York: W. H. Freeman, 2001); Elisabeth Crawford, *Nationalism and Internationalism in Science, 1880–1939: Four Studies of the Nobel Population* (Cambridge: Cambridge University Press, 1992); and István Hargittai, *The Road to Stockholm: Nobel Prizes, Science, and Scientists* (New York: Oxford University Press, 2002).

54. Hargittai, *The Road to Stockholm*, 18.

55. Max Talmey, "The Auxiliary Language Question," *Modern Language Journal* 23 (1938): 172–186, on 174; and Maryse Lafitte, "Quelques hypothèses sur la place du français et de l'anglais dans le monde actuel . . . ," in Chartier and Corsi, eds., *Sciences et langues en Europe* (2000): 179–191, on 184.

56. On simplicity, see, among many others, W. Brackenbusch, *Is English Destined to Become the Universal Language of the World?* (Göttingen: W. Fr. Kaestner, 1868), 5–6. On manliness, see Jespersen, *Growth and Structure of the English Language*, 2, 11.

57. Thierfelder, *Die deutsche Sprache im Ausland*, II:72.

58. Thierfelder, *Die deutsche Sprache im Ausland*, II:311.

59. Quoted in David C. Gordon, *The French Language and National Identity (1930–1975)* (The Hague: Mouton, 1978), 49.

60. Claude Truchot, *L'Anglais dans le monde contemporain* (Paris: Le Robert, 1990), 38; McCrum, *Globish*, 222.

61. Andrew W. Conrad and Joshua A. Fishman, "English as a World Language: The Evidence," in Joshua A. Fishman, Robert L. Cooper, and Andrew W. Conrad, eds., *The Spread of English: The Sociology of English as an Additional Language* (Rowley, MA: Newbury House, 1977): 3–76, on 7–8; Crystal, "The Past, Present, and Future of World English," 30–31.

62. C. W. Hanson, *The Foreign Language Barrier in Science and Technology* (London: Aslib, 1962), 2. See also D. N. Wood, "The Foreign-Language Problem Facing Scientists and Technologists in the United Kingdom—Report of a Recent Survey," *Journal of Documentation* 23 (1967): 117–130, on 117–118; W. J. Hutchins, L. J. Pargeter, and W. L. Saunders, *The Language Barrier: A Study in the Depth of the Place of Foreign Language Materials in the Research Activity of an Academic Community* (Sheffield: University of Sheffield, 1971); and John Gray and Brian Perry, *Scientific Information* (Oxford: Oxford University Press, 1975), 22.

63. Mario Pei, *One Language for the World* (New York: Biblo and Tannen, 1968),

45; and S. Frederick Starr, "English Dethroned," *Change* (May 1978): 26–31, with the reference to science on 30.

64. Nicholas Ostler, *The Last Lingua Franca: English until the Return of Babel* (New York: Walker, 2010), 14; Paul Cohen, "The Rise and Fall of the American Linguistic Empire," *Dissent* (Fall 2012).

65. Pei, *One Language for the World*, 104.

66. Conrad and Fishman, "English as a World Language," 13, 56; Robert Phillipson, *Linguistic Imperialism* (Oxford: Oxford University Press, 1992), 157.

67. On economic strength as a causal factor, see Ammon and McConnell, *English as an Academic Language in Europe*, 13.

68. William Grabe and Robert B. Kaplan, "Science, Technology, Language, and Information: Implications for Language and Language-in-Education Planning," *International Journal of the Sociology of Language* 59 (1986): 47–71.

69. Eugene Garfield, "English—An International Language for Science?," *Current Contents* (26 December 1967): 19–20; Kaplan, "The Hegemony of English in Science and Technology"; Montgomery, *Does Science Need a Global Language?*, 83.

70. Pamela Spence Richards, "The Soviet Overseas Information Empire and the Implications of Its Disintegration," in Bowden et al., eds., *Proceedings of the 1998 Conference on the History and Heritage of Science Information Systems* (1999): 206–214, on 210.

71. Péter Medgyes and Mónika László, "The Foreign Language Competence of Hungarian Scholars: Ten Years Later," in Ammon, ed., *The Dominance of English as a Language of Science* (2001): 261–286, on 263–264; Ammon, *Die internationale Stellung der deutschen Sprache*, 142–143; idem, "To What Extent is German an International Language?," in Patrick Stevenson, ed., *The German Language and the Real World: Sociolinguistic, Cultural, and Pragmatic Perspectives on Contemporary German* (Oxford: Clarendon Press, 1995): 25–53, on 36; Ferenc Fodor and Sandrine Peluau, "Language Geostrategy in Eastern and Central Europe: Assessment and Perspectives," in Maurais and Morris, eds., *Languages in a Globalising World* (2003): 85–98, on 87, 97.

72. Tatjana Kryuchkova, "English as a Language of Science in Russia," in Ulrich Ammon, ed., *The Dominance of English as a Language of Science: Effects on Other Languages and Language Communities* (Berlin: Mouton de Gruyter, 2001): 405–423, on 420–421.

73. Skudlik, *Sprachen in den Wissenschaften*, 212.

74. See, for a sampling of such views, Schoenfeld, *The Chemist's English*, 145; and Maurice Crosland, *Science under Control: The French Academy of Sciences, 1795–1914* (Cambridge: Cambridge University Press, 1992), 11. For the opposing position, see Michel Debré, "La langue française et la science universelle," *La Recherche* 7, no. 69 (July-August 1976): 956; and Hans Hauge, "Nationalising Science," in Chartier and Corsi, *Sciences et langues en Europe* (2000): 151–159, on 157.

75. Skudlik, *Sprachen in den Wissenschaften*, 22–23, 98.

76. Ammon, "To What Extent is German an International Language?," 45–46. See also Winfried Thielmann, *Deutsche und englische Wissenschaftssprache im Vergleich: Hinführen—Verknüpfen—Benennen* (Heidelberg: Synchron, 2009), 317; and Günter Bär, "Die nationalen Hochsprachen, z.B. Französisch und Deutsch, als Grundlagen der nationalen Kulturen in der Auseinandersetzung mit der Weltsprache Englisch," in Manfred Hättich and Paul Dietmar Pfitzner, eds., *Nationalsprachen und*

die Europäische Gemeinschaft: Probleme am Beispiel der deutschen, französischen und englischen Sprache (Munich: Olzog, 1989): 64–78.

77. Traxel, "Internationalität oder Provinzialismus?," 588.

78. G. A. Lienert, "Über Werner Traxel: Internationalität oder Provinzialismus, zur Frage: Sollten Psychologen in Englisch publizieren?," *Psychologische Beiträge* 19 (1977): 487–492.

79. Joshua A. Fishman, "English in the Context of International Societal Bilingualism," in Fishman et al., eds., *The Spread of English* (1977): 329–336, on 334–335; and also Alexander Ostrower, *Language, Law, and Diplomacy: A Study of Linguistic Diversity in Official International Relations and International Law*, 2 v. (Philadelphia: University of Pennsylvania Press, 1965), I:398. Ostrower's prime example was the contemporary (1965) ignorance of Russian on the part of American scientists.

80. Ulrich Ammon, "German as an International Language of the Sciences—Recent Past and Present," in Gardt and Hüppauf, eds., *Globalization and the Future of German* (2004): 157–172, on 164; Claude Hagège, *Contre la pensée unique* (Paris: Odile Jacob, 2012), 64.

81. Hans Joachim Meyer, "Global English—a New Lingua Franca or a New Imperial Culture?," in Gardt and Hüppauf, eds., *Globalization and the Future of German* (2004): 65–84, on 68.

82. Ulrich Ammon, "German or English? The Problems of Language Choice Experienced by German-Speaking Scientists," in P. H. Nelde, ed., *Sprachkonflikte und Minderheiten* (Bonn: Dümmler, 1990): 33–51; Miquel Siguan, "English and the Language of Science: On the Unity of Language and the Plurality of Languages," in Ammon, ed., *The Dominance of English as a Language of Science* (2001): 59–69, on 59.

83. Theo Bungarten, "Fremdsprachen als Barrieren in der Wissenschaft?," *Zeitschrift für Sprachwissenschaft* 4, no. 2 (1985): 250–258, on 255; Cristina Guardiano, M. Elena Favilla, and Emilia Calaresu, "Stereotypes about English as the Language of Science," *AILA Review* 20 (2007): 28–52, on 29.

84. Philippe Van Parijs, "Tackling the Anglophones' Free Ride: Fair Linguistic Cooperation with a Global Lingua Franca," *AILA Review* 20 (2007): 72–86.

85. Lewin and Jordan, "The Predominance of English," 436.

86. Liliana Mammino, "The Mother Tongue as a Fundamental Key to the Mastering of Chemistry Language," in Charity Flener Lovitt and Paul Kelter, eds., *Chemistry as a Second Language: Chemical Education in a Globalized Society* (Washington, DC: American Chemical Society, 2010): 7–42; Kwesi Kwaa Prah, *Mother Tongue for Scientific and Technological Development in Africa* (Bonn: Deutsche Stiftung für internationale Entwicklung, 1993).

87. Ralph Mocikat in Mocikat and Alexander Kekulé, "Soll Deutsch als Wissenschaftssprache überleben?," *Zeit Online* (28 April 2010), accesssed at www.zeit.de/wissen/2010-04/deutsch-forschungssprache. Werner Traxel, the disgruntled psychologist, made a similar point in "'Publish or Perish!'—auf deutsch oder auf englisch?," *Psychologische Beiträge* 21 (1979): 62–77, on 68.

88. Montgomery, *Does Science Need a Global Language?*, 116.

89. Pörksen, "Anglisierung—der dritte große Entlehnungsvorgang in der deutschen Sprachgeschichte," 13.

90. Skudlik, *Sprachen in den Wissenschaften*, 24.

Conclusion

1. Edward Sapir, *Language: An Introduction to the Study of Speech* (Mineola, NY: Dover Publications, 2004 [1921]), 184.

2. John Chadwick, *The Decipherment of Linear B*, 2d ed. (Cambridge: Cambridge University Press, 1992 [1958]).

3. H. Beam Piper, "Omnilingual," Project Gutenberg eBook, http://www.gutenberg.org/files/19445/19445-h/19445-h.htm, accessed 7 December 2008, on p. 11.

4. Piper, "Omnilingual," 42.

5. Piper, "Omnilingual," 44–45.

6. Piper, "Omnilingual," 46.

7. A point raised, inter alia, by Sabine Skudlik, *Sprachen in den Wissenschaften: Deutsch und Englisch in der internationalen Kommunikation* (Tübingen: Gunter Narr, 1990), 221; and Scott L. Montgomery, *Does Science Need a Global Language? English and the Future of Research* (Chicago: University of Chicago Press, 2013), 66.

8. David Crystal, *The Stories of English* (Woodstock, NY: Overlook Press, 2004), 508; Henry Hitchings, *The Secret Life of Words: How English Became English* (New York: Farrar, Straus & Giroux, 2008), 336; Nicholas Ostler, *The Last Lingua Franca: English until the Return of Babel* (New York: Walker, 2010), 62.

9. Examples are too many to count. For interesting instances of the first of these genres, see Walter M. Miller, *A Canticle for Leibowitz* (Philadelphia: Lippincott, 1960 [1959]); and Poul Anderson, *Orion Shall Rise* (New York: Timescape, 1983).

10. J. E. Holmstrom, "The Foreign Language Barrier," *Aslib Proceedings* 14, no. 12 (December 1962): 413–425, on 414.

11. John DeFrancis, *The Chinese Language: Fact and Fantasy* (Honolulu: University of Hawaii Press, 1984); Jing Tsu, *Sound and Script in Chinese Diaspora* (Cambridge, MA: Harvard University Press, 2010).

12. Montgomery, *Does Science Need a Global Language?*, 5.

13. For the ecology perspective, see Peter Mühlhäusler, "Language Planning and Language Ecology," *Current Issues in Language Planning* 1, no. 3 (2000): 306–367. Against, see Douglas A. Kibbee, "Language Policy and Linguistic Theory," in Jacques Maurais and Michael A. Morris, eds., *Languages in a Globalising World* (Cambridge: Cambridge University Press, 2003): 47–57.

14. K. David Harrison, *When Languages Die: The Extinction of the World's Languages and the Erosion of Human Knowledge* (New York: Oxford University Press, 2007), 6–7.

15. Mühlhäusler, "Language Planning and Language Ecology," 333; Winfried Thielmann, "The Problem of English as the *Lingua Franca* of Scholarly Writing from a German Perspective," in Anthony J. Liddicoat and Karis Muller, eds., *Perspectives on Europe: Language Issues and Language Planning in Europe* (Melbourne: Language Australia, 2003): 95–108, on 103.

16. Ulrich Ammon, "Die deutsche Sprache in den Wissenschaften," in Conseil de la langue française, Gouvernement du Québec, *Le français et les langues scientifiques de demain: Actes du colloque tenu à l'Université du Québec à Montréal du 19 au 21 mars 1996* (Quebec, Canada: Gouvernment du Québec, 1996): 209–220, on 219; Dieter Föhr, *Deutsch raus—Englisch rein: Vom Abdanken einer großen Kultursprache* (Berlin: Pro BUSINESS, 2008), 115.

17. Michel Debré, "La langue française et la science universelle," *La Recherche* 7, no. 69 (July-August 1976): 956. For a summary of French cultural attitudes toward English more broadly, see Brian Weinstein, "Francophonie: Purism at the International Level," in Björn H. Jernudd and Michael J. Shapiro, eds., *The Politics of Language Purism* (Berlin: Mouton de Gruyter, 1989): 53–79.

18. Claude Hagège, *Contre la pensée unique* (Paris: Odile Jacob, 2012), 237; Jean Darbelnet, "Le français face à l'anglais comme langue de communication," *Le français dans le monde* 89 (1972): 6–9, on 9.

19. Claude Truchot, *L'Anglais dans le monde contemporain* (Paris: Le Robert, 1990), 74, 128; David Graddol, *The Future of English?* (London: British Council, 1997), 8–9; Philippe Meyer, "The English Language: A Problem for the Non-Anglo-Saxon Scientific Community," *British Medical Journal* 2 (7 June 1975): 553–554. Resignation to the flood of English was already widespread at least by 1962: Pierre Burney, *Les langues internationales* (Paris: Presses universitaires de France, 1962), 71.

20. Laurent Lafforgue, "Le français, au service des sciences," *Pour la Science* (28 March 2005): 8.

21. Lawrence H. Summers, "What You (Really) Need to Know," *New York Times* (20 January 2012): ED26.

22. See "Inside Google Translate" at http://translate.google.com/about/index.html; and "How Does Google Translate Work?" at http://www.geekosystem.com/how-does-google-translate-work, both accessed on 28 September 2011.

23. Graddol, *The Future of English?*, 30. On the metaphor of "languages" in this context, see the excellent discussion in Jörg Pflüger, "Language in Computing," in Matthias Dörries, ed., *Experimenting in Tongues: Studies in Science and Language* (Stanford: Stanford University Press, 2002): 125–162.

24. For a general introduction, see Brian McConnell, *Beyond Contact: A Guide to SETI and Communicating with Alien Civilizations* (Sebastopol, CA: O'Reilly, 2001).

25. Paul Davies, *The Eerie Silence: Renewing Our Search for Alien Intelligence* (Boston: Houghton Mifflin, 2010), 183.

26. George Basalla, *Civilized Life in the Universe: Scientists on Intelligent Extraterrestrials* (New York: Oxford University Press, 2006), 200.

27. Hans Freudenthal, *Lincos: Design of a Language for Cosmic Intercourse, Part I* (Amsterdam: North-Holland, 1960), 11.

28. Freudenthal, *Lincos*, 13.

29. Freudenthal, *Lincos*, 21.

30. I. S. Shklovskii and Carl Sagan. *Intelligent Life in the Universe* (Boca Raton, FL: Emerson-Adams, 1998 [1966]), 430.

INDEX

Page numbers in italics refer to figures.